Sensor Technology in the Netherlands: State of the Art

Sensor Technology in the Netherlands: State of the Art

Proceedings of the Dutch Sensor Conference
held at the University of Twente,
The Netherlands, 2–3 March 1998

Edited by

A. van den Berg
and

P. Bergveld

MESA Research Institute,
University of Twente,
The Netherlands

SPRINGER-SCIENCE+BUSINESS MEDIA, B.V.

A C.I.P. Catalogue record for this book is available from the Library of Congress.

ISBN 978-94-010-6103-2 ISBN 978-94-011-5010-1 (eBook)
DOI 10.1007/978-94-011-5010-1

Printed on acid-free paper

Preface

The 1998 Dutch Sensor Conference held on March 2-3, 1998, at the University of Twente in Enschede, the Netherlands is the third in a series of biannual meetings. The series of conferences is initiated by the Dutch Technology Foundation (STW) in order to stimulate the industrial application of scientific sensor research. The main reason for this has been the fact that in spite of a very long and outstanding reputation on high quality sensor research, few products have been marketed in the Netherlands up to now. This sensor conference not only offers academic and industrial researchers an opportunity to present the latest results of research and development on sensors, actuators and microsystems, but in addition offers designers and developers from industry inspiration for new products and gives them the chance to meet young researchers.

In this second MESA Monograph the Proceedings of the third National Sensor Conference are collected. It contains both a collection of the latest research and development results from all major Dutch centers for sensor research, as well as several contributions on the aspects of sensor commercialization, as presented at the conference as (invited or submitted)oral or poster contributions. Thus, it provides an excellent overview of the state of the art of Dutch Sensor Technology in 1998, which explains the title of this Monograph volume. In addition we invited two Swiss researchers to present their view on sensor development in the field of biochemical sensor systems and the utilization of new micromechanical sensing principles for chemical sensor(arrays).

At this place I would like to acknowledge the contributions of the programme committee, the local organizing committee, STW, the sponsors, many MESA members and, last but not least the contributing authors.

Albert van den Berg
Conference chairman.

Programme Committee

A. van den Berg (University of Twente (MESA), chairman)
P. Bergveld (University of Twente (MESA))
W. Olthuis (University of Twente (MESA))
R. Wolffenbuttel (Delft University of Technology (DIMES))
G. Meier (Delft University of Technology, DIMES)
H. Leeuwis (3T BV)
A.J. Kölling (Texas Instruments)

Sponsors

TEXAS INSTRUMENTS — Texas Instruments

3T — 3T BV

Twente MicroProducts

Contents

MEASUREMENT STRATEGIES FOR OBJECT IDENTIFICATION

Paul P.L. Regtien
University of Twente, Department of Electrical Engineering
Laboratory for Measurement and Instrumentation
P.O.Box 217, 7500 AE Enschede, The Netherlands
email: p.p.l.regtien@el.utwente.nl

0. Abstract

Advanced automation of industrial processes such as product handling, assembly and inspection requires further development of systems for the identification of objects involved in such processes. Various measurement methods for obtaining information about the identity of objects, based on the detection of features by which the objects can be characterised, are reviewed. Examples of such features are shape (or particular geometric properties) and material properties. The measurement of geometric features is performed by a colour camera, or the combination of a black-and-white camera and structured light. Material properties are detected by eddy current sensors. Most of these methods are illustrated with examples taken from a research project about the recognition of electronic components on PCB's, for recycling purposes. Finally, some comments on the combination of sensor data (sensor fusion) to enhance the reliability of the identification process are given.

Keywords: object recognition; object identification; computer vision; machine vision; eddy current

1. Introduction

Identification of an object implies two essential elements: a set of characteristic properties describing the actual object and a set of models or object classes to which the object is being assigned unambiguously. The identification process comprises the measurement of the properties that describe the object's identity, the matching of the acquired data with the model data and finally, the assignment of the object under observation to a particular class, according to a specified criteria. Looking to a cat (the measurement, data acquisition), we could identify this cat as an animal, or just a cat, or a Persian cat, or our left neighbour's cat, depending on the required level of classification and the availability of specific data. Obviously, going from a higher class down to a lower or subclass, the amount of information required for a proper classification increases, so does the complexity and processing time of the identification process. The more particular the class, the more specific should be the detection system.

Key problem in identification is the definition of the characteristic properties for a certain class. Each class should be described by a unique set of properties. Such properties need be detectable by a proper set of sensors (detectors). So, the detector or measurement system should be tailored to those features of the object that enable adequate identification.
Features may appear in a variety of modalities. We can identify a person upon observing his face, appearance, voice, particular clothes, or even a characteristic gait or sound of footsteps.

In the recognition of persons, many features are involved, which can hardly be described completely by a machine. They appear in a large number of combinations within a frame of almost infinite variability. In this paper we only discuss the identification of man-made objects. The number of particular characteristics is low compared to that of natural entities and phenomena.

An example of identification of man-made objects is the recognition of buildings from aerial photographs [1]. This is an example of object recognition by remote sensing: the available measurement data is restricted to images from a number of different optical bands, but all of limited resolution.

An important application of (man-made) object recognition is product inspection: failures in a product should be detected automatically, to alleviate human exertion, to obtain a more objective and reliable criterion for rejection of faulty products, and to safe money. Also in inspection systems faults should be accurately defined first, while the detection system should be designed such that it is able to detect and classify those particular faults.

Figure 1. Yarn package [AKZO-Nobel]; to be inspected on various faults: yarn failures (broken fibres, nooses, fluffs, pokes), shape faults, cone damage, transfer tail, contamination, all within specified tolerance bands. The inspection station consists of a number of detectors, each for a particular set of faults.

This example illustrates the importance of a proper description of features to obtain an acceptable inspection result: neither correct packages might be rejected nor faulty ones accepted. Although faults can be considered as man-made objects, we will not take them into account further in this paper, and narrow down the discussion to complete objects only.
Further, we focus on a restricted set of features: geometric properties (shape) and material properties (electric, magnetic).

2. Retrieval of geometric properties from 2D grey-tone images.

The most common way to characterise an object is by observing its shape. Most automatic identification systems use one or more cameras to generate an image of the object. This image (or a set of two images or even a sequence when 3D information is required) is analysed by some image processing algorithm [2], using the intensity distribution in the image. The shape of the object is extracted from particular patterns in light intensity in the image.

Image processing alone is not enough for a proper recognition. First of all, specified conditions for getting a proper image must be fulfilled: an illumination that yields adequate contrast and no disturbing shadows; a camera set-up with a full view on the object in the scene and with a camera that has a sufficiently high resolution, not to loose relevant details. Obviously, a 2D image only shows a certain aspect of the object, never a complete view (self-occlusion). In case of more than one object, some of them could be (partially) hidden behind others (occlusion), a situation that makes the identification much more difficult.

Even in the most favourable situation, the image alone does not always reveal enough information for a correct identification result. At least we need a model of the imaging process itself: position and orientation of the camera(s), camera parameters like focal length, position of the light source(s) with respect to the object and camera, since all these aspects determine the properties in the image from which features are being extracted. Although the pose (position and orientation) of the object in the scene could be derived from the available information and a priori knowledge of the imaging system, we will disregard in this paper the possibility of pose determination, and restrict to identification only.

Many algorithms have been developed to extract particular features from an image, that is built up of thousands of samples (in space and time) having grey-tone values or just black and white (binary images). The image is searched for particular combinations of adjacent pixels such as edges, from which region boundaries are derived. Noise in the image may disturb this process, and special algorithms are developed to reduce noise effects. Finally, an image results that reflects in a way at least some geometrical aspects of the object.

We show two examples of object recognition from 2D images. The examples differ in available a priori information. In the first case, we just have two images (stereo) of a number of objects; from each object we have a complete and exact geometrical model. The goal is to identify all the objects in the scene (figure 2). Clearly, a single image only would lead to ambiguity: differently shaped objects can have identical projections. The use of stereo images can eliminate such an ambiguity.

Figure 2. Example of the identification of objects (toys). A geometric model (wire frames with corner points) is available; positions and orientations in the scene are unknown. Identification is performed by a matching algorithm, using geometric hashing. The white points indicate the result of the identification process.

The original images are searched for corner points (on the crossings of edges). The result is a large number of points (some hundreds), many of them being false alarms, due to shadows, highlights or noise in the image. To identify the objects, all possible aspects of each model are compared (matched) with all the possible combinations of points in the scene. Using a special search algorithm (geometric hashing), we end up with a hypothesis about the most probable objects [3]. The result is shown in figure 2. Obviously, not all the objects have been identified; note also that some objects that are partly occluded are identified correctly.

In this identification process, both images from the stereo set-up have been used. This yields two sets of candidate corner points. Each point in one image corresponds with one single point in the other, however, it is not evident which one. This general problem in stereo vision, the corresponding problem [4], is solved here by using knowledge about the position of the cameras. As a matter of fact, many other methods have been studied to obtain the required information from a set of stereo images [5].

The second example illustrates a completely different strategy for obtaining 3D shape information from 2D images. The method is based on "structured light", a kind of optical coding. A set of lines with known spacing is projected onto the object of interest (figure 3). From the observed shift and rotation of the projected lines with respect to the lines on the reference plane (the background), height information can be achieved. To obtain a complete range image, a sequence of grids is projected, each with different (binary related) spacing.

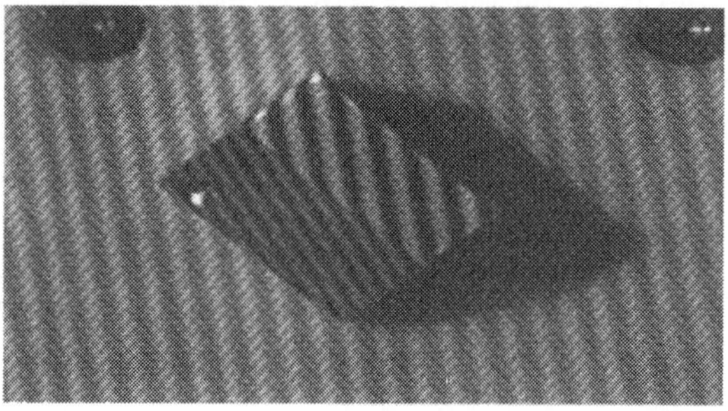

Figure 3. Principle of the structured light method to obtain 3D information from a 2D image. Only one grid out of a set of 8 is shown here. Shadowed areas are eliminated by illumination from two different angles.

The method yields a range image representing geometric properties of the object. In order to identify the object, this range image should be matched to a known model of the object. Again, the application dictates the amount and type of a priori information. In the application of the components on a PCB, the components of interest are modelled by a set of superquadrics, mathematical functions with five parameters that describe the size and shape of the bodies. The surface points of a superquadric satisfy the equation

$$\left\{\left(\frac{x}{a_1}\right)^{\frac{2}{\varepsilon_2}}+\left(\frac{y}{a_2}\right)^{\frac{2}{\varepsilon_2}}\right\}^{\frac{\varepsilon_2}{\varepsilon_1}}+\left(\frac{z}{a_3}\right)^{\frac{2}{\varepsilon_1}}=1$$

The parameters a_i define the size and ε_i the shape of the body. The range images as observed by a camera are fitted to the available models; the best fit results in a hypothesis about the identity of the component that is subjected to the test [6,7]. Some results will be discussed in section 5 of this paper, together with examples of colour images and high-resolution greytone images.

3. Object identification based on tactile imaging

A tactile imager provides shape information based on physical contact with the object under test. Basically, a tactile sensor consists of a matrix of pressure sensors, each responding to the local force that is exerted by the object. The output is a spatially sampled pressure image, with a resolution that is mainly determined by the pitch of the sensor grid. Many principles for the construction of tactile sensors have been proposed, during the last three decennia [8].

The most popular types are those based on elastomeric, piezoresistive materials, preferably those that are available in continuous sheets. Such materials allow the realisation of high-resolution tactile imagers. Disadvantages of this type of sensors are the strongly non-linear force-resistance relation, poor reproducibility and hysteresis. The resulting image data (figure 4) are not accurate and not stable. However, even with inaccurate pressure data, object identification can be performed, by using a priori knowledge about the characteristics of the objects.

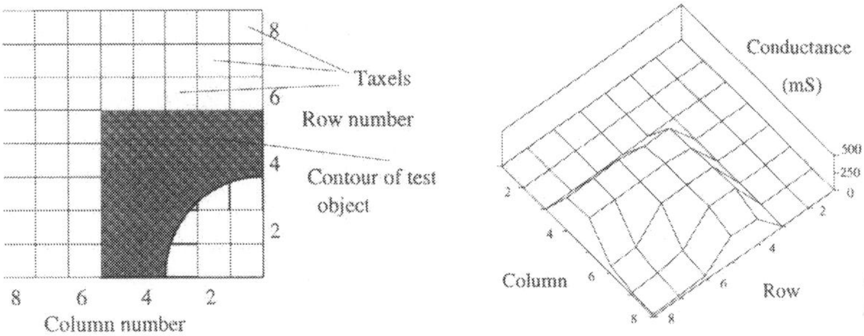

Figure 4. left: positioning of an object on a 8 by 8 tactile sensor; right: sensor response.

The quality of resistive tactile matrix sensors is also limited by mechanical and electrical cross-talk. The former is caused by the mechanical stiffness of the elastomeric layer, resulting in broadening of the point spread function describing the sensor's transfer characteristic. In order to minimise wiring, selection by rows and wires is preferred, although this strategy might cause electrical cross-talk as is illustrated by the equivalent electric circuit of figure 5.

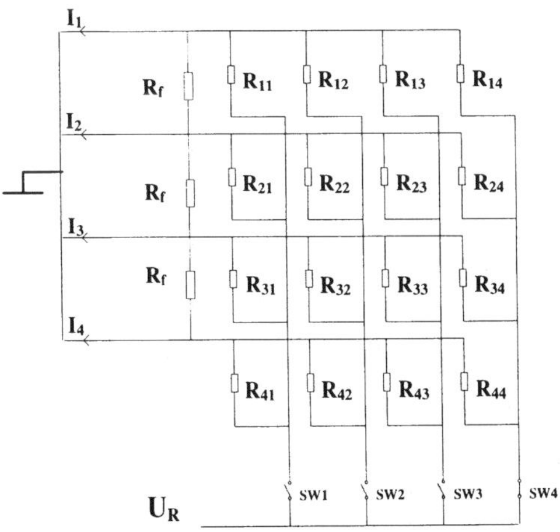

Figure 5. Simplified electronic model of a 4x4 tactile matrix sensor based on a piezoresistive sheet; taxel selection is performed here by connecting a reference voltage through a multiplexer to only one column, and the individual measurement of all row currents.

In this diagram, the taxel resistances are represented by the parameters R_{ij}. The resistances R_f account for the resistance between adjacent rows. In fact, all these resistances are distributed over the sensitive sheet, but for simplicity they are modelled as lumped elements. Taxel R_{11} is selected by closing switch SW1, and simultaneous measurement of current I_1. Obviously, there is a subnetwork of all other resistances in parallel to R_{11}, so the measurement of R_{11} is disturbed by the resistance values of all non-selected taxels. Elimination of this effect is achieved either by (virtual) grounding or active guarding of the non-selected rows and columns. The method is rather effective, but requires complex additional electronics.

Despite poor performance of the individual sensor elements, it appears possible to use tactile matrix sensors for identification of objects. A 16x16 tactile sensor based on highly conductive rubber has been tested successfully in a three-finger robot for grip optimisation in autonomous control with dextrous grippers [11]. In this experimental set-up, a neural network has been trained to identify three classes of objects, based on the characteristic properties of the contact surface: spheres (point contact), cylinders (line contact) and flat objects. Moreover, the system is also able to localise the objects in the gripper, and for cylindrical and flat objects to determine their orientation.

4. Identification based on material properties.

Among the various features of an object, the kind of material of which it is composed is a property that can be employed for identification, when the acquisition of suitable optical images fails. In this section, specificity in bulk material is the feature that will be looked for. Again, various modalities can be candidate to extract the required information, such as thermal capacity, mass density, compliance or electric conductivity, of which in particular the latter can be implemented in a rather simple way. The method we discuss here is based on the occurrence of eddy currents, induced in the material by an alternating magnetic field brought in the vicinity of the object. It should be noted that eddy currents flow mainly in the outside layer of the object, so identification occurs on the basis of the material just in the outer shell of the object.

Basically, an eddy current sensor consists of a coil supplied with an alternating current, resulting in a magnetic field protruding outside the sensor body. When a conductor is present in the region of the external field, free charge carriers (electrons) experience Lorentz forces and will move around in a rather disordered manner by non-homogeneities of the material. These eddy currents counteract the field of origin, resulting in a reduction of the magnetic flux in the coil, which in turn lowers the self-inductance L. The strength of the eddy currents increases with the conductivity of the object, which makes the sensor material specific. Obviously, the output of this sensor also varies with the distance between object and sensor, which is the basis of the well-known contactless proximity sensor.

Ferromagnetic objects that are subjected to the eddy-current sensor will raise the coil's self-inductance, due to a lower magnetic resistance (reluctance) of the magnetic circuit. Hence, the eddy current sensor can distinguish between objects with different conductivity and different magnetic permeability.

The eddy current sensor together with the object can be modelled by a simple transformer circuit, in which the object acts as the secondary coil of the transformer with a material-dependent resistive load. With this model, the equivalent impedance at the primary terminals of the transformer can be derived:

$$R = R_1 + \frac{k^2 L_1 \omega^2 L_2/R_2}{1+\omega^2 (L_2/R_2)^2}$$

$$L = L_1 - \frac{k^2 L_1 \omega^2 (L_2/R_2)^2}{1+\omega^2 (L_2/R_2)^2}$$

Here, k is the coupling factor, accounting for the distance to the object, whereas L_2 and R_2 reflect the object's ferromagnetic and resistive properties, respectively. R_1 and L_1 are the resistance and self-inductance of the sensor coil. From the simultaneous measurement of the real and imaginary parts of the sensor impedance, and using this model, material specific information can be obtained (parametrized by the ratio L_2/R_2), and independent of the gap width between object and sensor head.

In its most simple form, the eddy current sensor can be used in a single-point measurement of the material feature: after a training (calibration sequence) with the complete set of objects it is possible to distinguish between, for instance, steel, aluminium and copper objects, and between ferro- and non-ferro materials [10]. When combined with a scanning mechanism, the eddy current sensor can be used to obtain a conductivity image of objects. Figure 6 shows a picture of a piece of PCB with different components on it; the goal is to identify these components on the basis of differences in the materials, as shown in the impedance images in the same figure.

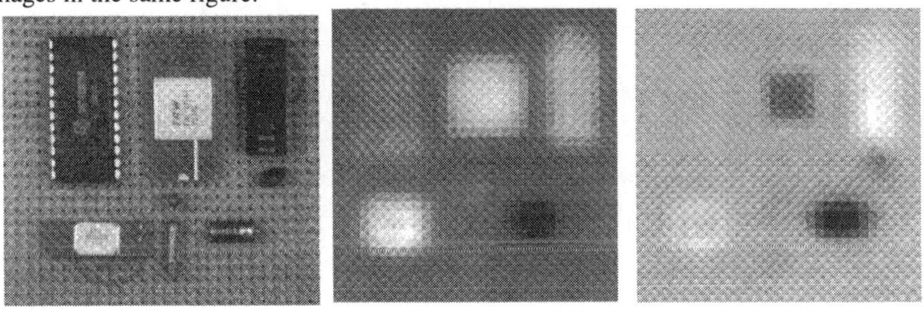

Figure 6. Left: grey-tone reproduction of a colour image of a test PCB with various components. Middle and left: impedance images taken by a scanning eddy-current sensor, showing the real and imaginary parts of the impedance, respectively.

5. Identification of electronic components on PCB's: a case study

Recycling of waste materials contributes to reduction of environmental problems. A prerequisite for recycling is the separation of the primary waste streams into more homogeneous material flows. In this example we discuss the recycling of printed circuit boards (PCB's) from outdated electronic equipment (TV sets, computers). A PCB might contain components that could be reused (memory chips) or that are harmful for the environment (batteries). We describe here a project in which a system is being developed for the identification of components on waste PCB's.

The system consists of four identification units: a range imager based on structured light, a high-resolution grey-tone camera, a colour camera and an eddy current sensor. The PCB under test passes subsequently these stations. Each station comes up with a hypothesis about certain components on the PCB, restricted in the prototype system to IC's (of various types), electrolytic capacitors and batteries. The range imager identifies the components on the basis of 3D shape. Characteristic details in a high-resolution image are checked in the second station, for instance the presence of an array of small bright points, indicating an IC. The colour camera system uses the spectral characteristics in the image of the PCB, to obtain a proper segmentation of the regions representing the objects of interest [12]. The segmentation process is further improved by using a priori knowledge about the component's shape, whereas the object recognition is performed using the technique of inexact attributed graph matching [13]. Finally, the eddy-current sensor is used here as an additional test for suspicious components that are identified with insufficient certainty [14].

Figure 7 shows a picture of a complete PCB, taken with a standard black-and-white camera, and using diffuse illumination. Figure 8 is a height map of the same PCB, as obtained from the range imaging system (using structured light).

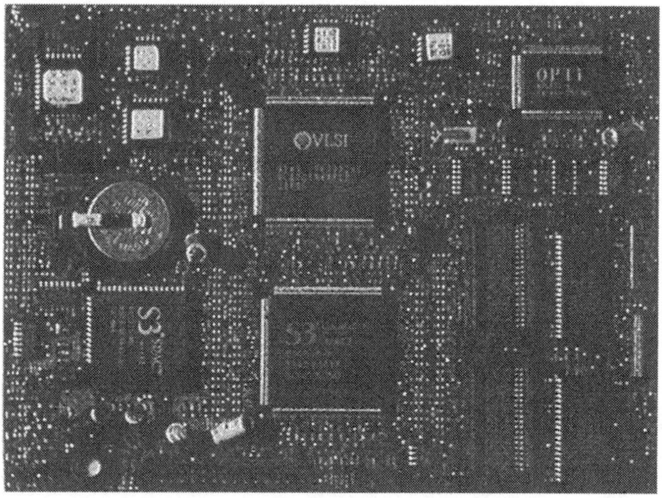

Figure 7. Example of a PCB. A variety of electronic components (IC's, capacitors), some empty sockets, and left-middle a battery.

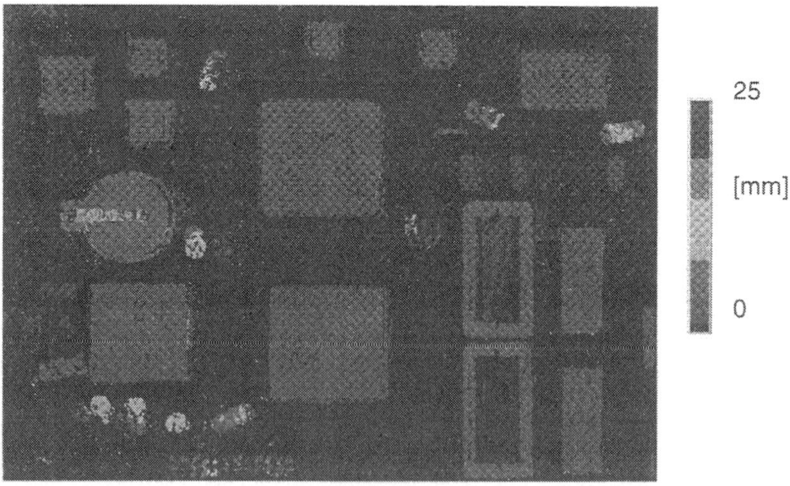

Figure 8. Range image (height map) of a PCB; the tilted, upstanding capacitors are clearly visible.

Figure 9 illustrates the difference in the images obtained from a standard area-scan camera (right) and a high-resolution line-scan camera (left). A complete high-resolution image is obtained by moving the PCB in front of the line-scan camera. The image is used for component identification based on specific details, for instance the pins of the IC's (by template matching) and the characters printed on the components (OCR).

Figure 9. Detail from (left) a high-resolution line-scan camera (1728 pixels per line) and (right) a standard black-white camera image.

Each of the first three stations generates a set of hypotheses about the possible identity of the components on the PCB, and to each of these hypotheses an uncertainty level has been connected. The uncertainty about the identity of an object could be reduced by combining the outputs of the individual systems in such a way that the total uncertainty is lower than that of the individual systems. This is not a trivial process, because the uncertainty is different for the various identification systems, and not all objects are identified by all systems. For instance, the range imager can hardly distinguish between empty and occupied IC-sockets, and the colour camera has some difficulties in detecting electrolytic capacitors. In this application the hypotheses from the first three stations are combined using Dempster-Shafer theory of data fusion [15]. This method takes into account the differences in levels and classes of uncertainty.

6. Conclusions

In this paper we reviewed a number of methods for the identification of (man-made) objects. Evidently, the quality of the identification (the certainty of the generated hypothesis about the object's identity) increases when more a priori information about the object is available and actually used in the recognition process. All systems discussed here and probably most others as well suffer from the fact that they can only detect a limited number of features (if they are able at all to detect them). It is shown that identification upon a single feature or a single modality will not result in a sufficiently high certainty. The key solution to this problem is the use of a multi-sensor system, that is designed to detect various features of different nature. A proper combination of the individual results (sensor or data fusion) should enhance the overall quality of the identification.

Acknowledgement

The author likes to thank E.van Dop, N.H.Kroupnova, H.A.L.van Dijck and H.Kuiper for their contribution to the research on the various subjects described in this paper, SENTER for financial support of the recycling project and AKZO for the financial support of the package inspection project.

References

1. L.J Spreeuwers, K.Schutte, Z.Houkes, A model driven approach to extract buildings from multi-view aerial imagery. Automatic Extraction of Man-Made Objects from Aerial and Images (II), Ascona, Austria (1997), pp. 109-118, ISBN 3-7643-5788-6.
2. F. van der Heijden, Image Based Measurement Systems, Wiley & Sons, West Sussex, England (1994), ISBN 0-471-95062-9
3. H.A.L. van Dijck, M.J. Korsten, F. van der Heijden, Robust 3-dimensional object recognition using stereo vision and geometric hashing. ICIP'96 International Conference, Lausanne, Switzerland (1996), ISBN 0-7803-3672-0
4. O. Faugeras, Three-dimensional Computer-Vision: a Geometric Viewpoint, The MIT Press, 1993
5. R.M.Hoogeveen, M.J.Korsten, The use of silhouettes in 3D scene recognition and pose estimation, 9th Scandinavian Conference on Image Analysis. Theory and Applications of Image Analysis II, Uppsala, Sweden (1996), pp.183-196, ISBN: 981-02-2448-6
6. E.R. van Dop, P.P.L. Regtien, Object recognition from range images using superquadric representations, Proceedings of the IAPR Workshop on Machine Vision Applications, Keio University, Tokyo, Japan (1996), pp.267-270
7. E.R.van Dop, P.P.L. Regtien, Volumetric segmentation of range images for printed circuit board inspection, Proceedings of the International Society for Optical Engineering, Automated Optical Inspection for Industry, Beijing, China (1996) pp.687-694, ISBN 0-8194-2300-9
8. P.P.L.Regtien, Tactile imaging, Sensors and Actuators A, 31 (1992), pp. 83-89
9. P.P.L.Regtien, E.G.M Holweg, A tactile matrix with neural network for object recognition and pose estimation, XIV Imeko World Congress, Tampere, Finland (1997), Volume IX b, pp. 37-42, ISBN 951-96042-8-6
10. N.H.Kroupnova, Z. Houkes, P.P.L. Regtien, Application of eddy-current imaging in multi-sensor waste separation system, Proceedings of the International Conference & Exhibition on Electronic Measurement & Instrumentation, ICEMI'95, Shanghai, China (1996), pp.196-199
11. E.G.M.Holweg, Autonomous control in dextrous gripping, PhD thesis, Delft University of Technology, 1996
12. Kroupnova, N.H., Gorte, B., Method for multi-spectral images segmentation in case of partially available spectral characteristics of objects, Proceedings of the Machine Vision Applications in Industrial Inspection IV, International Society for Optical Engineering -SPIE-, San Jose, California, USA (1996), pp.210-218, ISBN: 0-8194-2039-5
13. N.H. Kroupnova, Recognition of objects from multi-sensor data. PhD thesis University of Twente, 1997, ISBN 90-9006902-X
14. N.H.Kroupnova, Z.Houkes, P.P.L.Regtien, Model-based parameter estimation from eddy-current images and application in a multi-sensor waste separation system. Proceedings of the 1996 National Sensor Conference, Delft, The Netherlands (1996), pp.135-138, ISBN: 90-407-1321-9
15. G.Shafer, A mathematical theory of evidence, Princeton University Press, 1976

A SMART 2.5D IMAGE SENSOR CONTAINING A PSD ARRAY AND AN ON-CHIP MULTIPLEXER

Michiel de Bakker, Piet W. Verbeek, Gijs K. Steenvoorden
Pattern Recognition Group, Department of Applied Physics
Delft University of Technology
Lorentzweg 1, 2628 CJ Delft, the Netherlands

0. Abstract

In the present paper we will discuss a smart optical sensor that has been designed for the purpose of sheet of light range imaging. The sensor consists of an array of Position Sensitive Detectors (PSD's). The PSD's have a length of 20mm and a width of 20μm. The pitch in the array is 28μm. This PSD array is used to record the cross sectional line of a sheet of laserlight with an object. Three dimensional information is computed using triangulation. A complete surface can be measured by sweeping the sheet of light over the object. On-chip electronics deal with low-pass filtering and multiplexing of the sensor signals.

Keywords: Position Sensitive Detectors, range imaging

1. Introduction

For the purpose of quality inspection and computer vision a smart image sensor is being developed containing an array of Position Sensitive Detectors (PSD's) and on-chip (bipolar) electronics [1,2]. A PSD is a lateral photodiode that delivers two photocurrents from which the centre of gravity of an incident light distribution can be determined [3,4,5,6]. The image sensor is capable of capturing range data at 2MHz at a 12 bits resolution. This is approximately 100 times faster than using a standard CCD image sensor for the same purpose, and 10 times faster than other reported application specific devices [7,8,9,10].

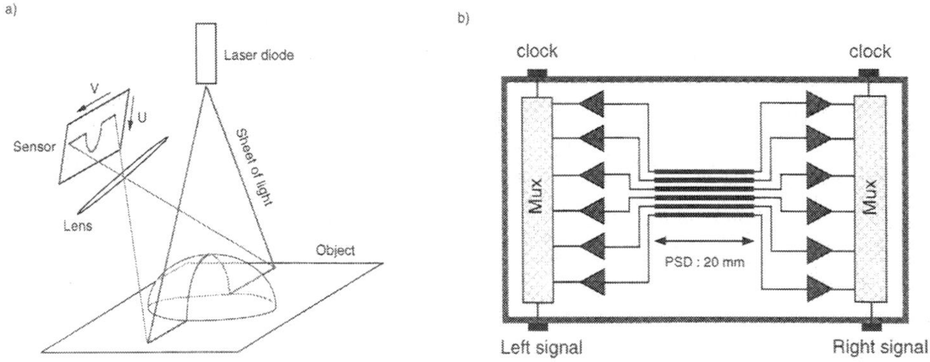

Figure 1 a) Sheet of light range imaging b) smart PSD sensor with 6 PSD's.

The technique used to measure range data is called sheet-of-light range imaging. A light-sheet is projected onto an object. The cross-sectional line of the light-sheet and the object is recorded using an image sensor, and world co-ordinates are computed by means of a triangulation computation [see Fig. 1a]. The co-ordinates (U,V) in the image plane are measured as follows: the cross-sectional line is sampled in the V-direction with a 28μm pitch. For each discrete V-position the centre of gravity of the light distribution is measured with a PSD. These centre of gravity positions are the U co-ordinates in the image plane. If a complete surface has to be recorded, the object has to be moved perpendicular to the light sheet (e.g. on a conveyor belt).

The main reason for the on-chip electronics is to reduce the number of bonding-pads by on-chip multiplexing. Apart from supply voltages, the only contact with the outside (off-chip) world is a clock-signal "in" and a demultiplexed signal "out". The on-chip electronics consists of :

- pre-amplifiers that low-pass filter the sensor-signals at 16 kHz (videospeed)
- multiplexer implemented in ECL-technology

Fig. 1b shows a sketch of the on-chip signal-processing. Each PSD has two pre-amplifiers: one for the left, and one for the right photocurrent. The multiplexer selects the PSD signal that has to be connected to the output channel.

Further signal processing proceeds with off the shelf electronics. The first step is variable gain amplification. This is necessary in order to deal with fluctuations of light intensity. If the AD-converter has to be driven at an input level with acceptable resolution, the low light level values have to be amplified stronger than high intensity values. Next, AD conversion is performed at 12 bits and 2 MHz. The next step is that a correction has to be performed for each separate on-chip preamplifier. The preamplifiers have to be corrected for both gain and offset values. Finally the centre of gravity (i.e. the U co-ordinate) and corresponding range value have to be computed. This is done using Digital Signal Processors (DSP's).

2. Prototype device

A test-device has been produced at the Delft Institute for Micro Electronics and Submicrontechnology (DIMES) [see Fig. 3]. In this test-device there are three interesting features. First of all there is the combination of a PSD-array with on-chip electronics. The second new feature is that we have produced PSD's with a 2 cm length in a process where the standard die size equals 1 by 1cm. This has been achieved by using 2 masksets: one containing the electronics, and one containing solely the PSD-array. By positioning the masks very accurately the technicians at DIMES were able to make PSD's exceeding the standard die length, for which we have measured a linearity that was better than 1:1000 (this will be shown in Sec. 2.1). Third, a couple of PSD's have been fabricated that were very close to the saw-line of the die. The idea here was that we want to be able to place 2 PSD-arrays sufficiently close together in order to make a larger array, where the 28μm pitch in the V-direction [see Fig.1] is maintained. This way the total chip-surface of one device can be kept within margins that are acceptable regarding the yield-stress of the IC-process used.

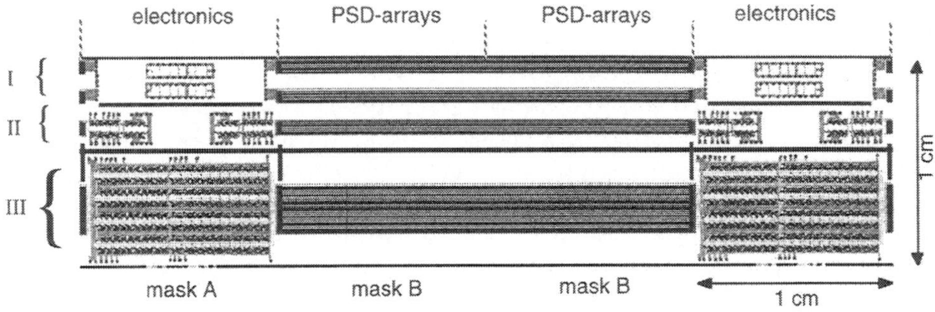

Figure 2 construction of 2cm PSD's with readout electronics using four adjacent masks:
area I separated PSD arrays and electronics
area II array of 6 PSD's connected to the on-chip electronics
area III array of 80 PSD's connected to the on-chip electronics

In the remainder of this paper we will discuss two important properties for the PSD array we have measured. We will also provide some examples of the sensor signals of the first prototype device containing 6 PSD's and readout circuitry.

2.1 Linearity

The most important characteristic (and also one of the strong properties) of PSD's is their high linearity. The centre of gravity of light can be derived from the sensor signals using [6]:

$$U_{cog} = \frac{L_{psd}}{2} \frac{I_{right} - I_{left}}{I_{right} + I_{left}} \qquad (1)$$

In Eq.1 U_{cog} is the location of the centre of gravity (cog), L_{psd} is the length of the Position Sensitive Detector, and I_{left} and I_{right} are the photocurrents to the left and right contact, respectively. The factor $L_{psd}/2$ makes sure that the origin (U=0) is located at the centre of the PSD.

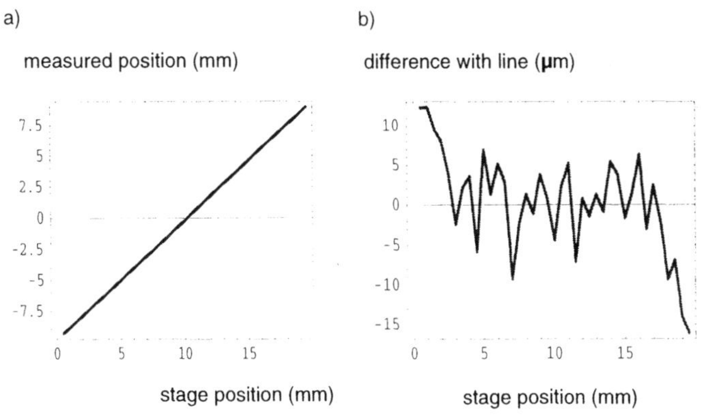

Figure 3 Linearity measurements: a) Measured position vs. translation stage position
b) difference of Figure 3a with a linear fit.

For Eq. 1 to hold, it is most important that the PSD layer is perfectly homogeneous. We would like to know if linearity is maintained if we produce 2cm PSD's by aligning 2 masks. We have measured the linearity by placing a laserdiode on a translation stage (with a position accuracy of 10μm). In Fig. 3a the measured position (derived from the PSD signals using Eq. 1) is depicted against the translation stage position. In Fig 3b the difference of Fig. 3a with a linear fit through the data is shown. From Fig. 3b we learn that for the inner 2 cm of the PSD there is no significant deviation from linearity, and that the accuracy corresponds to the accuracy of the translation stage. At the outer regions the PSD delivers centre of gravity values that deviate from linearity. Since this deviation is known it can be accounted for in a calibration procedure.

2.2 Electrical crosstalk

A possible cause for problems could be electrical crosstalk between two neighbouring PSD's. If charge carriers that have been generated at PSD_n can diffuse toward PSD_{n+1} or PSD_{n-1} this could degrade the resolution in the V-direction. In order to check for possible crosstalk problems we performed a sensitivity scan in the V-direction. The PSD-chip was mounted on a microscope stage and through one of the eyepieces of the microscope a laserbeam was focused on the PSD-chip. The size of the laserspot was approximately 8μm. In Fig. 4 the results of this scan perpendicular to the PSD's is depicted.

From the data of Fig. 4 can be seen that electrical crosstalk is not a serious hazard: the amount of signal at the edge of the metal layer (i.e. V=-20μm or V=18μm) is less than 1% of the intensity at the centre of PSD_n. This can only become a serious problem in the case of large fluctuations in both intensity and position. One must also be aware that the graph in Fig. 4 is the convolution of the metal layer mask and the (Gaussian) laserspot. This implies that the presence of photosignal at V<-20μm and V>18μm can also be caused by light coming from the edges of the laserbeam.

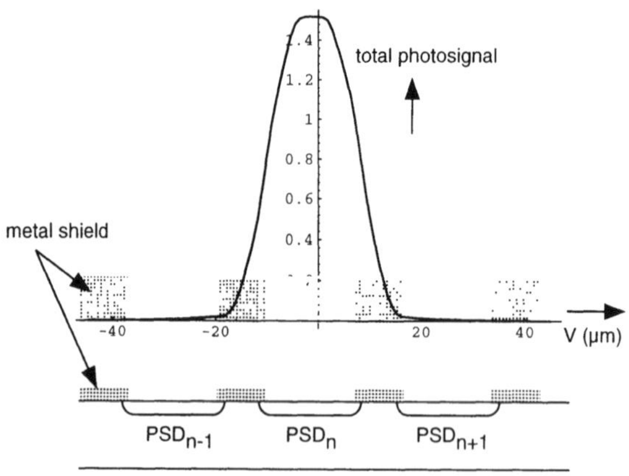

Figure 4 Electrical crosstalk: a sensitivity scan in the V direction. Total photosignal after iv conversion, and a cross-section of the PSD-array.

2.3 The sensor signals

A prototype sensor containing six PSD's and the on-chip circuitry has been tested. The output of the sensor signals, as displayed on an oscilloscope, are shown in Fig. 5 for three situations. Fig. 5a represents the situation that no light falls onto the sensor. The other two scope outputs are the result of a light stripe that has been projected at the right side of the PSD's [Fig. 5b] and at the centre of the PSD's [Fig. 5c]. The upper scope signals correspond to the right PSD signals and the lower scope signals correspond to the left PSD signals. The scope signals must be interpreted as follows. The signal level indicated with a "0" is the state where no sensor signals are connected to the output channel. This zero state is used for triggering the AD-converter. The next six signal levels (indicated with 1-6) correspond to the six PSD's that are time-base multiplexed to the output channel. In this example the readout frequency was 100kHz.

In Figs 5b and 5c the first PSD captures the highest lightintensity (i.e. $I_{left}+I_{right}$). The total amount of light decreases for PSD's 2 to 6. The centre of gravity can be computed from the scope signals using Eq. 1, after correction for preamplifier offset and gain. The offset values of the preamplifier can be derived from Fig. 5a. In Fig. 5c the left and right signals are approximately equal leading to a centre of gravity near U=0 ($I_{left}=I_{right}$ in Eq. 1). It can also be observed that the total amount of light is higher in Fig. 5c than in Fig. 5b (compare $I_{left}+I_{right}$ for PSD_1). This is indicated with a larger iso intensity contour for Fig. 5c.

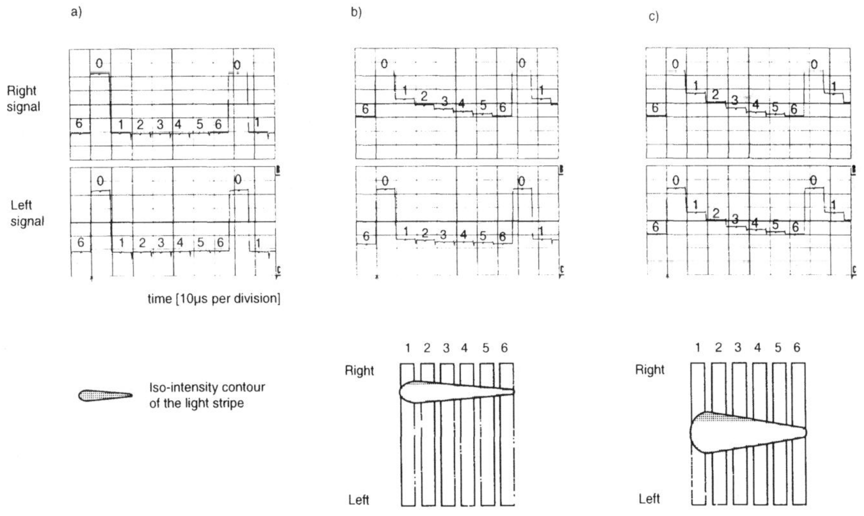

Figure 5 Sensor signals a) no light signal: the preamplifier offset values b) projected light stripe at right side of PSD c) projected light stripe at centre of PSD. The iso intensity contours indicate the position and the intensity of the light stripe.

3. Conclusions

In the present paper we discussed a smart image sensor, specifically designed for sheet of light range imaging. We discussed a prototype with 6 PSD's and on-chip circuitry consisting of preamplifiers and a multiplexer. It was shown that aligning two masks in order to obtain larger sensors did not influence the linearity of the PSD's. We also showed that electrical crosstalk was no problem in the sensor array. Finally we explained the working of the prototype sensor and the sensor signals.

Acknowledgements

This research is sponsored by the Dutch Technology Foundation (STW).

References

1. P.W. Verbeek, F.C.A. Groen, G.K. Steenvoorden, M. Stuivinga, "Range cameras at video speed based on a PSD-array or on CCD-interpolation", Proceedings of the IASTED Int. Symp. on Robotics and Automation, 1985, pp. 67-69.
2 M. de Bakker, P.W. Verbeek, F. van den Ouden and G.K. Steenvoorden "High Speed Acquisition of Range Images", Proc. of the 13th ICPR, track C, 1996, pp 293-297.
3 J.T. Wallmark, "A New Semiconductor Photocell Using Lateral Photoeffect", Proc. IRE, April 1957, vol. 5, pp. 474-483.
4 G. Lucovsky, "Photoeffects in Nonuniformly Irradiated p-n Junctions," Applied Physics 31 (6), 1088-1095 (1960).
5 D.J.W. Noorlag, Lateral-photoeffect position sensitive detectors , Delft University of Technology, 1982.
6 W.P. Connors, "Lateral Photodetector Opterating in the Fully Reverse-Biased Mode," IEEE Trans. on Electr. Devices ED-18 (8) (1971).
7 Paul J. Besl, "Active, Optical Range Imaging Sensors", Machine Vision and Applications, 1, 1988, pp. 127-152.
8 Anders Åstrom, Smart Image Sensors, Ph.D. Thesis Linköping University, Linköping, 1993.
9 T. Kanade, A. Gruss and L. Richard Carley, "A Very Fast VLSI Rangefinder", Proc. of the Int. Conf. on Robotics and Automation, April 1991, pp. 1322-1329.
10 J.-A. Beraldin, "Digital Three-dimensional Imaging in the Infrared at the National Research Council of Canada", SPIE Vol. 2269 Infrared Technology XX, 1994, pp.208-225.

ADAPTIVE OPTICAL SYSTEM BASED ON A MICROMACHINED ADAPTIVE MIRROR

Gleb Vdovin and Serhat Sakarya
Laboratory of Electronic Instrumentation, Delft University of Technology,
P.O.Box 5031, 2600 GA, Delft. Fax: +31-15-2785755,
Email: gleb@ei.et.tudelft.nl

0. Abstract

An implementation of an adaptive optical system based on a 15-mm 37-channel silicon micromachined adaptive mirror [1] is reported. The control loop consists of a beam quality sensor – in our case a brightness sensor placed in the focus of a positive lens – and electronics which includes a single 12-bit ADC to input the quality parameter into the control computer and a 37-channel high-voltage 8-bit DAC to control the shape of the mirror.

Keywords: adaptive optics, micromachined adaptive mirror

1. Introduction

This paper describes an experimental setup that uses a micromachined adaptive mirror (see Fig. 1) to compensate for various forms of aberration in a collimated light beam. The system is controlled by an optimization program that uses high-voltage 8-bit DACs to change the voltages on the mirror actuators. Feedback is derived from a brightness sensor that has been placed in the focus of a positive lens (see Fig. 1).

This experimental control system, based on discrete components, is reconfigurable for different control criteria, although a particular implementation of the control electronics (such as the high-voltage amplifiers) can be integrated into silicon. This integration will allow for more control channels and therefore for higher correction quality.

In our experiments we searched for optimization schemes securing robust improvement of the beam quality for a random initial aberration. We have compared different control algorithms with respect to their speed of convergence, global optimization robustness and the achievable beam quality.

2. Algorithms

Correction of the phase aberration is optically equivalent to global optimization of the beam brightness, which is a nonlinear function of 37 variables. The control algorithm should be able to converge within a reasonable number of iterations given that the hardware allows for up to 1000 changes of the mirror figure per second. We have experimented with genetic algorithms (GA) and a simple gradient algorithm.

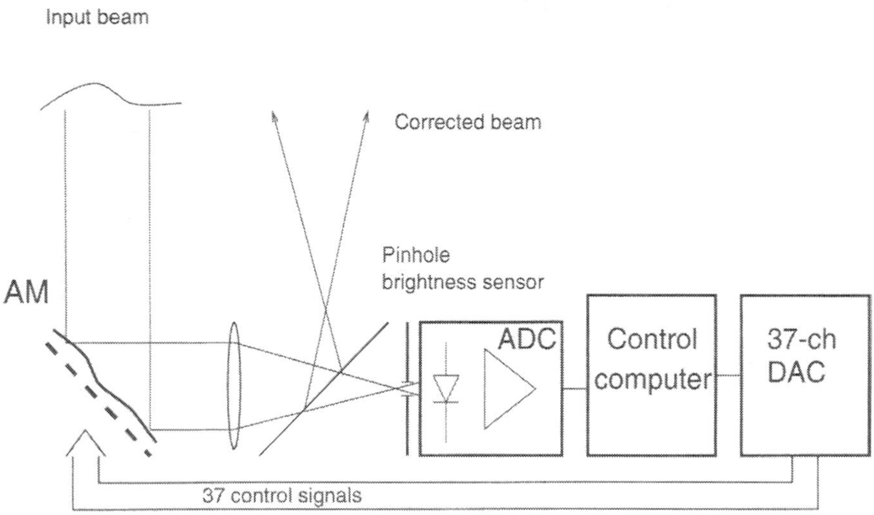

Figure 1: 37-channel micromachined AM (top) and experimental setup for brightness optimization (bottom).

2.1. Genetic algorithms

Genetic algorithms are inspired by Darwin's theory of survival [2]. In our application we begin the process of optimization with a randomly formed "population" of "individuals" (system states), each represented by a 37-dimensional vector of mirror control voltages. This randomly formed population is genetically improved during the optimization process. The quality of each individual is measured by its fitness value, which can be conveniently taken equal to the feedback value. While a GA is slow at the final stage of the optimization process, they are very efficient at finding the global region of the function optimum starting from a set of random system states.

A degenerate form of random search, performed at the first stage of genetic optimization by randomly initializing the mirror control voltages is further continued by reproducing the individuals into following generations using three genetic operations: selection, crossover and mutation.

Crossover involves combining two individuals into one new individual (i.e. "mating" two individuals) in such a way that characteristics of both parents are present in the child. This operation determines the direction of the entire process. The chance of being selected for mating is directly proportional to the brightness value. Finally, mutation may be applied to random individuals to assure genetic diversity of the population. Crossover is done by randomly selecting portions of bits from each of the parents – see table 1 for an example.

	Act. 1	Act. 2	...	Act. 37
Parent 1	00110101	10111010		11010100
Parent 2	11001010	01001001		01001101
Child	00101100	01111001		11010100

Table 1: Random crossover of two selected 37-channel 8-bit parents

Instead of mutating random individuals, we take the worst individual (i.e. lowest brightness value) of every iteration and replace it with a completely random one, while the best individual is simply being copied into the next population. Experiments show that this approach increases the speed of the algorithm and prevents a potential degeneration of the population.

2.2. Gradient optimization

While the GA is potentially very fast with regard to finding promising regions of optimization space, once there, the method is very slow at finding the (local) optimum. This is where it is wise to switch to gradient optimization:

$$V_i^+ = V_i + \alpha \Delta V_o / \Delta V_i \tag{1}$$

where i is the channel number, ΔV_i is a small probe change introduced to the channel control voltage, ΔV_o is the measured change of the feedback value (beam brightness) and α is an application specific coefficient. During one iteration of the algorithm, this operation is performed on all 37 channels.

Optimization (1) performs fairly well in the proximity of an optimum, but does not converge to the global optimum if started from a random state. For reasons such as these, we used a combination of this algorithm with the GA, where the GA finds a good quasi-optimum position and the optimization algorithm quickly finds the best vector.

3. Experiments

When the pinhole is moved in the plane orthogonal to the optical axis – see Fig. 1, the system will attempt to maximize the power through the pinhole, introducing the necessary amount of tilt in the reflected beam, by shaping the mirror surface. We have shown that real-time program control of beam tilt is possible after the system is calibrated to produce consequently four different diffraction-limited beams with different tilt components. Arbitrary tilt is achieved by storing the control vectors for calibrated components in the computer memory and by linear interpolation in the field defined by these four pre-calibrated vectors. This makes possible a simple projection by rapidly scanning the beam between the image points (see Fig. 2). Micromachined adaptive mirrors have negligible hysteresis, permitting for quick and precise one-step program control of the surface shape.

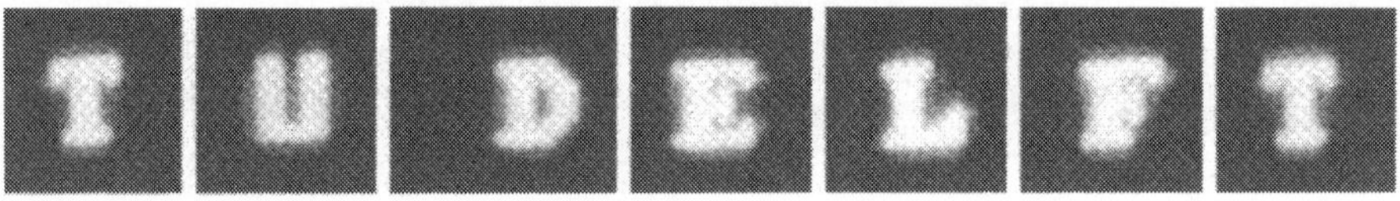

Figure 2: Examples of a diffraction-limited focal spot scanning.

We obtained a good efficiency of correction for aberrated light beams with diameters of up to 10mm. In particular, quick programmable correction of tilt in the range of $5 \cdot 10^{-3}$rad and optimization - based correction of spherical aberration, coma, astigmatism and higher order aberrations were demonstrated. For instance, Fig. 3 illustrates the quality of correction for a light beam passed through a wavy plexiglas plate.

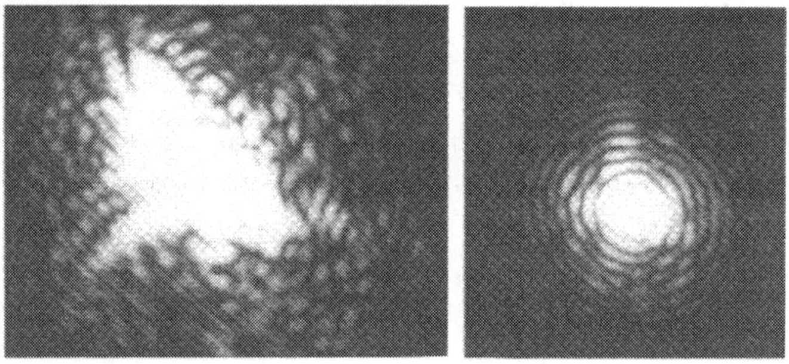

Figure 3: Focal distribution of the aberrated beam before (left) and after (right) correction

4. Conclusions

An adaptive optical system with a micromachined adaptive mirror was built and tested. We demonstrated good performance in correction of random aberrations with low spatial frequencies and for tilt correction in the range of approximately $5 \cdot 10^{-3}$rad. The approach can be used for quick correction of tilt, defocus and quasi-static random optical aberrations of optical systems. Potential applications include laser microscopy, precision scanning through aberrated media, correction of image quality in systems with limited field of view (space instruments) and correction of human eye aberrations in visual science.

References

[1] G. Vdovin, S. Middelhoek and P.M. Sarro, Technology and applications of micromachined silicon adaptive mirrors, Optical Engineering **36**, 1382–1390 (1997)

[2] C. Darwin, On the origin of species by means of natural selection, or the preservation of favoured races in the struggle for life, Murray, London (1860)

THE INTEGRATED COMPASS WATCH MICROSYSTEM

Ronald J. W. T. Tangelder, Hans G. Kerkhoff
MESA Research Institute, University of Twente
P.O. Box 217 7500 AE Enschede, The Netherlands
email : ronaldt@ice.el.utwente.nl

0. Abstract

An electronic compass watch microsystem has been designed and tested. The microsystem uses two micro-machined fluxgate magnetic sensors, and the angle calculation is based on the pulse position method. The compass has been designed to have an accuracy of one degree. The analogue and digital circuitry of the system fit on a single Sea-of-Gates array of 200k transistors. Together with the sensors and passive components it can be combined on a single Multi Chip Module.

Keywords: micro systems, fluxgate sensors, mixed-signal design and testing

1. Introduction

A number of magnetic sensor systems for navigational use have been presented until now [1-5]. However none of them have been shown to function as a complete compass system on a single chip or Multi Chip Module(MCM). The main problem is the requirement of sensors and mixed analogue and digital circuitry on a single silicon substrate.

With the introduction of micro-machined fluxgate sensing elements with electroplated permalloy cores [6] this problem has been partly solved. These sensors are particularly suitable for applications in the earth's magnetic field range as they combine high sensitivity with good integrability [5, 6]. They can even be produced in a CMOS compatible process or can otherwise be used in an MCM implementation. Their operating principle ensures robust and reliable operation of a portable compass system. Above mentioned properties are very appropriate for a fully integrable electronic compass system.

2. Global system overview

An overview of the total system is given in Figure 1. The electronic compass functions by measuring the magnetic field in a horizontal plane in two perpendicular directions. The angle to the magnetic north is calculated by taking the arctangent of the division of the two measurants.

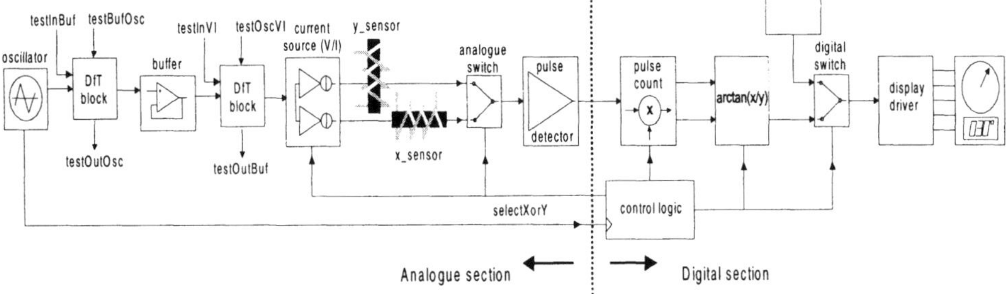

Figure 1: Overview of the electronic compass system

The system comprises of a analogue front-end, which excites the sensors with a triangular waveform and converts the resulting sensor output to measurable digital signals. The digital back-end performs the angle calculation and LCD control.

It uses a multiplexing technique by exciting only one sensor at a time. This reduces both momental power consumption and chip area since only one oscillator is needed. The supply voltage is currently 5 Volts, but can be scaled down to 3.5V.

Both digital and analogue parts are implemented on a Sea-of-Gates Array (SoG) [7] (see Figure 2). The SoG consists of 4 quarters, each with about 50k pmos/nmos pairs. It is mainly intended for digital applications, but can very well be used for analogue designs too (see [8, 9]). Capacitors can be made by putting the second metal layer above the first one, but will consume a considerable amount of area. Very large capacitors and resistors should be realised on the substrate of the MCM.

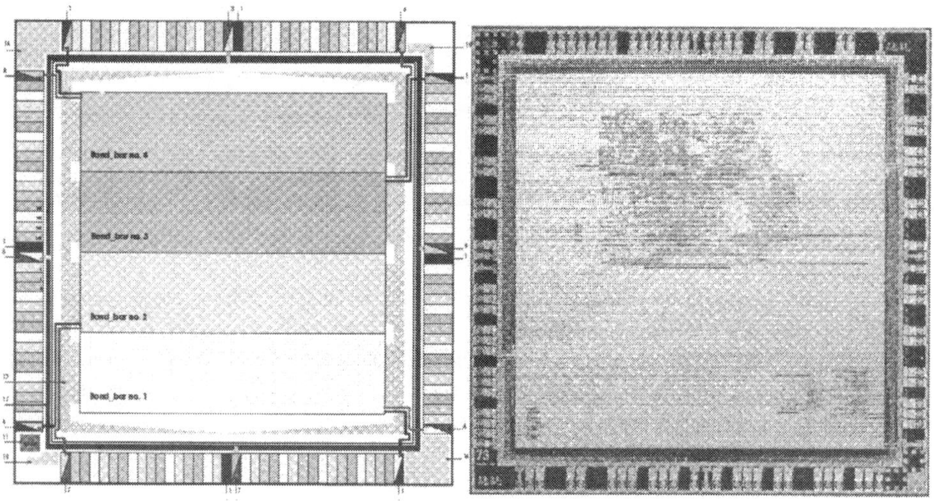

Figure 2: The Sea-of-Gates Array and the realisation of the mixed signal chip

The digital part of the integrated compass occupies 3 quarters and the analogue part 15 % of a quarter. Since each quarter has a separate power supply, we have used two different power supplies for both the digital and analogue parts. To the analogue part Design for Test (DfT) blocks have been added to be able to test the analogue macros separately using the MisMatch mixed signal test environment [10, 11].

The SoG, the two micromachined sensors and required passive components will be combined on a single MCM, equipped with boundary scan test structures [12].

2.1 Fluxgate sensors

For fluxgate sensors several operating principles can be used. Most common is the second harmonic measurement [13, 5, 6]. We, however, use the pulse position method [13], which results in a very simple communication between the analogue and digital part [3]. This will be explained in subsection 3.2.

2.1.1 Operating principle

A fluxgate sensor is a form of transformer, which is deliberately driven into saturation periodically with a symmetrical excitation field (H_{exc}) (see Figure 3a, solid line). If an external magnetic field H_o is applied (the earth's magnetic field in this case), the sensor remains saturated for a longer period in one direction and shorter in the other direction (see Figure 3a, dashed line). The resulting change in induction voltage (V_{ind}) in the pickup coil is a measure for the external magnetic field. If a triangular wave excitation field is applied, the induction voltage consists of pulses which shift in time due to the external field.

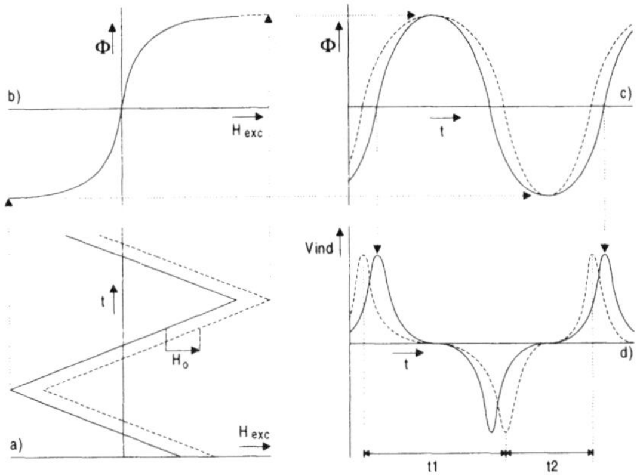

Figure 3: Pulse position operating principle of a fluxgate sensor, $H_{exc} \equiv I_{exc}$ and $V_{ind} = d\phi/dt$

The electrical model of the sensor is shown in Figure 4. The non-linear behaviour of the coil has to be described in the magnetic domain. The ELDO macro model *nlcore2* [14] uses the parameters B_{sat}, μ_o, and H_c to model the magnetic behaviour. It is based on the core model described in [15]. The complete sensor can be modeled by using this macro model as a two port between the magnetic nodes.

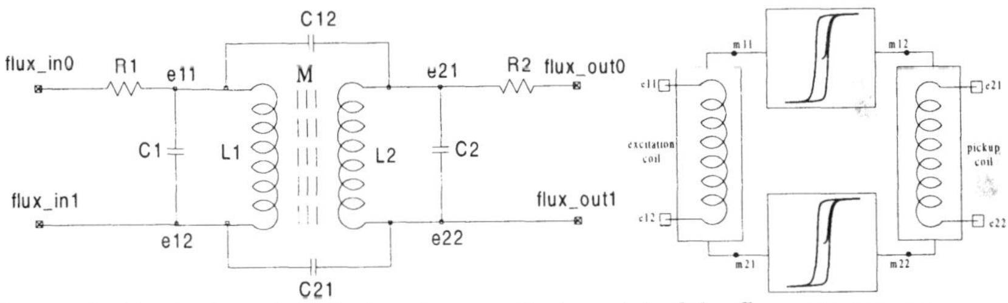

Figure 4 : Equivalent electrical and magnetical model of the flux gate sensor (m11 to m22 are magnetic nodes, e11 to e22 are electrical ones)

An ideal sensor should reach saturation with an applied field with the same magnitude as the earth's magnetic field (H_K=10 Oe). Measurements were carried out on an actual fluxgate sensor [6] and showed that it reached saturation at 15 times the magnitude of the earth's magnetic field. Therefore, currently a discrete miniaturised fluxgate sensor has been used both in the simulation and the actual system. The ideal micromachined sensor is not available yet, but will be realised in near future.

The resulting voltage over the pickup coil and the actual voltage over the excitation coil, after applying a triangular excitation current of 12 mA peak to peak with a frequency of 8 kHz, have been measured and are displayed in Figure 5. The first measurement is without a field. In the second measurement a field is applied. The pulse shift is clearly visible. Notice also the change in impedance of the excitation coil, when saturation is reached.

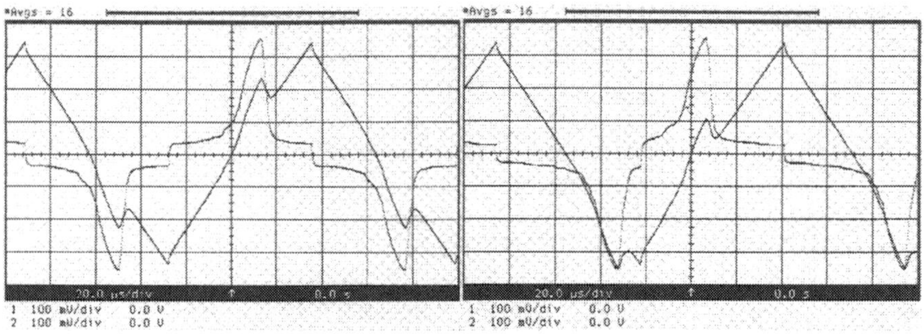

Figure 5: Real fluxgate sensor data, without and with a field applied.

2.1.2 Sensor structure

The micromachined fluxgate sensor is based on a two-layer metalisation silicon process technology and ferromagnetic films. A permalloy core is sandwiched between both metal layers as illustrated in Figure 6. This layout results in a transformer-like structure with a excitation coil and a pick-up coil. The fabricated micromachined fluxgate sensing elements shown [6] each have a different number of windings.

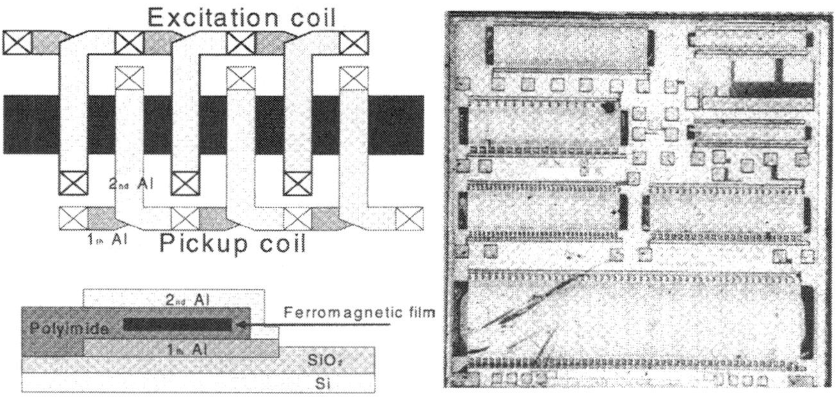

Figure 6 : Structure and realisation of fabricated micro-fluxgate sensing element

3. Analogue section

The analogue section consists of two main parts. The excitation current source will be discussed first and next the pulse position detector is dealt with.

3.1 The excitation current source

The current source consists of three different types of analogue macros being a triangular waveform generator or oscillator, a buffer and two VI-converters to drive the two sensors. Also two DfT blocks have been added in between these macros in able to test them separately. The current source generates a triangular wave with an amplitude of 12 mA peak to peak, which is sufficiently high to drive the fluxgate sensors into saturation (see Figure 3a, b and c). The frequency of this generated waveform is 8 kHz. The triangular waveform oscillator uses a capacitance of 10 pF, which is still easily realisable at the SoG. An external resistor of 12.5 MΩ, which is realised on the substrate of the MCM, is connected to it.

Best sensitivity is obtained when the applied magnetic field is twice the saturation field. The linearity of the waveform is not very essential but the dc-offset is. The sensors have a high series resistance, which requires the use of a balanced differential output. With the supply voltage at 5 Volt, sensors with a resistance as high as 800 Ω can be driven. The resistive character of the sensors is used to linearise the excitation current sources.

3.2 The pulse position detector

The output of the sensor consists of voltage pulses which are detected in the second part of the scheme (see Figure 3d and Figure 5). Their position in time with respect to each other is measured by detecting both the falling edge of the positive pulse and the rising edge of the falling pulse. The pulse position detector processes a logic 1 after the falling edge of the positive pulse, which changes to a digital 0 after the rising edge of the negative pulse, and vice versa.

This CMOS compatible binary signal is sampled by a high speed digital up-down counter. The fraction of time in a period at which the output of the pulse detector is high is a direct indication of the field component measured. Since the analogue output consists only of one digital compatible signal, a complicated AD-converter is not necessary, which would have been the case for methods based on second harmonic measurements.

4. Digital section

The digital control logic has two main functions. It enables the analogue section and the digital high speed up-down counter only when they are needed and it controls the multiplexing of the two sensors. The first approach contributes to diminish the power consumption further

The pulse count part contains a high-frequency (4.194304 MHz) up-down counter, which transforms the output of the pulse detector into two integer values x and y, each indicating the field component of the x- and y-sensor.

The arctangent part gets an x- and an y-value from the up-down counter and computes arctan(x/y), using a cordic-like algorithm [16, 17]. It uses only 8 cycles to calculate the direction with an accuracy of one degree. The calculation method is insensitive to local variations of the magnitude of the earth's magnetic field, which is necessary since the magnitude varies between 25μT in South America and 65μT near the South Pole.

The pulse count part and the arctan part can be modified easily to compute the direction with an arbitrary precision. However, there will always be a bottle neck in the previous parts as the sensitivity of the fluxgate sensor and the analogue section are limited.

The digital part contains also common watch options as added features. The display driver selects either the direction or the time.

5. Simulation and test verification

Simulations have been carried out with the Compass Design Automation Tools and Anacad ELDO for the analogue and mixed-signal simulations. For the system simulation a macro model of the fluxgate sensor has been used as shown in Figure 4. Simulations indicate that an accuracy within one degree is possible. The layout has been realised with the Ocean Design System [18]. The resulting mixed signal chip has been tested on a Microlex mixed signal tester, using the MisMatch tool [10, 11]. To be able to evaluate the complete MCM, first the new sensors have to be fabricated.

6. Conclusions and future work

A miniaturised electronic compass watch system has been designed. The analogue and digital parts fit on a single Sea of Gates array and have been both realised and tested. The compass watch system combines sensing devices and analogue and digital circuitry in a single system by means of a Silicon based MCM. The ideal sensor is not available yet but the system is designed to broad specifications so it can operate with these new fluxgate sensors.

Acknowledgements

We would like to thank the students of the D1-project, as well as G. Diemel and E.A.M. Klumperink who had some nice ideas about the analogue circuitry. S. Kawahito provided us with his first versions of micromachined fluxgate sensors, which helped us to determine the feasibility of the system in an early stage. Finally N. Engin, H. Speek and M. Weusthóf offered us their support with the MisMatch test environment and the test system.

References

1. J. Peters, "Automotive Navigation Using a Magnetic Flux-Gate Compass", IEEE Trans. Veh. Tech., VOL VT-35, No. 2, May 1986, pp 41-47
2. Sprotte, et al, "CMOS Magnetic-Field Sensor System", proc. ESSCIRC 1993, 1993, pp 254-257
3. S. Hollema, et al, "Compass Watch, Report on the design of an Integrated Compass Watch", D1 project report, ICE, 1994, Twente University, The Netherlands
4. Malcovati, et al, "A Magnetic Sensor with Current-controlled Sensitivity and Resolution", Sensors and Actuators, A46-47, 1995, pp 284-288
5. Gottfried-Gottfried, et al, "A Miniaturized Magnetic Field Sensor System Consisting of a Planar Fluxgate Sensor and a CMOS Readout Circuitry", Transducers '95, Sensors and Actuators, 289 - A12, 1995, pp 229 - 232
6. Kawahito, et al, "Micro-Fluxgate Magnetic Sensing Elements using Closely Coupled Excitation and Pickup coils"', Transducers 1995, Sensors and Actuators, 290 - A12, 1995, pp 233 - 236
7. Frehe, "The division of the Sea of Gates Chip into 4 quarters", Internal report, Technical University Delft, the Netherlands
8. de Haan, "Analogue Circuit Design on Digital Sea-of-Gates Array", Ph.D. Thesis, University of Twente, September 1995, the Netherlands
9. van Dongen and V.Rikking, "Advanced Analogue Circuit Design on a Digital Sea-of-Gates Array", proc. ED&TC 1994, pp. 70-74, Paris, France
10. Engin, H.G. Kerkhoff, R.J.W.T. Tangelder, H. Speek , "Automatic Test Plan Generation for Mixed-Signal Circuits", 3rd IEEE International Mixed Signal Testing Workshop, June 3-6, 1997, Seattle, USA, pp. 126-131
11. Engin, H.G. Kerkhoff, R.J.W.T. Tangelder, H. Speek , "A tool for Automatic Test Plan Generation for Mixed Signal Circuits", IEEE Design, Automation and Test in Europe Conference, February 23-26, 1998, Paris, France
12. Oliver, H.G. Kerkhoff, "Test Structures on MCM Active Substrate: Is it Worthwhile", proc. ED&TC 1996, pp. 126-128, Paris, France
13. Ripka, "Review of fluxgate sensors", Sensors and Actuators A, 33, 1992, pp 129-140
14. "ELDO Electrical circuit simulator-version 4.1.x - reference manual", ANACAD Computer Systems, 1992
15. Chan, A. Vladimirescu, Xiao-Chun Gao, P. Liebmann, J. Valainis. "Nonlinear transformer model for circuit simulation", IEEE Trans. on Computer aided design, VOL 10, No. 4, April 1991, pp. 476-482.
16. Spaniol, "Arithmetik in Rechenanlagen", Teubner Studienbücher Informatik, 1976
17. Tangelder, G. Diemel, H.G. Kerkhoff, "Smart Sensor System application : an integrated compass", IEEE European Design and Test Conference, March 17-20, 1997, Paris, France, pp. 195-199
18. Groeneveld and P. Stravers, "Ocean: the Sea-of-Gates Design System Manual", 1994, Delft University of Technology, The Netherlands

NON-DESTRUCTIVE FOOD-STERILITY AND PACKAGE-QUALITY TESTING OF ASEPTICALLY PACKED FOOD PRODUCTS BY IMPEDANCE MEASUREMENTS

Stoyan N. Nihtianov and Gerard C. M. Meijer

Electronics Research Laboratory, Faculty of Electrical Eng., Delft University of Technology, Mekelweg 4, 2628 CD Delft

0. Abstract

An impedance method for simultaneous non-destructive package-quality and food-sterility testing of aseptically packed food products is presented. The method is intended for 100% on-line testing of TBA (Tetra Brik Aseptic) food containers.

An electric model has been developed for the measured impedance. A relation is derived between the resistive component of the measured impedance and the conductivity of the food. The impact of local disintegration of the inner thermoplastic layer of the packaging material (local leakage) on the measurement impedance is analyzed.

The experimental results prove the rapidity and the sensitivity of the method. They also show that we can even roughly determine the size of the leakage area and the distance between its location and a reference point.

Keywords: food-sterility, package-quality, testing, impedance

1. Introduction

The destructive sterility testing of a part of the production lot currently used by food-stuffs producers cannot meet the contemporary demands of improved product safety. It entails the loss of large amounts of ready-to-use food but still does not ensure consumer safety. The alternative is an intensive on-line non-invasive sterility testing of the already packed food product. [1]

Unfortunately, the intensive non-invasive sterility testing of aseptically packed food products is a much more sophisticated procedure than the existing invasive testing procedures. There is a number of reasons for this: the signals that carry information about the sterility of the packed food should be extracted from the food container, without opening it; the results must be independent of the tolerances of the tested packs in shape, size, weight, etc. and from alterations of the test conditions; the testing method must be applicable in an industrial environment; the speed of the testing procedure has to be compatible with the speed at which the packages are produced; no changes of the packaging technology that could deteriorate the reliability and the productivity and could increase the production costs, are acceptable.

The sterility of the packaged food, even when guaranteed by a 100% sterility-testing procedure within a comparatively short period of time after its production, cannot ensure a

This work is supported by the Dutch Foundation for Technical Sciences (STW), project Del.66.4369, and the companies: Tetra Brik, Tetra Pak Research & Development and Unilever.

long shelf-life. What additionally is needed is a high-quality food container. It must be tight enough to prevent food spoilage caused by interaction with the surrounding non-sterile environment. To ensure this, an inspection of the tightness of the packaging container is carried out. For TBA (Tetra Brik Aseptic) food containers, the control is principally focused on the innermost thermoplastic layer of the multiple-layered packaging laminate, since any possible disintegration of this layer (local leakage) may entail the contents reaching the barrier layer (aluminum foil) or the fibre layer, in which event the other barrier properties of the laminate are lost, even if no actual liquid leakage occurs through the laminate. [2]

This paper presents a novel technique for simultaneous non-destructive package-quality and food-sterility testing of aseptically packed food products in TBA containers by impedance measurements.

The proposed method is rather insensitive to the variations in the shape of the container. The experimental results prove the rapidity and the sensitivity of the method. They also show that we can even roughly determine the size of the leakage area and the distance between its location and a reference point.

2. Food-sterility testing

In order to sense the impedance alterations of the packed food, we have to pass an electric current through the walls of the TBA food container and through the food itself. It has been proved in [3] that this is not possible to do by using external electrodes, as one of the intermediate layers of the multi-layer packaging material is a very good conductor (aluminium foil) and prevents external electromagnetic fields from entering the package. To solve this problem, a small-surface electrode is fixed on the food-facing side of the multiple-layer packaging material during its manufacture. The part of a second electrode is given to the conductive Al-foil intermediate layer of the packaging material. Fig.1 shows the current distribution (with dashed lines) and the potential zones in the food product, as simulated by a Maxwell Electrostatic simulator.

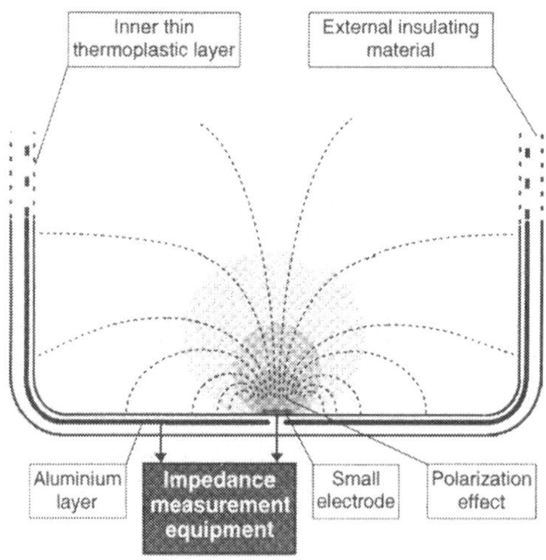

Fig.1. A part of a cross-section of a TBA food container with electric field lines and potential zones displayed, when voltage is applied between the small electrode, fixed on the inner surface of the container wall, and the aluminium layer.

The field below the small electrode can be modeled with a fixed shunt capacitor (not displayed in Fig.1). The electric field extends inward the container rather than along the wall. This is valid for a very wide range of the conductivity σ and the relative dielectric constant ε' of the liquid. The potential drop associated with the current flow mainly occurs in the vicinity of the small electrode. The current through the inner thermoplastic layer has its highest density in the immediate neighborhood of the small electrode. With increase of the distance from the electrode the current density drops rapidly. These are the reasons why this method appears to be quite insensitive to the variations in shape of the container, especially when they take place far away from the small electrode.

Figure 2 presents the equivalent electric circuit of the setting of Fig.1. The value of the capacitance C_{bulk} depends on the dielectric properties of the liquid and the measurement setting. Our experiments with the method, described in [4], proved that for various liquid food products its value is normally a few pF and its influence on the measured impedance can be sensed at frequencies above 10 MHz.

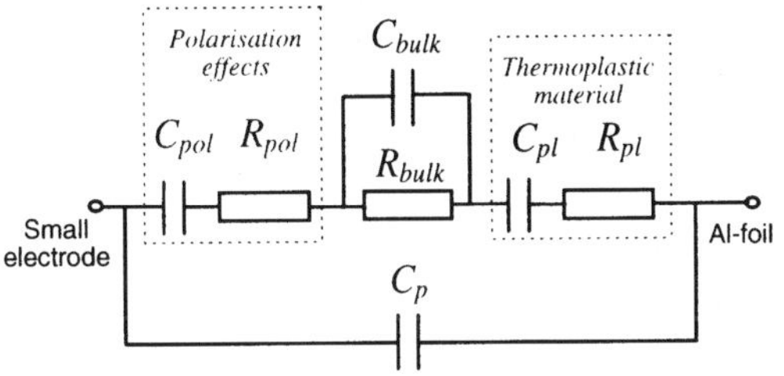

Fig.2. Equivalent electric circuit of the measured impedance, where: C_{pol} and R_{pol} represent the polarisation effect at the surface of the small electrode which is in galvanic contact with the liquid; C_{pl} and R_{pl} represent the impedance of the plastic material between the Al-foil and the liquid; C_{bulk} represents the dielectric properties of the food; R_{bulk} is the bulk resistance of the liquid; C_p is a shunt capacitance.

It is important that the parasitic shunt capacitance C_p (see Fig.3), created between the small electrode and the aluminium foil through the thermoplastic material, is kept, also, as small as possible. If we neglect the influence of C_{bulk} and C_p and we use a simplified series model for the measured impedance Z_Σ, its active component R_Σ equals the sum of R_{pol}, R_{bulk} and R_{pl}.

It is well known that the values of R_{pol} and R_{bulk} are dependent on the conductivity σ of the liquid [4]. The experimental results proved that there is a similar dependence of R_{pl} (the active component in the series model of the thermoplastic layer impedance) on σ, that gives rise to the absolute sensitivity of this method, when compared with other methods. On our opinion, the reason is the redistribution of the current density through the thermoplastic material, due to the spatial redistribution of the current lines within the container when the conductivity of the contained liquid varies.

3. Package quality testing

Local leakage causes redistribution of the current flow in the packaged food. This effect is very well expressed at low frequencies, at which a comparatively low-impedance "tunnel" is created between electrode 1 and the leakage area (part of the aluminium layer, which is in galvanic contact with the food), as shown in Fig.3. The current density in the different areas of the cross-section is depicted with different intensity of the gray color.

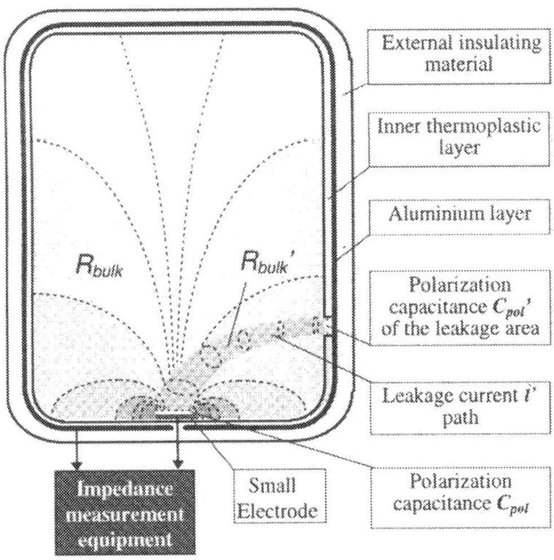

Fig.3. A cross-section of a TBA food container with current lines, when a voltage is applied between the small electrode and the aluminium layer. An impedance "tunnel" caused by a local leakage is also shown.

To model this effect we add an additional "leakage" branch (R_{bulk}', R_{pol}', C_{pol}') to the equivalent electric circuit in Fig.2 (see Fig.4). For simplicity, the parasitic components C_p and C_{bulk} are not depicted in Fig. 4 as their influence on the value of the measured impedance in the low frequency range is negligible.

For a great variety of liquid food products, the polarization capacitance

C_{pol}' has a much higher value than C_{pl}, in the frequency range up to 10 kHz. This is true even for a very small surface of the leakage area, so that the following correlation is valid: $[(R_{food}'+R_{pol}')/(R_{food}+R_{pl})]<(C_{pol}'/C_{pl})$. It can easily be shown that, in this case, with a decrease in frequency there is an increase in the relative value i'/i of the current through the leakage branch (R_{food}', R_{pol}', C_{pol}'). This is why, at low frequencies, this branch has greater impact on the measured impedance. At the same time the resistance of the leakage branch R_{bulk}' has always higher value than the resistance R_{bulk}. As a result, in the case of local leakage, both the equivalent resistance R_{Σ} and the equivalent capacitance C_{Σ} of the measured total impedance between electrode 1 and the aluminium layer rise in value.

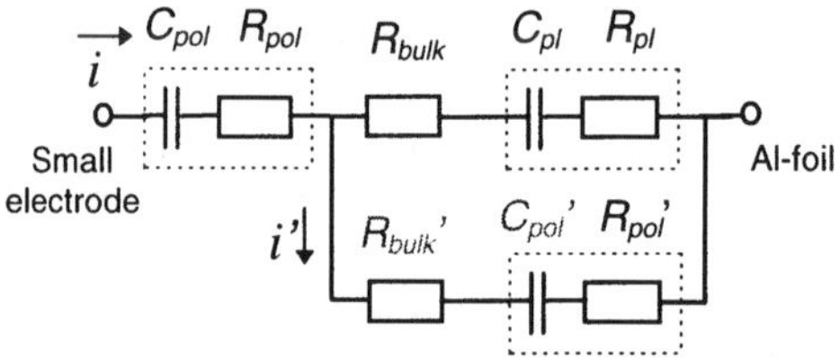

Fig.4. The simplified equivalent electric circuit of the measured impedance from Fig.3 with additional "leakage" branch, comprising: R_{bulk}' - resistance of the food to the current i', which flows through the leakage branch; C_{pol}' and R_{pol}'- representing the polarisation effect at the surface of the local leakage area.

At higher frequencies the influence of the capacitors on the measured impedance drops, and with this the influence of the local leakage branch becomes much smaller, because $(R_{food}'+R_{pol}')>>(R_{food}+R_{pl})$. An additional contribution to this effect is the frequency dependence of the polarization capacitance C_{pol}', which decreases in value with the increase in frequency. In fact, at frequencies above 100 kHz, at which the sterility testing is carried out, the effect of local leakages is already negligible.

4. Experimental results

The impedance measurements were carried out by HP 4194A Impedance Analyzer. Pilot 1litre TBA packages with saline in them were used. A small surface electrode was fixed on the inner side of the package wall. To present the results, a series R_{Σ}-C_{Σ} model of the impedance was used.

The measured capacitance C_{Σ} showed much lower dependence on the conductivity changes of the saline solution than R_{Σ}. That is why when we test the sterility of the food we prefer to extract the required information from the measured resistive component of the impedance R_{Σ}.

To assess the sensitivity of the described sterility testing method, we used a reference method to measure the conductivity of fluids, described in [5]. The theoretical relation between the conductivity σ and the measured resistance R by this method is: $R=1/4\pi\sigma r=0.282/(\sigma.\sqrt{S})$, where $S = 4\pi r^2$ is the surface of the metal sphere. To define the transfer function $R_{\Sigma}(\sigma)$ that best fits the results of our experiments, we used polynomial curve-fitting. Eight values of the conductivity σ were used, uniformly spread over the range: 0.05 S/m to 2.5 S/m. With the least-squares method a second-order relation was generated that repeats the experimental results with maximum tolerance of ±1%, according to the equation

$$R_{\Sigma} = \frac{A}{\sqrt{S}\sigma} - \frac{B}{\sqrt{S}\sigma^2} + C. \tag{1}$$

In equation (1) A, B and C are constants for a given frequency. The constant C represents the resistive component of the impedance of the thermoplastic layer, when the conductivity of the food σ has an infinite value. The surface S of the small electrode is used as a parameter. The constants *A*, *B* and *C* change their values with frequency, which makes the sensitivity of the method frequency dependent.

Table 1 presents the values of the constants *A*, *B* and *C* for a few frequencies at which we obtained the highest sensitivity.

TABLE 1

Frequency / Constant	50 kHz	100 kHz	200 kHz	500 kHz
A	0.389	0.387	0.384	0.385
$B \times 10^3$	0.507	0.58	0.942	2.61
C, Ω	1.77	0.85	0.455	0.16

As it was already mentioned above, the potential drop associated with the current that passes from electrode 1 to the aluminium foil is largely in the vicinity of electrode 1. This is why the conductivity of the food in this area of the package has the greatest impact on the measured impedance. The simulated large changes of the package shape at distance of more than 30 mm away from electrode 1 (the dimensions of the package were: 62mm x 95mm x 165mm) did not show noticeable influence on the measurement result.

We tested experimentally how a local leakage through the inner thermoplastic layer of a one-liter TBA food container filled with drinking water changes the values of R_Σ and C_Σ. The surface of electrode 1 was 70 mm^2. Two types of leakage were simulated - a pin-hole with a surface less than 1 mm^2 and another leakage with a surface 70 mm^2. Both leakages were created first at a distance of 5 mm from electrode 1 and then at 90 mm. Figure 5 shows the frequency dependence of the parameters R_Σ, C_Σ and $\varphi_\Sigma = \mathrm{arctg}(-1/(R_\Sigma . C_\Sigma))$ of the measured impedance Z_Σ .

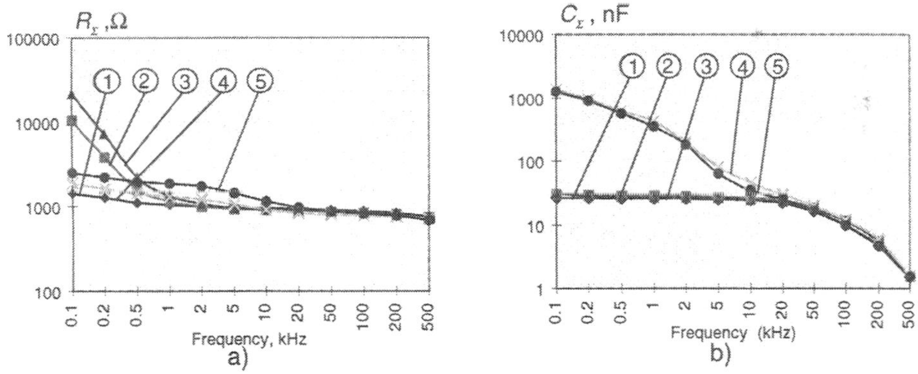

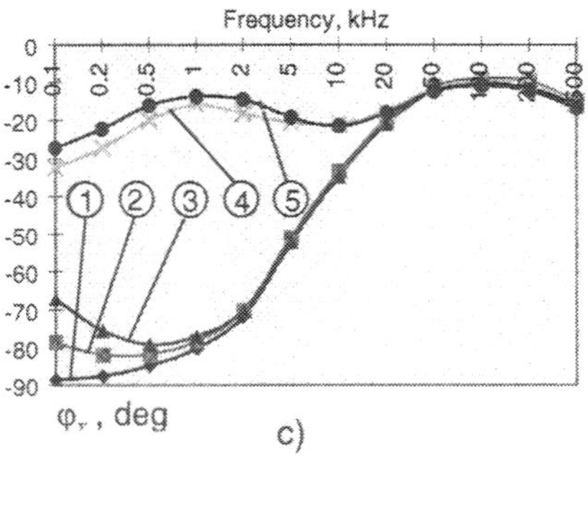

Fig. 5(a, b, c). Dependence of R_Σ (a), C_Σ (b) and φ_Σ (c) on the presence of local leakage in a one liter TBA food container versus frequency. Five different situations have been simulated: 1) no leakage; 2) a pin-hole leakage at a distance of 5 mm from electrode 1; 3) a pin-hole leakage at a distance of about 90 mm from electrode 1; 4) a 70 mm^2 leakage at 5 mm distance from electrode 1; 5) a 70 mm^2 leakage at a distance of approximately 90 mm from electrode 1.

From the experimental results we can conclude that at low frequency the measured capacitance C_Σ is largely dependent on the surface of the leakage area and also slightly dependent on the distance to electrode 1. The measured resistance is both dependent on the surface of the leakage area and on the distance to electrode 1. From the combined information about R_Σ and C_Σ we may have an idea about both the surface of the leakage area and how far away it is from electrode 1. The phase shift φ_Σ of the measured impedance combines both the sensitivity of R_Σ and C_Σ to leakage, and that is why it can be considered to be the best parameter to detect leakage within very wide range of its size and location.

To use the measured values of R_Σ and C_Σ as criteria for the existence of leakage in the package, we need information about the conductivity σ of the food. This information can be obtained by a similar measurement at a higher frequency, at which the effect of local leakage on the measured impedance is negligible. It can easily be seen from Fig.5 that at frequencies higher than 50 kHz the five curves almost merge into one single curve. When the leakage leakage detection is combined with sterility testing of the packaged food at, let's say, frequency 100 kHz , the required information about the conductivity σ of the food can be automatically obtained.

5. Conclusions

A novel technique for simultaneous non-destructive food-sterility and package-quality testing of aseptically packed food products by impedance measurements is presented. It is intended for 100% on-line quality testing of TBA food containers. The physical effects on which the method is based are discussed. An electrical model has been developed for the measured impedance and a relation is derived between the active component R_Σ of the measured impedance and the conductivity of the food. The impact of local leakage through the inner thermoplastic layer of the packaging material on the measured impedance in the low frequency range has been investigated.

The results obtained with this method proved to be *independent* of large changes in the package shape, provided that they are not in the immediate vicinity of the small-surface electrode. The method does not require significant *changes* in the packaging technology.

References

1. H. Gestrelius, T. Mattila-Sandholm and R. Ahvenainen, Methods for non-invasive sterility control in aseptically packaged foods, Trends in Food Science and Technology, Vol.5, 12 (1994), 379-383.

2. St. N. Nihtianov and G. C. Meijer, Non-destructive Package Quality Testing of Aseptically Packed Food Products With Impedance Measurements, to be published in: Proc. The Fourth International Conference On Electronics, Circuits, and Systems, ICECS'97, Cairo, December 15-18, 1997.

3. Stoyan N. Nihtianov and Gerard C.M. Meijer, Non-invasive Impedimetric Sterility Testing of Aseptically Packed Food Products, Proc. Anniversary Scientific Conference: "Fifty Years Technical University - Sofia", Fourth Edition of The National Scientific Conference "Electronic Engineering", Sozopol 1995, Book 1, 52-59.

4. St. N. Nihtianov and G. C. Meijer, "Indirect Conductivity Measurement of Liquids in Flexible containers", *Proc. Instrumentation and Measurement Technical Conference, IMTC/97*, Ottawa, Canada, May 19-21, 1997, *Vol.2*, pp. 919-922.

5. H. A. Haus and J. R. Melcher, "*Electromagnetic Fields and Energy*", Prentice Hall, New Jersey, USA, 1989, ISBN 0 13 249020 X, pp.260-261.

TRANSDUCTION PRINCIPLES AND INTEGRATION OF CHEMICAL SENSORS INTO A MICROMECHANICAL ARRAY DEVICE

R. Berger,[a,b] H.P. Lang,[a,b] J.P. Ramseyer,[b] F. Battiston,[b] J.H. Fabian,[c,*] L. Scandella,[c] C. Andreoli,[a] J. Brügger,[a] M. Despont,[a] P. Vettiger,[a] E. Meyer,[b] H.-J. Güntherodt,[b] Ch. Berger,[a] J.K. Gimzewski[a]

[a]IBM Research Division, Zürich Research Laboratory, CH-8803 Rüschlikon, Switzerland
[b]Institute of Physics, University of Basel, CH-4056 Basel, Switzerland
[c]Laboratory for Micro- and Nanostructures, Paul Scherrer Institute, CH-5232 Villigen PSI, Switzerland

0. Abstract

Ultraprecise measurement tools are indispensable components for most scientific engineering enterprises and for manufacturing process controls. For such applications, micromechanical devices are potentially very promising because a variety of chemical and physical effects can be transduced into a mechanical response on the nanometer scale. Here we discuss micromechanical cantilever-based devices designed and operated to attain ultimate limits in sensitivity and speed, which is achieved by minimizing the sensor size. Such sensors have been used to analyze an entire range of transduction mechanisms down to the nanogram, picoliter, femtojoule, and attomole level. In particular, we use microengineering technologies to design and fabricate a Nanotechnology olfactory sensor (NOSE) device with a volume of only ≈1 mm^3. A new sequential readout technique, the fabrication process of cantilever sensor arrays, and the data acquisition electronics components used for sensor experiments are described in detail.

Keywords: Cantilever array, micromechanical sensor, chemical nose, nanotechnology

1. Introduction

The invention of the scanning force microscope (SFM) [1] linked the use of micromechanics and microscopy at the atomic scale. The extreme sensitivity of micromechanical devices coupled with nanoscale science techniques extends their use well beyond imaging surface properties. Thermal, mass, magnetic, electrical, chemical, stress, and flow signals—among others—can be detected at ultimate limits by transduction into micromechanical motion [2]. Nanoscale science techniques allow measurements of cantilever responses with a resolution of ≈1 pm over a wide dynamic range. In terms of size, the smaller the sensor geometry scales, the less material is necessary to transduce a signal into a micromechanical deflection. Typically our cantilevers are ≈500 μm long, ≈100 μm wide, and ≈1 μm thick, which results in an actuated mass of only a few nanograms. Fabricated with microtechnology, the devices realize both ultrahigh sensitivities and short response times well beyond what is achievable with standard techniques. Here, we briefly review three fundamental signal domains, mass, heat

*Also at Institute for Experimental and Applied Physics, University of Kiel, D–24098 Kiel, Germany

and **stress**, and discuss how they can be transduced into a micromechanical response (Fig. 1). In particular, the focus of the paper is on the micromechanical fabrication processes as well as the data acquisition systems and techniques to read out cantilever sensor arrays.

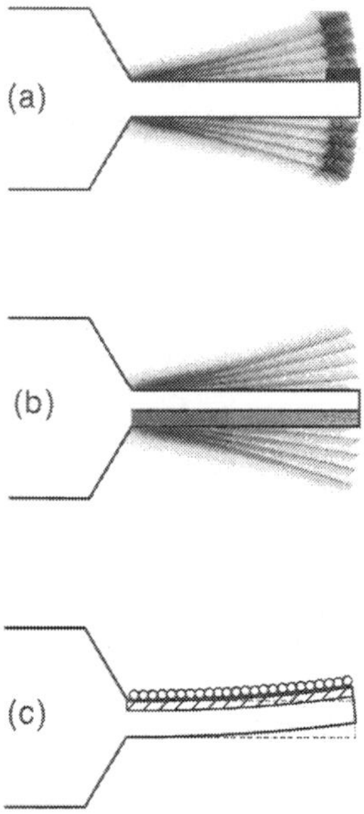

Figure 1. Examples of basic operational modes of a micromechanical cantilever-based sensor. (a) In the *thermogravimeter* application the cantilever is excited to vibrate at its resonance frequency, which is determined by the mass of an attached specimen. (b) The *calorimeter* device can measure permanent and transient temperature variations via the bimetallic effect by detecting the bending of the cantilever beam. (c) Surface stress changes arising from specific molecular adsorption on one side cause a permanent bending of the cantilever beam.

2. Mechanical transduction of chemical processes

2.1 Gravimetric sensor

Cantilever sensors can be applied to monitor mass changes when they are vibrated to resonance, as schematically outlined in Fig. 1a. The cantilever shown is operated as a microbalance to weight and monitor the change of a sample's mass. For a quantitative analysis, two types of coating with a sensitive layer have to be distinguished: continuous coating (CL) on the entire cantilever surface and local coating within a defined area on the cantilever, e.g. towards its apex (AL). When molecules adsorb in a CL configuration, the mechanical properties, such as the stiffness, of the specific sensing overlayer upon adsorption can interfere with the mass change induced resonance frequency shift. Experiments using a gelatine-coated sensor have been shown to be very sensitive to humidity, where adsorption of water increases the resonance frequency of the sensor [3–5], which is in contradiction to increasing mass. This crosstalk to mass loading can be minimized by restricting the sensitive area to a narrowly defined location on the cantilever surface. The mass change, Δm, in an AL configuration is determined by

$$\Delta m = \frac{K}{4\pi^2}\left(\frac{1}{f'^2} - \frac{1}{f^2}\right) \quad , \tag{1}$$

where f is the resonance frequency before adsorption, f' is the resonance frequency during adsorption, and K the sensor spring constant [6]. Restricting the area of the sensitive layer reduces the amount of molecules adsorbed, which results in a smaller mass change and hence a smaller resonance frequency shift of the sensor. To increase the sensitivity, porous material can be attached to the apex of the sensor.

Zeolites are nanoporous materials that are selective to molecular shapes [7] and are used in catalysis [8]. For this purpose, single zeolite crystals with edge lengths of 40 μm $\times$ 25 μm $\times$ 80 μm were attached to the apex of the cantilever, e.g. to sense humidity by measuring the resonance frequency [9]. Given that zeolites have varying pore sizes and adsorption properties, the devices have potential as gas analyzers.

The sensitivity of a cantilever to mass loading is mainly determined by the excited cantilever resonance frequency. Theoretical estimates based on commercially available cantilevers have indicated that their use as microbalances allows sensitivity to mass loadings down to a minimum detectable mass of $\approx 10^{-16}$ g [10].

2.2 Calorimeter

The performance of the cantilever sensor is extended to that of a calorimeter by depositing a layer onto the beam (Λ), which has a different expansion coefficient than the cantilever material (κ). By following the deflection of the apex of the sensor, tiny temperature changes can be detected via the "bimetallic" effect, as outlined in Fig. 1b. Applications of these sensors were first demonstrated by using the device as an ultrasensitive calorimeter involving heat changes at the femtojoule level for investigating surface catalytic reactions [11]. This was achieved by coating the sensor with a thin Pt layer, which catalyzes the exothermic conversion of H_2 and O_2 to form water vapor in high vacuum [11]. The vertical deflection, z, of the sensor is given for a heating at the apex by

$$\Delta z = \frac{t_\kappa + t_\Lambda}{t_\Lambda^2 K} \frac{(\alpha_\kappa - \alpha_\Lambda) l^3 P}{w(\lambda_\kappa t_\kappa + \lambda_\Lambda t_\Lambda)} , \qquad (2)$$

where K is

$$K = 4 + 6\left(\frac{t_\kappa}{t_\Lambda}\right) + 4\left(\frac{t_\kappa}{t_\Lambda}\right)^2 + \frac{E_\kappa}{E_\Lambda}\left(\frac{t_\kappa}{t_\Lambda}\right)^3 + \frac{E_\Lambda}{E_\kappa}\left(\frac{t_\Lambda}{t_\kappa}\right) , \qquad (3)$$

λ_κ, λ_Λ are the thermal conductivities, α_κ, α_Λ are the thermal expansion coefficients, t_κ, t_Λ are the thicknesses of the sensor layers, P is the power generated at the apex, w is the width, and l the length of the sensor [12]. Such calorimeters can be calibrated *in situ* by following the bending of the sensor for a known power transduced into heat at the sensor. Standard Si cantilevers coated with a 0.4-μm layer of Al were found to have a sensitivity down to the picojoule regime and a response time on the millisecond scale [11].

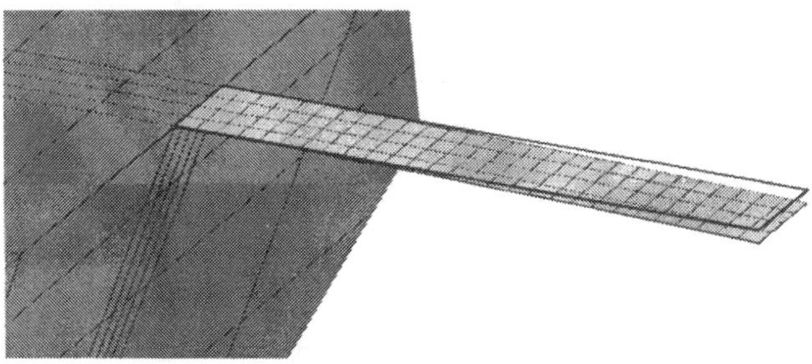

Figure 2. A FE model of the calorimetric sensor, which is $l = 465$ μm, $w = 75$ μm, $t_{Si} = 0.4$ μm and $t_{Al} = 0.6$ μm, $t_{Al}(t_{Si} + t_{Al})^{-1} = 0.6$. The sensor is bent towards the Si layer due to a uniform temperature increase of 2 °C.

Optimizing the material properties and the geometry of such sensors, one obtains a sensitivity in the attojoule range and a thermal response of milliseconds [13,14]. These micromechanical calorimetric devices are able to detect temperature variations as small as 10^{-5} K.

Finite-element (FE) simulations eliminate the need to perform detailed experimental optimization of device characteristics and help optimize material combinations to achieve the best device performance. The geometry of a Si-based cantilever calorimeter clamped rigidly at one end is shown as a FE model in Fig. 2. FE simulations [15] were conducted to find the maximum bending of the sensor as a function of the layer thickness ratio between Si and a deposited material for Al, polystyrol (PS), and polyethylen (PE) (Table I).

Table I. Expansion coefficients and Young's modules used for FE simulations. The thickness ratio between a Si cantilever and a deposited layer—either Al, PS, or PE— $t_\Lambda(t_{Si}+t_\Lambda)^{-1}$ is calculated, which produces the maximum deflection change Δz upon a temperature change. The latter value is then listed for a temperature increase of 2 °C.

Λ	α (10^{-6} K^{-1})	E (GPa)	$t_\Lambda(t_{Si}+t_\Lambda)^{-1}$	Δz (μm)
Si	2.33	170	0	0
Al	23.9	80	0.6	7
PS	75	3.2	0.87	24
PE	200	0.5	0.95	66

From the simulations one can see that choosing PE, which has an 8.3× higher expansion coefficient than Al, results in an approximately 9× higher deflection for the same temperature increase.

The principle of bimetallic cantilevers has been extended to allow photothermal spectroscopy with a sensitivity of 150 fJ on picogram layers of materials [16]. In addition, enthalpy changes associated with phase transitions in nanogram samples of alkanes were detected and determined quantitatively [17]. This technique extends the analysis of small quantities of material to the nanogram realm, which is inaccessible to standard thermal analysis, where size-dependent properties of samples can be studied [18,19]. Masses of the attached samples can be determined simultaneously from a resonance frequency analysis of the sensor as described in Subsect. 2.1. The use of these sensors is generally not limited to a specific medium; they can also be operated successfully in liquid environments [20].

2.3 Detection of surface stress changes

Compared to transient thermal signals in heat detection, stress in an overlayer causes a permanent bending of the cantilever (Fig. 1c). Stress changes arising from volume changes of attached samples, associated with phase transitions, have been detected [17]. The magnitudes of bending signals associated with structural phase transitions in alkanes were measured to be ≈20 times higher than that of bimetallic actuation arising from the transition heat. The surface stress change $\sigma_s = \sigma t_f$, due to an overlayer of thickness t_f, can be determined from the cantilever curvature radius R using Stoney's formula [21]

$$\sigma_s = \frac{E_\kappa t_\kappa^2}{6R(1-\nu_\kappa)} \quad , \tag{4}$$

where ν_κ is Poisson's ratio of the sensor material. This transduction mechanism requires specific coating of the cantilever on one side, which binds either only specific molecules or a general class of molecules. Such sensitive coatings can be tailored to bind molecules in a preferential and specific manner. Promising candidates for designer surfaces are based on alkanethiol molecules, which are known to self-organize into well-ordered, densely packed films [22–24]. Surface stress changes were measured *in situ* during the self-assembly of alkanethiols on gold using a micromechanical sensor [25]. Self-assembling monolayers (SAMs) of methyl-terminated alkanethiol, $CH_3(CH_2)_{n-1}\mathbf{CH_3}$, can be substituted by an ω-functionalized mercaptohexadecanoic-acid SAM, $CH_3(CH_2)_{n-1}\mathbf{COOH}$, on gold. The principle of ω-functionalized molecules allows the detection of specific molecules by docking a guest molecule at the monolayer–air interface via a lock-and-key mechanism [25–27]. When the sensor is used as an electrode in an electrochemical environment it is possible to determine the deposition and removal of a fraction of a monolayer of material using the principle of detecting surface stress changes [28,29].

In the limit of the smallest detectable bending of standard-sized, commercially available cantilever sensors, surface stress changes of 10^{-7} Nm^{-1} can be probed. This corresponds to a change of zepto (10^{-21}) molar quantities of dodecanethiol [25]. A further reduction of the sensor thickness into the nanometer regime [30] promises detection limits of even smaller quantities.

3. Integration of micromechanical sensors

Using the high sensitivity, short response time, and tiny size and the technique to fabricate cantilever-based sensors, Nanotechnology olfactory sensor (NOSE) arrays [31] of micromechanical cantilever sensors have been fabricated. The surface of each cantilever can be covered by specific layers, e.g. a polymer, an ω-functionalized SAM, or microporous materials. This enables one to detect different interaction strengths with analytes and consequently permits a complex analysis of a variety of signal domains. The various operating modes provide different information due to several transduction mechanisms of a given analyte with the receptor layer, such as mass, heat, or surface stress changes, as described in the previous sections.

3.1 Microfabrication process steps for cantilever arrays

The fabrication process of thin cantilever arrays for NOSE applications uses an n-type silicon wafer with an epitaxially formed, boron-doped p-layer on top of it (Fig. 3a). First, 0.3 μm of a silicon nitride film is deposited on the wafer backside using plasma-enhanced chemical vapor deposition (PECVD). On the topside, a photoresist layer is lithographically structured, and the cantilever patterns are etched into the silicon using reactive ion etching (RIE). The depth of this etch step is about 2 μm and traverses the p-layer (Fig. 3b). The backside is structured with resist using double-side alignment, and the patterns are then used to open the windows in the etch mask (Fig. 3c). The wafer is mounted in a single-side, wet-etch chuck [32] to protect its topside, and the n-type silicon is etched using 60 °C, 40% potassium-hydroxide solution (KOH). The etch progress is controlled visually and stopped at about 5–10 μm membrane thickness (Fig. 3d). The remaining p-silicon layer provides an excellent etch stop for the final wet chemical etching in EDP [33] and enables very thin cantilevers to be homogeneously processed on a full-wafer scale (Fig. 3e). After careful rinsing and drying, the individual chips can be removed from the wafer frame. The resulting cantilevers have a typical thickness of 1 μm, depending on the initial p-layer thickness and the EDP etch time.

Using complementary metal oxide semiconductors (CMOS) [34] technologies and processes, these cantilever-based micromechanical sensors can be fabricated together with microelectronic components on a single chip. This should enable low-cost seamless sensors to be fabricated for the integration of NOSE with computer technology.

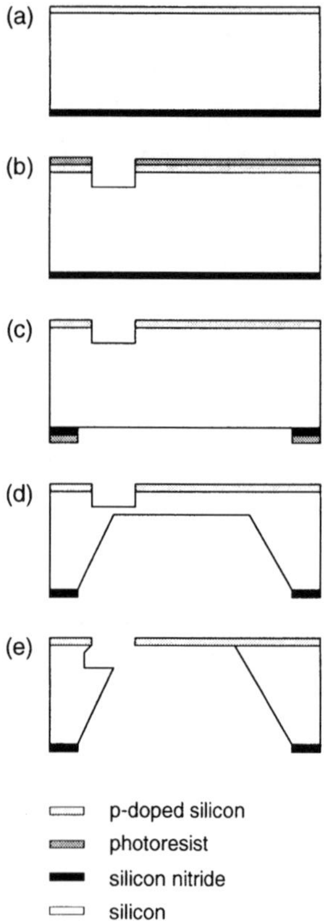

Figure 3. Fabrication sequence of cantilever array sensors for NOSE.

3.2 NOSE module

All operating modes and transduction principles outlined here can be employed simultaneously on a single micromechanical device to obtain "orthogonal" information. Orthogonality implies that signals from different analytes, although they produce similar responses in one operating mode, may produce a different response in another operating mode. Upon analysis of the response of various modes it is possible to gather the information required to distinguish between similar but non-identical chemical compounds.

The deflection of individual micromechanical sensors can be read out by using the beam-deflection technique, which is commonly used in SFM and is able to detect picometer deflections [35]. In our setup (Fig. 4) the light emitted from a light-emitting diode (LED) [36] is guided into a fiber ribbon (FR) consisting of eight individual multimode fibers [37]. A condensor-type focusing optics (L) situated at the entry window of the sampling chamber is used to collimate and focus the light onto the apex of each cantilever within the sensor array (AS). The reflected light impinges upon a position-sensitive detector (PSD) [38]. This causes a photo current, which is distributed between two opposite contacts corresponding to photo currents y_1 and y_2. Then the currents are converted into voltages, and the difference of both voltages are normalized by their sum.

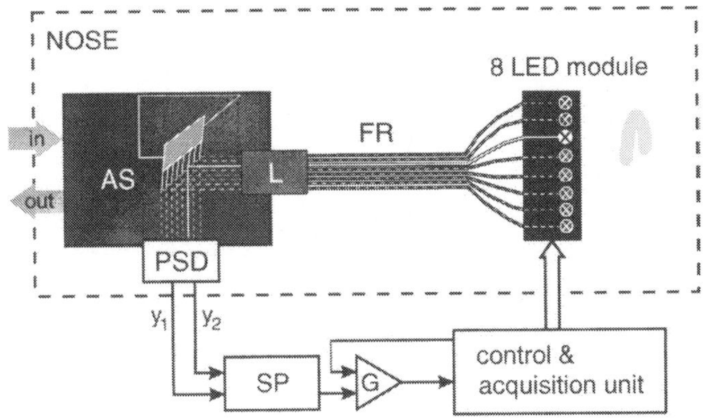

Figure 4. Schematic drawing of a NOSE module. Outlined are eight LEDs, one of which is switched on (the third from the top). The light is guided into a fiber ribbon (FR). Its end is connected to the lens system (L) of the NOSE chamber. The light of each LED is focused onto the corresponding cantilever of the array sensor (AS). After reflection on the AS, the light is collected by a position-sensitive detector (PSD) and converted into voltages by a signal preprocessor (SP). These components represent a NOSE module. The output signal is amplified (G) and fed into the control and acquisition unit, which adjusts the offset for optimal amplification of the signal (offset tune). 'In' and 'Out' arrows represent in- and outlets for the analyte.

3.3 Data acquisition

To read out a large number of cantilever sensor elements individually, time multiplexing was used [39,40]. The eight LED modules can be switched on and off individually in arbitrary time intervals by use of dedicated control electronics (Fig. 5). The light reflected by each cantilever within the array causes laterally separated light spots on the PSD. This causes an individual dc offset of the PSD signal, which can be eliminated by a compensation signal (offset tune) to achieve optimal amplification. The position of the incident light, and thus the cantilever deflection, can be determined. After analog–to–digital conversion, the 16-bit data is temporarily stored on a micro-controller board (μCB) and then transferred periodically to a personal computer.

Different NOSE modules can be accessed (Fig. 5) via memory mapping in the address space of a μCB. Each cantilever sensor represents a memory cell. This permits the use of standard memory accessing schemes to address each cantilever individually. Owing to the sequential readout of the sensors' deflection and the discrete time intervals of sampling, aliasing arises. This means that data must be low-pass filtered to cut off frequencies above $f_S/2n$ prior to acquisition in order to fulfill the Nyquist theorem. The term f_S denotes the sampling frequency of the analog–to–digital converter and n is the number of sensors. If data is subjected to additional arithmetic operations (e.g. subtraction of sensor responses) the temporal shift of the data points resulting from time multiplexing must be accounted for. To be able to subtract corresponding data points the curves must be interpolated. In a further step, signals from all sensor elements will be analyzed by employing neural network analysis.

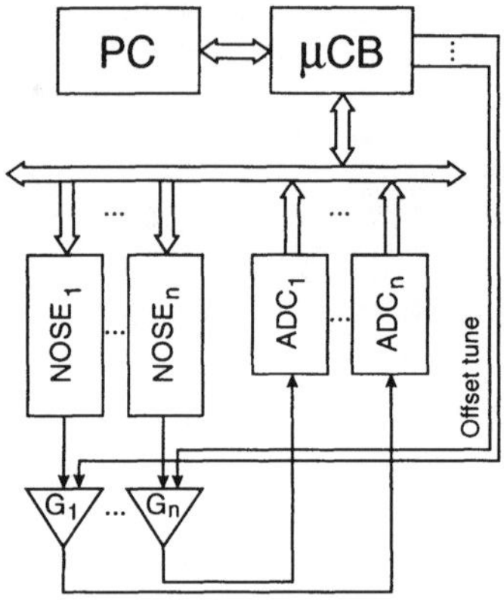

Figure 5. Several NOSE modules ($NOSE_1$, ..., $NOSE_n$) can be used in a NOSE system, each using its own amplifier (G_1, ..., G_n) and analog-to-digital conversion unit (ADC_1, ..., ADC_n). NOSE modules and ADCs are controlled via a bus by a microcontroller board (μCB). The μCB adjusts the offset of the amplifiers (G) and transfers data to a personal computer (PC).

Experiments with the NOSE module operated to detect surface stress changes have revealed that, compared to single sensors, mechanical disturbances from noise such as vibrations or turbulent gas flow can be more efficiently removed with array sensors by recording the differential responses between sensors and reference cantilevers [39,40]. This unique feature of arrays opens up the range of environments in which the sensor system can operate.

4. Outlook

The transduction principles presented here are by no means complete [2]. Further techniques, perhaps using more specifically designed microstructures, will extend the number and range of signal domains transduced into micromechanical deflection. In particular the shape of the micromechanical structure, which can be arbitrarily chosen, provides room for several orders of magnitude of device performance.

The advantage of such array devices in process control or online chemical and physical analysis is not only their small size ($\approx$1 mm^3) but also their low cost and insensitivity to external disturbances. This allows disposable sensor arrays to be integrated seamlessly into silicon technology for applications such as biochemical analysis or drug testing. Such devices operate in various media with high speed and sensitivity, and offer applications, for example, to oenology or fragrance design.

Acknowledgment

We gratefully acknowledge M. Baller, H. Biebuyck, E. Delamarche, W. Press, and P. Guéret for their support and discussions. In addition, we thank A. Koll, D. Lange, and H. Baltes for useful discussions. This work was supported by the Swiss National Science Foundation (NFP 36) and was partially funded by the Swiss Priority program MINAST Project 7.04 "NOSE".

[1] G. Binnig, C.F. Quate, and Ch. Gerber, Phys. Rev. Lett., 56 (1986) 930.
[2] R. Berger, H.P. Lang, Ch. Gerber, and J.K. Gimzewski, Microelectronic Engineering, 35 (1997) 373.
[3] T. Thundat, R.J. Warmack, G.J. Chen, and D.P. Allison, Appl. Phys. Lett., 64 (1994) 2894.
[4] E.A. Wachter, and T. Thundat, Rev. Sci. Instrum., 66 (1995) 3662.
[5] T. Thundat, E.A. Wachter, S.L. Sharp, and R.J. Warmack, Appl. Phys. Lett., 66 (1995) 1695.
[6] The spring constants were calculated from the cantilever geometry, which was determined from scanning electron micrograph images. For details see E. Meyer, R. Lüthi, L. Howald, M. Bammerlin, M. Guggisberg, and H.-J. Güntherodt, in Micro/Nanotribology and Its Applications, B. Bhushan (ed.), NATO ASI Series E: Applied Sciences, Vol. 330, Kluwer, Dordrecht, 1997, p. 193.
[7] T. Bein, Chem. Mater., 8 (1996) 1636.
[8] H. van Bekkum, E.M. Flanigen, and J.C. Jansen, Eds., Introduction to Zeolite Science and Practice, Studies in Surface Science and Catalysis, Vol. 58, Elsevier, Amsterdam, 1991.
[9] L. Scandella, G. Binder, T. Mezzacasa, J. Gobrecht, R. Berger, H.P. Lang, Ch. Gerber, J.K. Gimzewski, J.H. Koegler, and J.C. Jansen, submitted to Microporous Materials (1997).
[10] P. Oden, T. Thundat, and R.J. Warmack, submitted to Rev. Sci. Instrum. (1997).
[11] J.K. Gimzewski, Ch. Gerber, E. Meyer, and R.R. Schlittler, Chem. Phys. Lett., 217 (1994) 589.
[12] J.R. Barnes, R.J. Stephenson, C.N. Woodburn, S.J. O'Shea, M.E. Welland, T. Rayment, J.K. Gimzewski, and Ch. Gerber, Rev. Sci. Instrum., 65 (1994) 3793.
[13] E. Meyer, J.K. Gimzewski, Ch. Gerber, and R.R. Schlittler, in The Ultimate Limits of Fabrication and Measurement, M.E. Welland and J.K. Gimzewski (eds.), NATO ASI Series, Vol. 292, Kluwer, Dordrecht, 1995, p. 89.
[14] J. Lai, T. Perazzo, Z. Shi, and A. Majumdar, Sensors and Actuators A 58 (1997) 113.
[15] The ANSYS finite-element package was used to model the micromechanical cantilever. The cantilever was defined using a two-layered structural shell elements and was divided by uniform squares. The element size was reduced until convergent, consistent results were obtained (1120 elements for this FE analysis).
[16] J.R. Barnes, R.J. Stephenson, M.E. Welland, Ch. Gerber, and J.K. Gimzewski, Nature, 372 (1994) 79.
[17] R. Berger, Ch. Gerber, J.K. Gimzewski, E. Meyer, and H.-J. Güntherodt, Appl. Phys. Lett., 64 (1996) 40.
[18] W. Hemminger and G. Höhne, Calorimetry, VCH, Weinheim, 1984.
[19] S.L. Lai, J.Y. Guo, V. Petrova, G. Ramanath, and L.H. Allen, Phys. Rev. Lett., 77 (1996) 99.
[20] R. Berger, Ch. Gerber, J.K. Gimzewski, E. Meyer, and H.-J. Güntherodt, in Proc. Micro-Electro-Mechanical Systems (MEMS), Symposium on Heat Transfer – Thermomechanical Sensors, DSC-Vol. 59, ASME 1996, p. 49.
[21] G.G. Stoney, Proc. R. Soc. London Ser. A, 82 (1909) 172.
[22] R.G. Nuzzo, and D.L. Allara, J. Am. Chem. Soc., 105 (1983) 4481.
[23] C.D. Bain, E.B. Troughton, Y.-T. Tao, J. Evall, G.M. Whitesides, and R.G. Nuzzo, J. Am. Chem. Soc., 111 (1989) 321.
[24] G.M. Whitesides, C.T. Seto, and J.P. Mathias, Science, 254 (1991) 1312.

[25] R. Berger, E. Delamarche, H.P. Lang, Ch. Gerber, J.K. Gimzewski, E. Meyer, and H.-J. Güntherodt, Science, 276 (1997) 2021.

[26] K.D. Schierbaum, T. Weiss, E.U. Thoden van Velzen, J.F.J. Engbersen, D.N. Reinhoudt, and W. Göpel, Science, 265 (1994) 1413.

[27] B.-H. Huisman, R.P.H. Kooyman, F.C. van Veggel, and D.N. Reinhoudt, Adv. Mater., 8 (1996) 561.

[28] W. Haiss and J.K. Sass, J. Electroanal. Chem., 386 (1995) 267.

[29] S.J. O'Shea, M.E. Welland, T.A. Brunt, A.R. Ramadan, and T. Rayment, J. Vac. Sci. Technol. B, 14 (1996) 1383.

[30] J. Brugger, G. Beljakovic, M. Despont, N.F. de Rooij, and P. Vettiger, Microelectronic Engineering, 35 (1997) 373.

[31] This work is partially performed within the Swiss Priority Program in Micro- and Nano-System Technology (MINAST), No. 7.04 NOSE.

[32] J. Brugger, G. Beljakovic, M. Despont, H. Biebuyck, N.F. de Rooij, and P. Vettiger, submitted to Sensors and Actuators (1997).

[33] N.F. Railey, Y. Sugiyama, and T. van Duzer, J. Electrochem. Soc., 131 (1984) 162.

[34] H. Baltes and C.A. Leme, in Micro- and Nanosensor Technology/Trends in Sensor Markets, W. Göpel, J. Hesse, J.N. Zemel (eds.), Vol. 8, VCH, Weinheim, 1995.

[35] G. Meyer, N.M. Amer, Appl. Phys. Lett., 57 (1990) 2089.

[36] Optical transmitter LED module LSEST 02P, dominant wavelength $\lambda = 660$ nm, Fischer Elektronik, Nottebornstrasse 28, Postfach 1590, D–58511 Lüdenscheid, Germany.

[37] Europtics Ltd., Whiston, Merseyside, U.K., Assembly of fiber array of eight multimode fibers (core diameter 62.5 μm) and MT8/ST connectors from NTT corporation, Japan.

[38] Two-dimensional PSD 2L4, Laser Components, Olching/Munich, Germany.

[39] H.P. Lang, R. Berger, C. Andreoli, J. Brugger, M. Despont, P. Vettiger, Ch. Gerber, J.K. Gimzewski, J.P. Ramseyer, and E. Meyer, Appl. Phys. Lett. (1998) in press.

[40] H.P. Lang, R. Berger, F. Battiston, J.P. Ramseyer, E. Meyer, C. Andreoli, J. Brugger, P. Vettiger, M. Despont, T. Mezzacasa, L. Scandella, H.-J. Güntherodt, Ch. Gerber, and J.K. Gimzewski, Appl. Phys. A (1998) in press.

MEASUREMENTS ON SINGLE DNA MOLECULES

B.G. de Grooth, A. Agronskaya, Bennink, S.J.T. van Noort, K.O. van der Werf, J. Greve
Department of Applied Physics, University of Twente
P.O. Box 217, 7500 AE Enschede, The Netherlands
(e-mail: b.g.degrooth@tn.utwente.nl)

0. Abstract
All our understanding of how DNA works is based on evidence obtained with bulk methods. Moreover, information on the detailed conformation of DNA and interactions between DNA and proteins at a high resolution, have been obtained almost entirely under highly non-physiological conditions (vacuum, low temperature etc.). Recently several methods have become available that make it possible to study individual bio-molecules under physiological conditions. We will describe three of these methods and there application to DNA research: flow analysis of individual molecules, measuring mechanical forces of bio-molecules using optical tweezers and using an AFM to study the dynamics of DNA and DNA protein interactions.

Keywords: optical tweezers, DNA overstretching, denaturation, AFM, photolyase

1. Introduction

Probably DNA is the most important molecule for life on Earth. Although we precisely know how the genetic information is stored in DNA, little is known on the functional aspects of DNA. In a cell, the DNA is condensed by a factor of 10,000, in a very well organized way so that the important genes are still available for transcription. It is of great importance to understand how this is achieved. Numerous proteins are involved in this. Besides such structural aspects, various active processes take place on the DNA molecules, such as replication, gene transcription, damage recognition and damage repair. For all these processes, highly specialized proteins are active, and although our knowledge of these processes is rapidly improving, we are still at the beginning of our understanding. For an excellent introduction on this subject we refer to the book of Calladine and Drew [1]. DNA and DNA-protein interactions have been studied with a variety of methods, including various forms of spectroscopy, gel electrophoresis, and very importantly, X-ray diffraction. Although these techniques have yielded very important information, all these techniques are bulk techniques and thus represent the average properties of an ensemble of molecules. Recently, new technologies have been developed that enable us to study individual molecules under physiological conditions. In this article we will describe three examples that are currently studied in our laboratory.

2. Results and Discussion

2.1 Analysis of single molecules in flow

This technique is based upon a flow cytometer [2], that is optimized for the measurements of extremely weak fluorescent signals (see figure 1). The individual particles that will be measured are dispersed in liquid and are forced to flow in the middle of a small flow cell. This is achived using the principle of hydrodynamic focussing [3]. In the flow cell the particles are illuminated by a focused laser beam and the fluorescent light of the particles is measured. Individual DNA molecules can be measured in this way provided that they are labeled fluorescently with strong dyes such as TOTO-1 [4]. This intercalating dye stains DNA base-independently in a stoichiometric manner, so that the fluorescent signal is proportional to the length of the molecule. In figure 2 a typical example of a measurement is shown. In this case the sample consists of a particular DNA molecules, the so called λ DNA isolated from a *E. coli* bacteriophage, that is cut in a specific fragments using the enzyme Hind-III. This enzyme cuts DNA at places where the sequence is double AA. In this way fragments characteristic for the DNA are obtained. Thus the size distribution yields a 'fingerprint' of the original DNA molecule.

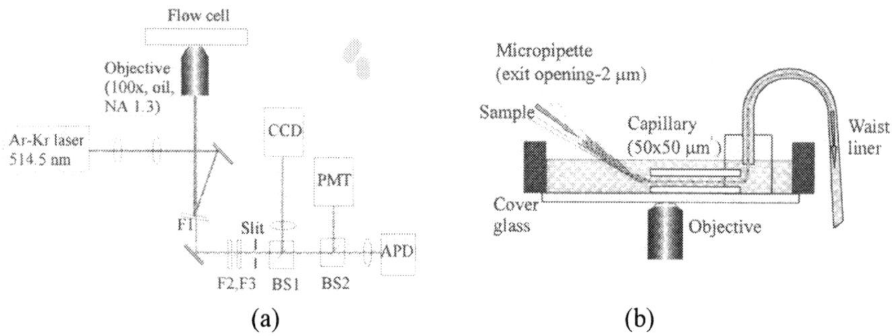

Figure 1. *Experimental set-up (a) to measure individual DNA molecules in a hydrodynamic focussing flow cell (b) CCD: image intensified CCD camera; PMT: photomultiplier; APD: avalanche photodetector.*

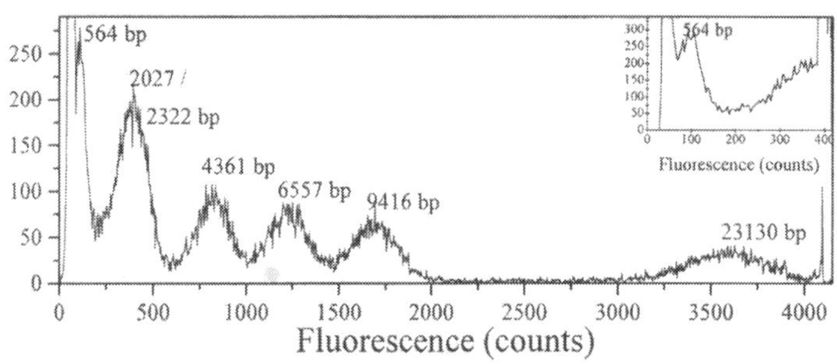

Figure 2. *Histogram of the fragments of λ DNA molecules after it was cut into pieces by the enzyme Hind-III.*

This method can be used for DNA sizing of DNA fragments ranging from 1.5 to 167 kbp (thousand base pairs). The lower limit is determined by the background signal from the water and can be improved by pre-bleaching the water, discriminating Raman scattering from fluorescence using time gated fluorescence measurements [5] and by decreasing the measuring volume. If the measurements can be automated and the sample handling can be improved, the method could be a alternative for gel electrophoresis with greatly improved analysis time, since the analysis can be performed in a few minutes. Extending this analysis to more parameters, e.g. measuring a specific DNA probe equipped with a fluorescent label that can be spectrally separated from the TOTO, could open a new area of DNA analysis including rare event detection of specific genes.

2.2 Measuring mechanical properties of individual DNA molecules

In 1986 Ashkin et al [6] showed that small micron sized beads could be trapped at a stable position using a tightly focused laser beam. The technique, referred to as optical trapping or optical tweezers, has been successfully applied on a variety of biological samples [7,8]. Recently, Smith and coworkers have used this technique to measure the mechanical stretching properties of single DNA molecules [9]. Here we describe similar experiments done in our laboratory (figure 3).

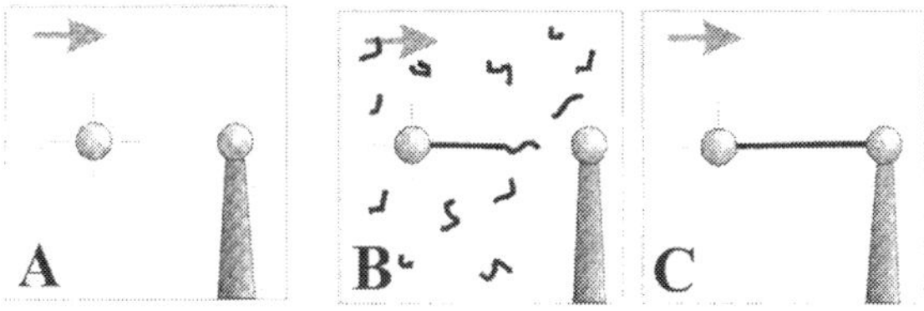

Figure 3. *Individual DNA molecules are connected to two different polystyrene spheres of 2 micron diameter using an optical trap and a glass micropipette with a 1 micron tip diameter. A: One streptavidin coated bead is trapped in the optical trap (indicated by 4 lines) and a second is hold by the glass micropipette. B: Buffer containing biotinylated DNA molecules is flown in the flow cell and one single molecule hits the trapped bead and attaches to it with one end. This can be detected by an increase of the force exerted on the trapped by due to the increased drag force. C: The second bead (on the pipette) is moved toward the open end of the DNA molecule to finish the system.*

The λ-DNA molecules which consist of 47,502 base pair are modified such that both ends of the molecules are biotinylated. Polystyrene beads with a diameter of 2 μm are flown into a small flow cell where a single bead is trapped using the optical trap. This bead is then transferred to the glass micropipette. A second bead is captured in the trap and the other beads are flown out of the flow cell. The buffer is now changed to a buffer containing the biotinylated DNA. A single DNA molecule in flow will hit the trapped bead and attach to it with one end. The next step is to 'fish' around with the bead on the glass micropipette, in order to get the open end attached to this bead. Now the system is ready to measure a force extension curve of an individual DNA molecule by moving the micropipette with respect to the laser trap position. The extension of the DNA molecule is determined as the distance between the beads using microscopic imaging. The force exerted on the DNA molecule can be obtained by measuring the small displacement of the trapped bead out of the laser trap. After calibration this directly yields the absolute force exerted on the molecule. The resulting curve (figure 4) shows an unexpected behavior of the elasticity properties of the DNA molecule. The behavior up to 18 μm is rather straight forward. Up to 16 micron the DNA is not completely stretched and the behavior can be described taking the entropy of the DNA molecule into account [10]: a fully stretched DNA molecule is subject to thermal fluctuations, which cause it to decrease it length. Therefore, a force is necessary to extent the DNA just to its full length of 16.4 μm. From this value up to 18 ì m, the DNA behaves as an elastic spring.

At a length of 18 micron, and a force of 65 pN, the molecule suddenly can be extended further to about 170 % of its normal contour length with only very minor effort. Stretching beyond this point the DNA eventually again behaves like a spring. The entire curve is not completely reversible. If we relax the same molecule again the force at certain lengths will be lower than when we stretched the molecule. This indicates that the conformation change of the molecule is not completely reversible. Different models for the structural conformation change has been proposed, but the exact nature of the overstretched conformation is not known yet [9,11].

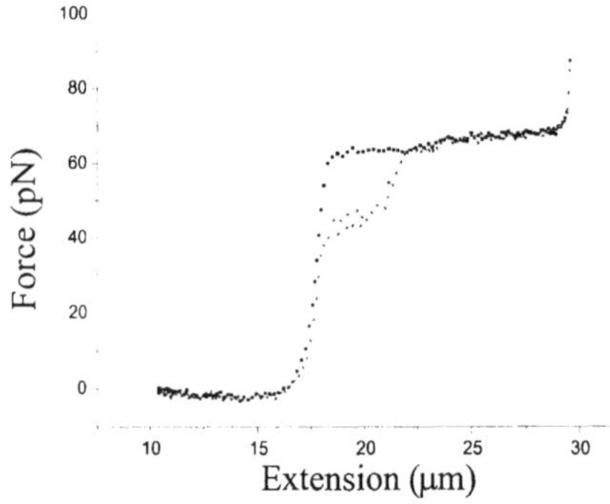

Figure 4. *Force extension curve of a double-stranded DNA molecule in buffer with 0.1 M NaCl. Black curve: Stretching. Gray curve: relaxing*

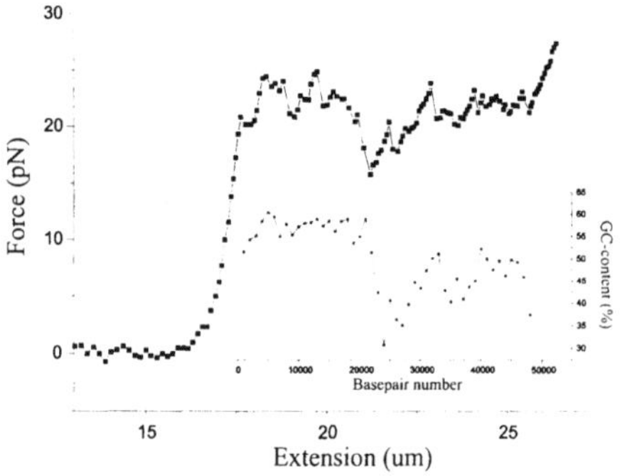

Figure 5. *Force extension curve of a double-stranded DNA molecule at low salt concentration. Insert: GC-content of λ-DNA as a function of position.*

If DNA molecules are brought into a buffer with a lower salt concentration, it is known that the DNA melting temperature decreases. The helix structure of the molecule is not stabilized anymore. We now build the system as described above and changed the buffer to distilled water in which no ions are present, and repeated the stretching of the molecule. The force-extension curve obtained is shown in figure 5. Up to a force of 20 pN the curve seems to be the same as in 100 mM Na Cl TE buffer, but after this the curve shows a very irregular behavior. This is interpreted as the actual dissociation ('melting') of the two strand of the DNA molecule. The 'melting' part of the curve seems very noisy, and arbitrary, but when we compare this with the GC-content of the DNA molecule, we observe a striking correlation (figure 5, insert).

2.3 Atomic Force Microscopy of DNA protein interaction

A very powerful method to study single molecules is the atomic force microscope, AFM. For biological applications, the ability to obtain images with a resolution about 20 nm seems to be extremely useful. However, under normal operation conditions, the tip will just move the biological molecules laterally across the surface. It was only after it was demonstrated that the AFM can also work with the much more gentle tapping mode in liquid that delicate biological samples could be studied [11]. We have shown that by careful optimizing the conditions of the sample and the instrument, the dynamic behavior of DNA molecules loosely bound to a mica surface, and the interaction with a individual protein molecules can be visualized with a series of AFM images. This is illustrated by the series of images below.

As AFM is a surface technique, the reactions have to proceed on a surface. In fact a paradoxical problem has to be overcome. To be able to image DNA reproducibly and with high resolution it is necessary that the DNA is attached to the surface. Otherwise the DNA molecules would be swept away when the tip taps on the molecules and the image would be severely distorted. However, to be available for interaction with other molecules, DNA must be free from the surface to avoid steric hindrance that might affect the reactions. Thus DNA should be immobilized just firmly enough to allow both imaging with AFM and interaction with proteins.

In this study formation of non-specific complexes of dsDNA with photolyase is presented. Photolyase is a well-studied enzyme responsible for the removal of thymine dimers in DNA. The crystal structure was resolved [12,13] and structural features have been related to results from photochemistry. Although the reaction has been extensively described biochemically, AFM can contribute to a better understanding of the dynamic aspects of the reaction such as the process of locating the damaged site.

Freshly cleaved mica discs (Ted Pella, Ca) were used as a substrate for immobilized DNA. Undamaged 500 bp dsDNA (Gensura, Ca), obtained by digesting plasmid DNA with EcoRI, was diluted to a final concentration of 2 ng/μl in a buffer containing 4 mM Hepes, 1 mM NaCl and 1 mM MgCl2, pH 6.5. The buffers were made in MilliQ filtered deionized water. Photolyase, obtained from Anacystis Nidulans, [13,14] was added to a final concentration of 0.12 ng/μl in the reaction buffer.

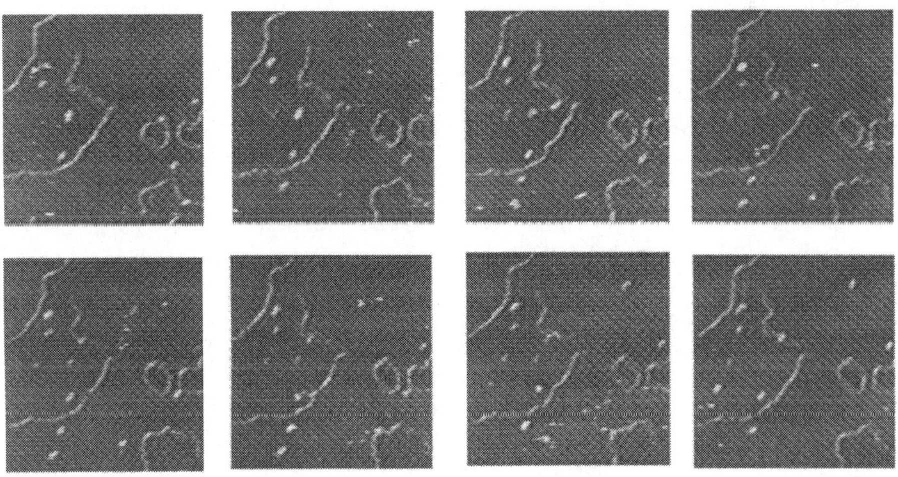

Figure 6. *Eight frames from a 30 minutes sequence of topography images obtained by tapping mode AFM at 1 frame per minute. The scan area was 0.5 x 0.5 μm^2, z-range 4 nm.*

After 10 minutes at room temperature 5 μl of the reaction mixture was deposited on mica. Directly after deposition the AFM was mounted over the sample and after a minute the liquid cell of the AFM was thoroughly rinsed with Hepes buffer, without DNA and photolyase. During sample preparation, mounting of the AFM and measurements, bio-molecules remain in buffer and are never dehydrated in order to keep them functional.

Figure ## shows 8 frames from a typical time series measurement of a photolyase DNA mixture. In this measurement the surface was scanned during 30 minutes at 1 frame per minute. During the 30 minutes of scanning no damage to the sample was visible, while the RMS roughness in the height of the atomically flat mica amounted less than 0.2 nm.

Acknowledgments
This work was supported in part by the Dutch organization of for Fundamental Research on Mater (FOM).

References

1. C.R. Calladine and H.R. Drew, Understanding DNA, Academic Press, San Diego, 1997
2. P.M. Goodwin, M.E. Johnson, J.C. Martin, W.P. Ambrose, B.L. Marrone, J.H. Jett, R.A. Keller, Rapid sizing of individual fluorescently stained DNA fragments by flow cytometry. Nucleic Acids Res 21 (1993) 803-806
3. V. Kachel, Fellner-Feldegg, E. Menke, Hydrodynamic properties of flow cytometery instruments. In: Flow Cytometry and Sorting, Second Edition, M.R. Melamed, T. Lindmo, M.L. Mendelsohn (eds). John Wiley & Sons, Inc., New York, (1990) 27-44
4. H.S. Rye, S. Yue, D.E. Wemmer, M.A. Quesada, R.P. Haugland, R.A. Mathies, A.N. Glazer, Stable fluorescent complexes of double-stranded DAN with bis-intercalating asymmetric cyanine dyes: properties and applications. Nucleic Acids Res 20 (1992) 2803-2812
5. R.A. Keller, W.P. Ambrose, P.M. Goodwin, J.H. Jett, J.C. Martin, M. Wu, Single-molecule fluorescence analysis in solution. Appl Spect 50 (1996) 12A-32A
6. A. Ashkin, J.M. Dziedzic, Optical trapping and manipulation of viruses and bacteria, Science 235 (1987) 1517-1520
7. S.C. Kuo, M.P. Sheetz, Optical tweezers in cell biology. Trends in Cell Biology 2 (1992) 116-118
8. S. Smith, Y Cui, C. Bustamante, Overstretching B-DNA: The elastic repsonse of individual doublestranded and singlestranded DNA molecules, Science 271 (1996) 795-799
9. C. Bustamante, J.F. Marko, E.D. Siggia, S. Smith, Entropic elasticity of lambda phage DNA, Science 265 (1993) 1599-1600
10. P. Cluzel, A. Lebrun, C. Heller, R. Lavery. J.-L Viovy, D. Chatenay, F. Caron, DNA: an extensible molecule, Science 271 (1996) 792-794
11. C.A.J. Putman, K.O. van der Werf, B.G. de Grooth, N.F. van Hulst and J. Greve, tapping mode atomic force microscopy in liquid, Appl. Phys. Lett. 64 (1994) 2454-2456
12. H.-W. Park, S.-T Kim, A. Sancar, and J. Deisenhofer, Crystal structure of DNA photolyase from Escherichia coli. Science 268 (1995) 1866-1872.
13. Tamada, T., K. Kitadoro, Y. Higuchi, K. Inaka, A. Yasui, P.E. de Ruiter, A.P.M. Eker, and K. Miki, Crystal structure of DNA photolyase from Anacystus nidulans. Nature Stuct. Biol. 4 (1997) 887891.
14. A.P.M. Eker, P. Kooiman, J.K.C. Hessels, and A. Yasui, DNA photoreactivating enzyme from the cyanobacterium Anacystis nidulans. J. Biol. Chem. 265 (1990) 8009-815.

NOVEL INTEGRATED OPTICAL REFRACTOMETER

G.J. Veldhuis and P.V. Lambeck
MESA Research Institute, University of Twente
P.O. Box 217, 7500 AE Enschede, The Netherlands
e-mail: g.j.veldhuis@el.utwente.nl

0. Abstract

A passive integrated optical waveguide refractometer based on bendloss changes in a spiral-shaped channel waveguide is proposed and fabricated in Si_3N_4/SiO_2-technology. The experimentally determined performance and the theoretical predictions, based on a simple model for so called 'Whispering Gallery Modes', agree very well. The minimum detectable refractive index change was measured to be $\Delta n=8\cdot 10^{-6}$.

Keywords: refractometry, integrated optics, sensors.

1. Introduction

Integrated optical sensors for measuring chemical or physical quantities show interesting prospects because of their high sensitivity, small size, low price and the absence of electro-magnetic interference. In the last decade many types of integrated optical sensors have been proposed and fabricated, where devices based on a Mach-Zehnder interferometer are generally considered as the most promising, due to their high sensitivity and intrinsic balancing [1]. Unfortunately, to exploit the full sensitivity potential, these devices often require on-chip modulation to tackle the phase ambiguity in the interference signal [2]. In this paper we introduce a novel passive waveguide refractometer [3], where refractive index changes induce a change in the bendloss of a spiral-shaped channel waveguide. Here no on-chip modulation is required. Measurements on a device fabricated in Si_3N_4/SiO_2-technology are presented and its performance is compared with theory.

2. Sensing principle

The losses of a bent optical waveguide increase strongly with a decrease of either the bending radius or the effective refractive index contrast defining the waveguide [4]. In the sensor we propose a single mode channel waveguide has been used which has a spiral shape as illustrated in figure 1a. The radius of curvature R of the waveguide decreases in the propagation direction so for a given effective refractive index contrast $\Delta N=(N_g-N_b)$, where N_g and N_b are the refractive indices of the waveguide and the background respectively, the losses will gradually increase in the propagation direction and at a certain point in the spiral all the light will be lost by the waveguide. The sensor principle is based on the shift of this point in the propagation direction caused by an increasing lateral refractive index contrast. In this way the depth of propagation of the light into the spiral is a direct measure for the refractive index contrast.

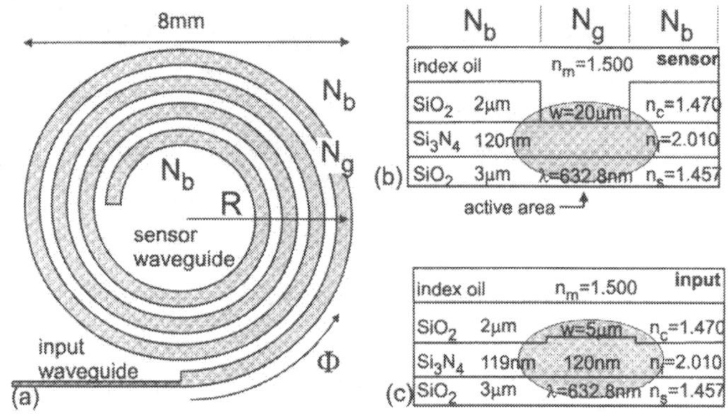

Fig. 1: Schematic representation of the refractometer, with a top-view of the spiral-shaped channel waveguide (a), a side-view of the sensor waveguide (b) and a side-view of the input waveguide (c).

3. Design

In literature simple expressions can be found for the attenuation, A, in bent waveguides which are in the so called Whispering Gallery Mode (WGM) regime [4]

$$A = \frac{2}{\pi}\frac{1}{\sqrt{100\Delta}}\cdot 10^{2.29-2.17R_n-0.58R_n^2} \quad (\mathrm{dB/rad}) \tag{1}$$

$$\text{with} \quad \Delta \equiv \frac{\Delta N}{N_b} \quad \text{and} \quad R_n \equiv \frac{N_b R}{\lambda}\cdot \Delta^{\frac{3}{2}} \cdot 1.137^{\Delta - 0.01}$$

The total power loss in the spiral as a function of the propagated angle, $\alpha(\Phi)$, can be calculated
by integrating (1)

$$\alpha(\Phi) = \int_0^{\Phi} \frac{2}{\pi}\frac{1}{\sqrt{100\Delta}}\cdot 10^{2.29-2.17R_n(\Phi)-0.58R_n^2(\Phi)}\cdot d\Phi \quad (\mathrm{dB}) \tag{2}$$

Using this expression the point were the light is lost by the waveguide, defined as $\Phi_{\alpha=20\mathrm{dB}}$, can be calculated.
The performance of the device will be determined by the choice made for the initial radius, R_0, and the angular dependency of the radius, $\delta R/\delta\Phi$, together with the waveguide cross section. From (1) it can be deduced that the sensitivity becomes higher with a larger bend radius. Nevertheless the bending radius will be limited by the device size we allow; in our case $1\mathrm{cm}^2$. From this we choose a initial radius R=4mm.The optimisation of $\delta R/\delta\Phi$ was done by evaluating a number of different spiral shapes and an attractive spiral shape was found to be

$$R(\Phi) = \frac{4}{1+\frac{\Phi}{25}} \quad (mm) \tag{3}$$

For obtaining a sensitive refractometer it is essential to find a waveguide geometry which converts a refractive index change in a sensing layer, Δn_m, efficiently into a change in the effective refractive index contrast, ΔN, i.e. we need a large value for $\delta\Delta N/\delta n_m$. This can be obtained by the waveguide illustrated in fig. 1b. If the refractive index of the index oil (in general the sensing layer showing the Δn_m) changes, only the effective refractive index of the waveguide, N_g, will change, whereas the refractive index of the background, N_b, will remain unchanged because here the optical field does not penetrate the oil. This means that for this type of waveguide optimizing $\delta\Delta N/\delta n_m$ is the same as optimizing $\delta N_g/\delta n_m$, which has been extensively studied earlier [5]. For our sensing device we chose to use a He-Ne laser with a wavelength λ=632.8nm and quasi-TE polarization. For the structure with the parameters as shown in figure 1b, made in the in our laboratory available Si_3N_4/SiO_2-technology [6], we then find $\delta\Delta N/\delta n_m=0.19$.

Using (2) and (3) together with the waveguide geometry (fig. 1b), we can calculate $\Phi_{\alpha=20dB}$, i.e. how far the light propagates into the spiral before the light is lost, as a function of the applied refractive index difference between the SiO_2-cladding and the index oil, n_m-n_c. This is the dotted curve in fig. 2. A straight-waveguide attenuation of 0.5dB/cm was incorporated in this analysis. It can be seen that $\delta\Phi/\delta n_m \approx 400\cdot 2\pi$ rad.

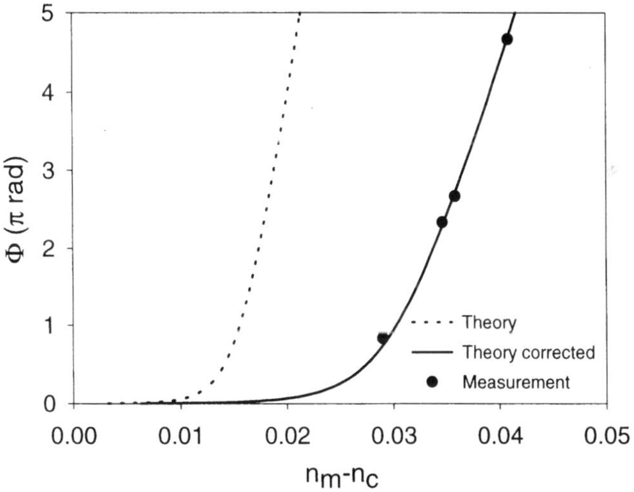

Fig. 2: Penetration angle of the light into the spiral ($\Phi_{\alpha=20dB}$) versus n_m-n_c. The dotted line is the theoretical curve, the solid line the theoretical curve for a residual SiO_2 layer of 40nm thickness; the dots are the experimental values.

4. Measurements and discussion

For demonstrating the feasibility of the sensing principle, measurements on the device were performed with a ccd-camera, allowing a simple determination of the point where the intensity of the light scattered by the waveguide, i.e. the light inside the waveguide, is reduced 20dB with respect to the intensity at the beginning of the spiral. The refractive index n_m was varied using several refractive index oils. Fig. 3 shows the penetration of the light into the spiral for four different values of n_m. The sensing capability of this device is clearly illustrated; the higher the refractive index of the oil the further the light penetrates the spiral (fig. 3a-d).

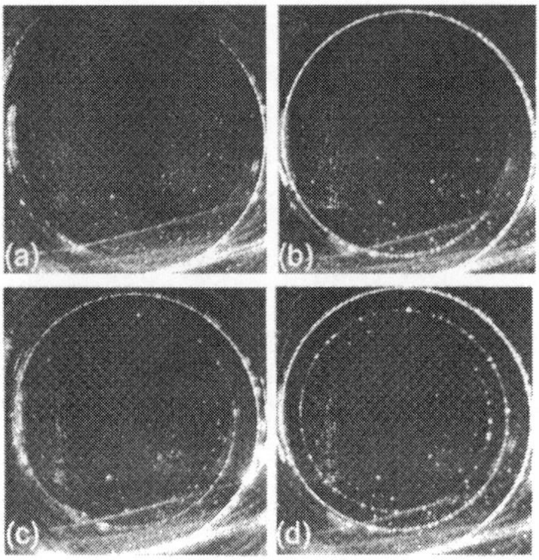

Fig. 3: Scattered light in the spiral for different values of n_m; n_m=1.4990 (a), 1.5047 (b), 1.5059 (c) and 1.5109 (d). Note that the pictures have an initial rotation of $\pi/12$ rad.

In fig. 2 the dots indicate how far the light penetrates the spiral ($\Phi_{\alpha=20dB}$) as a function of n_m-n_c. It is clear that the theoretical curve does not coincide with the experimentally found points. The reason for this was found by Scanning Electron Microscopy to be a 40nm thick residual layer of SiO_2 in the sensing waveguide area, apparently due to a too short etching time in the sensing waveguide defining etch step. The solid line in fig. 2 is the theoretical curve calculated for the real structure and it clearly agrees with the experimental points very well. This residual layer however results in a reduction of the slab sensitivity to $\delta\Delta N/\delta n_m$=0.10, resulting in $\delta\Phi/\delta n_m \approx 210 \cdot 2\pi$ rad.

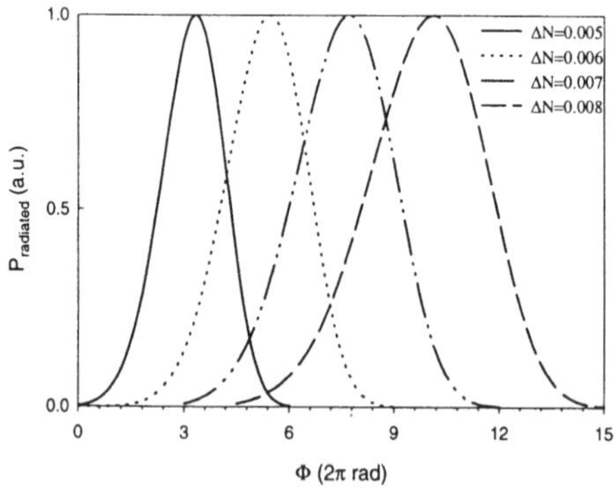

Fig. 4: Power radiated from the spiral versus the penetration angle, for different ΔN.

The resolution of the sensor can be evaluated by plotting the radiated power, $\delta P/\delta\Phi$, versus the penetration angle for different values of ΔN (fig. 4).

$$P_{radiated} = \frac{\partial P}{\partial \Phi} = -\frac{\partial}{\partial \Phi}\left[10^{-\frac{\alpha(\Phi)}{10}}\right] \qquad (4)$$

If we assume that the position of the peaks can be measured with an accuracy of 1/1000 of its FWHM, with a detector array or ccd-camera, a sensing resolution of $\Delta n = 8 \cdot 10^{-6}$ can be achieved, while for the originally designed structure a resolution of $\Delta n = 4 \cdot 10^{-6}$ can be calculated. This number can be improved by using a larger initial bend radius for the spiral. Also facilities for compensation of temperature effects can be incorporated. Furthermore the working point, i.e. the sensitivity region, can be adjusted to the specific application. Note here that SiO_xN_y layers can be made with a refractive index in the range $1.45<n<2.00$. Device to device reproducibility depends on technological limitations, which should be determined carefully.

5. Conclusions

In conclusion a novel highly sensitive passive integrated optical spiral-shaped waveguide refractometer was proposed and fabricated in Si_3N_4/SiO_2 technology. A resolution in the refractive index of $\sim 8 \cdot 10^{-6}$ is shown, and a value of $4 \cdot 10^{-6}$ is calculated for an even not completely optimized structure. The experimental performance of the device agrees very well with the simple model derived for WGM waveguide bends. The sensor concept is shown to be flexible (the range can be adjusted to practical specifications) and is expected to be well applicable in making useful integrated optical sensing systems.

Acknowledgements

The authors like to thank Rene Heideman for the valuable discussions on the subject. This research was financially supported by the Dutch Innovative Research Program (IOP) Electro-Optics.

References

[1] R.G. Heideman, G.J. Veldhuis, E.W.H. Jager and P.V. Lambeck, S&A B **35-36,** 234 (1996).
[2] P.V. Lambeck, R.G. Heideman and T.J. Ikkink, Med. & Biol. Eng. & Comp. **34**, 145 (1996).
[3] G.J. Veldhuis and P.V. Lambeck, Applied Physics Letters **71**, 2895 (1997).
[4] M.K. Smit, E.C.M. Pennings and H. Block, J. Lightwave Technol. **11**, 1737 (1993).
[5] O. Parriaux and G. J. Veldhuis, J. Lightwave Technol. (to be published).
[6] K. Worhoff et al., Proc. SPIE **3099**, 257 ((1997).

TEMPERATURE DEPENDENCE OF A THERMAL ACCELEROMETER

U.A. Dauderstädt *, P.M. Sarro **, P.J. French *
*Delft University of Technology, Department of Electrical Engineering, Laboratory for Electronic Instrumentation, Mekelweg 4, 2628 CD Delft, The Netherlands,
U.A.Dauderstadt@ET.TUDelft.NL
**DIMES, Delft, The Netherlands

1. Abstract

Silicon accelerometers are increasingly used in the automotive industry, but also applications for 3D mice or so-called gesture controlled remote controls are feasible. For such applications, small, low-cost sensors are needed which do not only meet certain requirements for sensitivity, working range or response time but can also work in a large temperature range as occurs in automobiles. In this paper, an accelerometer is described, which is based on a thermal principle. This accelerometer works at frequencies up to 300 Hz; it has an output signal of about 20 mV/g and shows very low temperature dependence and drift for a thermal device.

Keywords: thermal accelerometer, silicon accelerometer, thermopiles.

2. Introduction

Most commercially available silicon accelerometers are based on the piezoresistive or capacitive principle. Piezoresistive accelerometers have a relatively simple structure and can be fabricated by standard processes. Thus, they are well suited for low-cost low-performance applications. However, the necessity of calibration for temperature dependence and offset, which can be done by using thick film resistors on a ceramics substrate, makes them less suited for use as smart sensors. For applications, which require higher performance, capacitive accelerometers are used. These sensors can be combined with electronics on the same chip more easily, but they require rather sophisticated circuitry for the signal read-out and non-standard fabrication techniques, which makes them rather expensive.
In 1991, for the first time an accelerometer using a thermal principle was introduced [1]. This paper reports on the temperature dependence of this sensor, which is very low for a thermal device.

3. Principle

The thermal accelerometer, which is shown in Figure 1, is based on heat transfer. A seismic mass is positioned above a heat source. If the device is subjected to an acceleration a, an inertial force

$$F = ma \qquad (1)$$

acts on the seismic mass m and the mass will be displaced by an amount ξ

$$\xi = \frac{F}{k} = \frac{ma}{k} \qquad (2)$$

In the Equation (2), k is the spring constant of the suspension of the seismic mass.

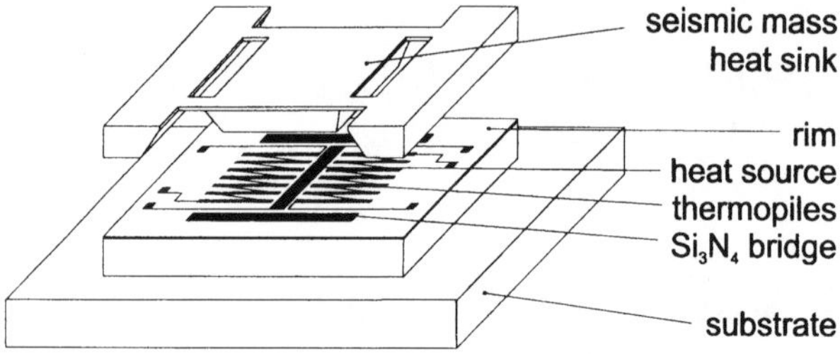

Figure 1: Schematic drawing of the thermal accelerometer. (The chips containing the seismic mass and the thermopiles are drawn apart to make the thermopiles visible. In the complete devices, these chips are glued or bonded together.)

The distance between the mass and the heat source then changes by ξ. Because of the temperature difference between the mass and the heat source, heat flows from the source to the mass. This flow is large if the distance to the mass is small and vice versa. A large heat flow causes the heat source to decrease in temperature, while for a small flow the heat source remains hot.

The total heat transfer from the heat source to the heat sink of the thermal accelerometer has three components: radiation, heat convection and heat conduction. Because of the small dimensions of the device (the distance between the heat source and the heat sink is only several micrometers) and the relatively small temperature difference between the heat source and the heat sink (several Kelvin), radiation and heat convection take only a very small part of the total heat transfer compared to the effect of heat conduction and can therefore be neglected.

By considering only the effect of heat conduction, the temperature difference between the heat source and the heat sink (the seismic mass) can be approximated by:

$$\Delta T = T_{HS} - T_0 = \frac{P}{K^*\lambda}\tanh(\lambda L) \tag{3}$$

with

$$\lambda = \sqrt{\frac{2K_{air}W(G_t + G_b + \xi)}{K^*(G_t + \xi)G_b}} \tag{4}$$

where T_{HS} is the temperature of the heat source, T_0 is the ambient temperature and P is the heating power. The variables G_t, G_b and W are explained in Figure 2. The constant K^* describes the thermal properties of the silicon nitride bridge and the thermopiles and can be written as

$$K^* = K_{SN}WD + n(K_{Al}D_{Al}W_{Al} + K_{PS}D_{PS}W_{PS}) \tag{5}$$

where K_i are the thermal conductivities, D_i and W_i are the width and thickness of the polysilicon and aluminum wires respectively and n is the number of thermocouples. The indices *SN*, *Al* and *PS* represent the materials silicon nitride, aluminum and polysilicon, respectively.

Polysilicon-aluminum thermopiles are used to determine the temperature difference ΔT. The output voltage of the thermopiles can then be given by

$$V_{OUT} = n\alpha\Delta T \tag{6}$$

where α is the Seebeck coefficient and n is the number of thermocouples. Polysilicon-aluminum thermocouples have a relatively large Seebeck coefficient of about 0.2…0.4 mV/K depending on the doping. Thus, with a ΔT of several Kelvin and about 40 thermocouples, output voltages of 80…100 mV can easily be obtained.

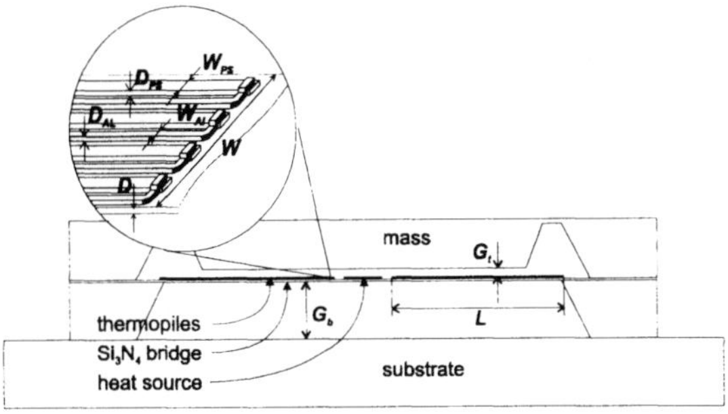

Figure 2: Cross section of the complete sensor and clipping of the silicon nitride bridge, which explain the variables used in the Equations (4) and (5).

Earlier presentations have shown that this accelerometer has a large output signal and can be fabricated using completely IC-compatible processes. Due to its small size it has a relatively short response time for a thermal device and can be used at frequencies up to 300Hz ([2],[3]). However, it might be expected that a thermal device would show a strong temperature dependence. In this work, this parameter is investigated.

4. Temperature Dependence

The temperature dependence of a complete thermal accelerometer is shown in Figure 3. The output voltage of the device depends linearly on the ambient temperature. The output is about 118 mV at room temperature and decreases by about 0.12 %/K.

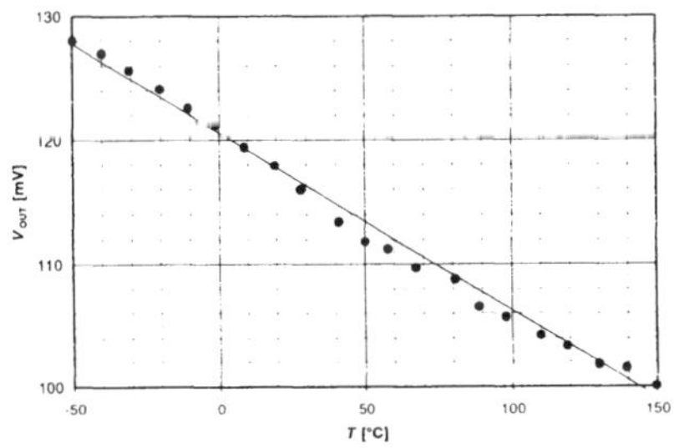

Figure 3: Output voltage of the Thermal Accelerometer versus ambient temperature (a=1g, Heating Voltage: $V_H = 2$ V).

Since every parameter, which is used to describe the behavior of the device, is temperature dependent, it is not easy to explain why the temperature characteristics of this device are as shown in the figure. However, there are some aspects, which have a larger influence on the accelerometer's behavior than others.

First of all, the temperature dependence of the thermal conductivity of air is an important parameter. The air gap between the heat source and the heat sink forms a thermal resistance, which changes when the seismic mass moves. This leads to a change in the heat flow between the heat source and the heat sink. But this resistance also changes when the thermal conductivity of the air changes.

The influence of this effect has been calculated by using Equation (3) and the data for the thermal conductivity of air has been taken from [4]. Figure 4 shows the result of this calculation. Within the considered temperature range, the thermal conductivity of air increases with temperature. As a result of this, the temperature difference ΔT decreases nearly linearly. The calculated temperature difference is about 9.7 K at room temperature and decreases by a factor of about 0.06 %/K.

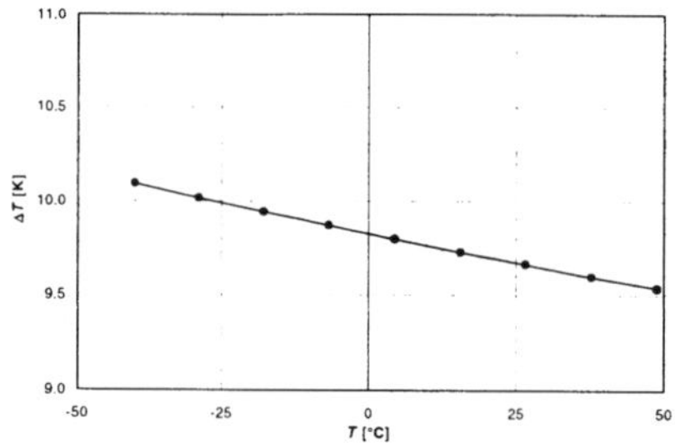

Figure 4: Calculated temperature difference between the heat source and the seismic mass of the Thermal Accelerometer versus ambient temperature. The calculations are based on the data given in [4].

In Figure 5, the output voltage of a single thermopile without the seismic mass versus temperature is shown. The output voltage is about 185 mV at room temperature and decreases by 0.08 %/K. The curve is linear in the temperature range of –30…145 °C. The temperature dependence here is considerably smaller than that of the complete device because the influence of the heat flow through the air gap has been eliminated.

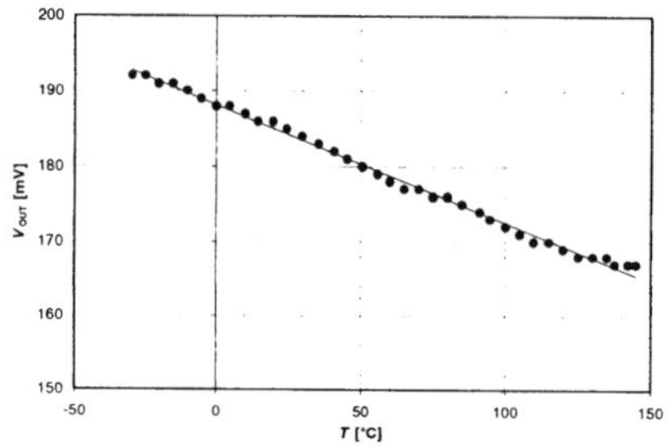

Figure 5: Output voltage of a single thermopile versus ambient temperature (Heating Voltage: $V_H = 2$ V).

The second important effect is the temperature dependence of the electrical conductivity of polysilicon. In the previous measurements, a constant voltage has been applied to the heating resistor. The heating power is then given by

$$P_H = \frac{V_H^2}{R_H} \tag{7}$$

Since the electrical resistivity of highly doped polysilicon has a positive temperature coefficient, the resistance of RH increases with increasing temperature resulting in a decrease in heating power. As shown in Equation (3), the temperature difference between the heat source and the heat sink depends linearly on the heating power. Thus, this temperature difference decreases with raising temperature leading to a decreasing of the output voltage.
There are two ways to eliminate this effect. The first possibility is it to use polysilicon with a lower doping concentration. By choosing the correct doping concentration, the temperature coefficient of the polysilicon can become zero or even negative. The disadvantage of this is that the Seebeck coefficient would also decrease if the doping concentration were lowered. But since the Seebeck coefficient of polysilicon is rather large this would be acceptable.

The other possibility is it to keep the heating power constant. In a smart sensor, this is rather difficult to do and thus the first option would be preferable. To test this hypothesis, the resistance was monitored and the voltage calculated to yield the desired heating power. The result of this measurement is shown in Figure 6. Now, the temperature dependence is no longer linear. The output voltage is about 115 mV at room temperature and the temperature coefficient ranges from -0.05 %/K at 0 °C to 0.12 %/K at 100 °C. The total change of the output voltage is now less than ± 1.2% in this temperature range.
In this measurement the effect of the temperature dependence of the electrical conductivity of polysilicon has been eliminated. The remaining effects are the temperature dependence of the Seebeck coefficient and of the thermal conductivity of air and polysilicon. Since the resistors of the thermopiles and the silicon nitride bridge are very thin, they have a rather large thermal resistance and the amount of heat flowing through them is small compared to the amount of heat that flows through the air gap to the seismic mass. Thus, the influence of this heat flow is rather small at any temperature.
The temperature dependence of the Seebeck coefficient has been investigated by F. Völklein et. al. [5]. The Seebeck coefficient increases with temperature, which would, in turn, cause the output voltage of the thermopiles to increase with temperature. In the actual devices, this effect is almost compensated by the temperature dependence of the thermal conductivity of air as Figure 6 shows.

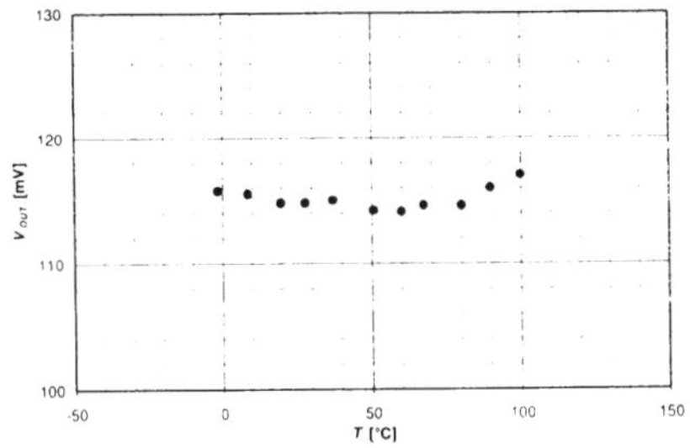

Figure 6: Output voltage of the Thermal Accelerometer versus ambient temperature (a=1g, Heating Power: P_H = 4 mW).

In conclusion it can be said that the thermal accelerometer has temperature dependence, which is rather low for a thermal device. This is because of the fact that the thermopiles do not measure an absolute temperature but a temperature difference, which is created on the chip. Thus, if the ambient temperature increases, the absolute temperature of the heat source increases as well and the difference remains (almost) constant.

5. Discussion and Conclusions

Table 1 gives an overview of some parameters of the thermal accelerometer in comparison with a commercially available device. The performed measurements show that the thermal accelerometer as presented in this paper has a very low temperature dependence. The temperature range of this device is within the limits for automobiles as given by MacDonald [6]. It has been shown how these temperature characteristics can even be improved by using lower doped polysilicon. This and the high output signal, the rather good frequency response and the possibility to fabricate the sensor by using standard IC processes make the thermal accelerometer an interesting alternative to other commercially available sensors.

Table 1: Characteristics of the Thermal Accelerometer in Comparison with a commercially available device

	ICS 3021	Thermal Accelerometer	
Range	2 g	(0.4… 0.8) g	
Frequency response	(0 … 350) Hz	(0 … 300) Hz	
Sensitivity	15 mV/g	(9 … 50) mV/g	
Zero acceleration output	(5 … 25) mV	≈ 100 mV	
Supply voltage	(5 … 10) V	(1 … 2) V	
Supply Power		(1 … 4) mW	
Temperature coefficient	1 %FS(±)	± 5 %FS 0.12 %/K	Constant heating voltage (V_H = 2 V)
		< ± 1.2 %FS (-0.05 … 0.12) %/K	Constant heating power (P_H = 4 mW, T_0 = (0 … 100) °C)
Operation temperature	(-55 … 125) °C	(-50 … 150) °C	

6. References

[1] R. Hiratsuka, D. C. van Duyn, T. Otaredian and P. H. S. de Vries, A novel accelerometer based on a silicon thermopile, Technical Digest 6th Int. Conf. Solid-State Sensors and Actuators (Transducers '91), San Francisco, CA, USA, June 24-27, 1991, 420-423.

[2] U. A. Dauderstädt, P. H. S. de Vries, R. Hiratsuka, P. M. Sarro, Silicon Accelerometer based on thermopiles, Sensors and Actuators A46 1-3 (1995), pp. 201-204.

[3] U. A. Dauderstädt, P. H. S. de Vries, R. Hiratsuka, J. G. Korvink, P. M. Sarro, H. Baltes, S. Middelhoek, Simulation Problems of a thermal Accelerometer, Sensors and Actuators A55 (1996), pp. 3-6.

[4] Handbook of Chemistry and Physics, 67th Edition, 1986-1987, p. E-2.

[5] F. Völklein and H. Baltes, Thermoelectric Properties of Polysilicon Films Doped with Phosphorus and Boron, Sensors and Materials, 3, 6 (1992), pp. 325-334.

[6] G. A. MacDonald, A review of low cost accelerometers for vehicle dynamics, Sensors and Actuators A, 21-23 (1990) 303-307.

NEARLY QUANTUM-LIMITED SQUIDS FOR A GRAVITATIONAL WAVE ANTENNA

J. Flokstra, E. Bartolomé, G.K. van Ancum*, M.J. van Duuren and H. Rogalla

Low Temperature Division, University of Twente, P.O. Box 217, 7500 AE Enschede, The Netherlands , * present address, KPN, Leidschendam
e-mail: j.flokstra@tn.utwente.nl

0. Abstract

Nearly quantum-limited Superconducting QUantum Interference Devices (SQUIDs) have to be used in resonant mass gravitational wave antennas to reach a displacement sensitivity of the order of 10^{-21} m/√Hz in a bandwidth of 100 Hz around the resonance frequency of about 900 Hz. The design of these SQUIDs will be described. We show that a good coupling between the inductive readout circuit and the SQUID can be obtained by choosing a relatively large hole size of the washer-type SQUID configuration.

Keywords: SQUIDs, quantum limit, gravitational wave antenna

1. Introduction

In the Netherlands an initiative has been taken to build a resonant mass gravitational wave antenna under the project name GRAIL. Groups from the Universities of Leiden, Amsterdam, Eindhoven and Twente and the high-energy physics center NIKHEF (Amsterdam) constitute the GRAIL team. The Twente group will be responsible for the readout system that has to detect the very small displacement of the spherical mass antenna after the passage of a gravitational wave. The antenna will have a diameter of 3 meters and a mass of about 110 tons. The operation temperature of 10 mK will be reached with a powerful dilution refrigerator of 100 μW cooling power at that temperature. The vibration damping of the support system of the sphere will be 320 dB. A gravitational wave will give an extremely small quadrupole distortion of the antenna dimensions. In order to be able to detect gravitational waves, the displacement sensitivity of a readout device should be of the order of 10^{-21} m/√Hz in a bandwidth of 100 Hz around the resonance frequency (900 Hz) of the sphere. This sensitivity is a factor of 100 better than that of existing antennas [1]. It is expected that with this extreme sensitive system, gravitational waves can be detected for the first time in history.

In this paper we will describe the motion sensor we aim for in the GRAIL design. For reaching the extreme displacement sensitivity superconducting sensors that operate near the quantum limit are necessary.

2. The overall system

A schematic picture of the readout system for the displacements of the spherical antenna is given in figure 1. The system consists of a mechanical transducer, a superconducting inductive network and a superconducting sensor (SQUID). The mechanical transducer consists of two masses with a resonance frequency very close to that of the sphere. When the vibrational energy is transferred to the last mass an amplitude gain of more than 100 can be obtained. The last mass is superconducting and its displacement modulates the self inductance of the coil facing this mass. The superconducting inductive circuit carries a persistent current and the modulation of the coil inductance induces a small ac current in the loop due to the principle of flux conservation. The ac current is sensed by the most sensitive sensor for magnetic flux that exists, being the SQUID. In total six sensors will be placed on the sphere in order to be able to determine the polarization and direction of the gravitational wave.

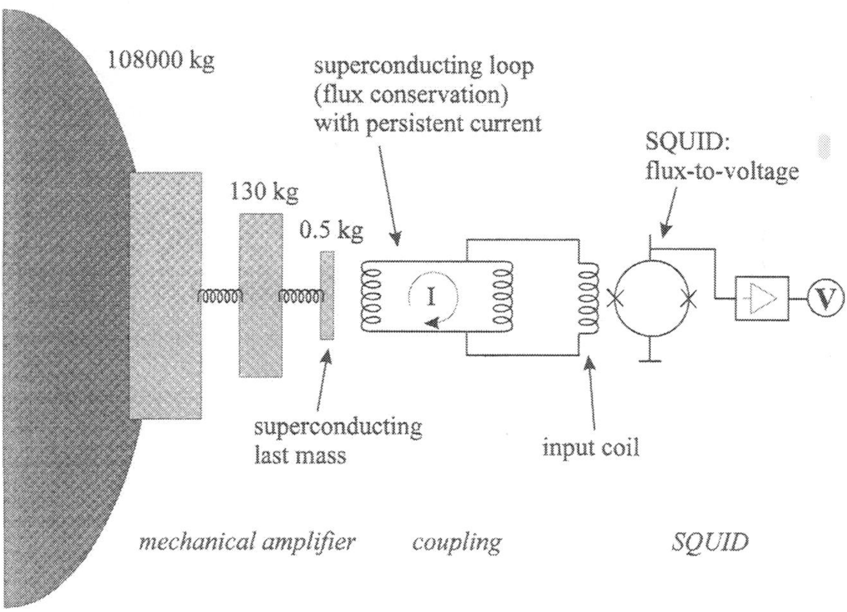

Fig. 1 - A schematic picture of the displacement readout system of the gravitational wave antenna.

The basic scheme of a dc SQUID is depicted in figure 2a. The two Josephson junctions, with a critical current I_0 and a self-capacitance C, are connected in parallel by a superconducting loop with inductance L_{sq}. To eliminate hysteresis on the current *vs.* voltage characteristic each junction is shunted by an external resistor R_s.

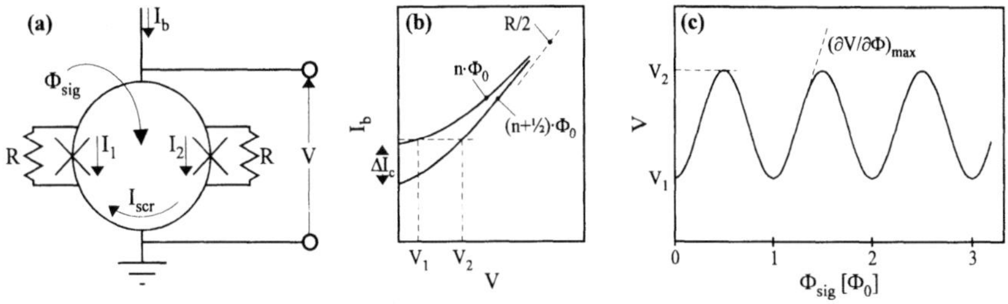

Fig. 2 - Scheme (a), I-V (b) and V-Φ (c) characteristics of a dc SQUID.

When the dc SQUID is biased with a constant current I_b, the voltage V across the SQUID is a function of the signal flux applied through the input coil. If the magnetic flux Φ_{sig} is varied , the voltage V oscillates with a period Φ_0 (one flux quantum) between the values V_1 and V_2, as shown in figures 2b and 2c, while the I-V characteristic oscillates between the two extreme values represented in figure 2b. Thus it is possible to measure the change in the voltage across the SQUID induced by a small change in the magnetic flux threading the SQUID loop. The SQUID plays then as a flux-to-voltage transducer with a transfer function $V_\Phi=(\partial V/\partial\Phi)_I$, which attains its maximum value at a flux around $(2n+1)\Phi_0/4$, n being an integer. However, in most applications the non-linear V-Φ_{sig} characteristic needs to be linearized. The dc SQUID is then operated as a null detector in a feedback loop (the so called "Flux Locked Loop") that provides a linear response. The dc SQUIDs made in the LTD of Twente are washer-type ones fabricated in thin film technology.

The main source of white noise in resistively shunted dc SQUIDS is the Johnson noise in the two resistive shunts R_s. The flux noise spectral density S_Φ, which determines the smallest detectable change in magnetic flux is found to be [2]:

$$S_\Phi = \frac{S_V}{(\partial V/\partial\Phi)^2} \approx 9\cdot\frac{2k_B T L_{sq}^2}{R},$$

where T is the operating temperature of the SQUID, and R is the parallel of the shunt R_s and the intrinsic resistance of the junction R_n.

The energy resolution ε is a convenient parameter to compare the sensitivity of SQUIDS with different inductances:

$$\varepsilon / 1\,\mathrm{Hz} = \frac{9k_B T L_{sq}}{R}.$$

For standard dc SQUIDs, ε is of the order of $10^4\hbar$. For many SQUID applications [3] this value is adequate. However, in order to reach the required sensitivity for a gravitational antenna, SQUIDs with an energy resolution below 10 $\hbar$ are needed.

3. Nearly quantum-limited SQUIDs

In order to design nearly quantum-limited devices suitable for the gravitational antenna, it is necessary to put the former expression of the energy resolution in terms of externally variable parameters of the dc SQUID. The inductance of the SQUID L_{sq} is considered to be mainly that of the washer, and the contribution of the slit is in first approach neglected: $L_{sq} \approx L_w = 1.25\mu_0 D$, where D is the dimension of the SQUID hole in the washer. The intrinsic resistance of the junction R_n is approximately $R_n \approx \frac{\pi}{4}\frac{2\Delta/e}{I_0}$, where $2\Delta/e$ is the gap voltage (~2.8 mV for a Nb based junction) and I_0 is the critical current through it. The capacitance per unit area of junction is supposed to be constant $C/A \approx 0.03$ pF/μm^2. Also the critical current density must be kept constant $J_0 = I_0/d^2 \approx 40$A/cm^2 (d being the junction size), as all the SQUIDs must be fabricated on the same wafer. Some extra requirements must be fulfilled in the design. First, to minimize the energetic noise the screening parameter $\beta_L = 2L_{sq}I_0/\Phi_0$ should be $\beta_L = 1$. In addition, to ensure the non-hysteretic behavior of the SQUID, the Mc Cumber parameter $\beta_c = 2\pi I_0 R^2 C/\Phi_0$ should be $\beta_c \leq 1$. Finally, the noise parameter, relating the Josephson coupling energy and the thermal fluctuations, should be: $\Lambda = I_0\Phi_0/2\pi k_B T > 10$ for the safe operation of the SQUID.

Taking into account these relations it is possible to write the energy resolution as a function of the temperature T, d and D:

$$\varepsilon/\hbar = 0.453\ T(K) d(\mu m) D(\mu m)^{1/2},$$

T and D fulfilling the operational condition:

$$\Lambda = 0.016 \times 10^6\ T(K)^{-1} D(\mu m)^{-1} > 10.$$

The energy resolution for a fixed temperature of 4.2K is represented in figure 3.

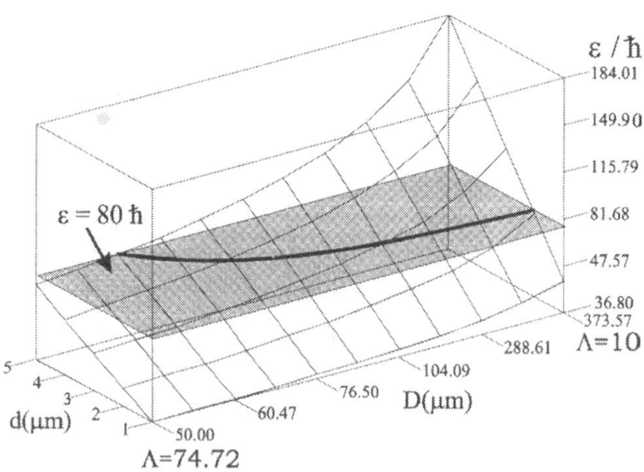

Fig. 3 - The theoretical energy resolution ε of a washer-type dc SQUID as a function of the junction size d and the washer hole D for $J_0 \approx 40$ A/cm^2, $\beta_C = 1$ and T = 4.2 K.

The above condition at this temperature implies that the SQUID hole size should be: D<374μm. Although in principle there is no lower limit for D, the smaller the SQUID hole size, the larger the number of turns necessary to couple the SQUID, Many problems will then arise. As the theoretical energy resolution is linear with T, to have nearly quantum limited SQUIDs with $\varepsilon \approx 2\hbar$ at 0.1K it is enough that their energy resolution at 4.2K is $\varepsilon \approx 80\hbar$. The intersection of the $\varepsilon \approx 80\hbar$ constant plane with the energy resolution surface defines a line on which a series of SQUIDS, (d,D) characterized, can be chosen. We conclude that SQUIDs with a relatively large hole size can be prepared that reach the quantum-limit at 0.1K. The large hole facilitates the coupling to the SQUID. It is found that junctions with (sub)micron dimensions are then needed. The construction of a number of such devices is at present under process.

4. System aspects

We have calculated the equivalent displacement noise $X_{noise}(m/Hz^{1/2})$ for the GRAIL antenna, according to reference [4]. The results for a 3-mode transducer, i.e. a mechanical transducer consisting of two masses (m_2, m_3) in addition to the antenna mass (m_1) are presented in figure 4. The effective antenna mass is 3×10^4 kg. The other two masses were varied in calculations, obeying $m_3=m_2^2/m_1$. The mechanical Q-factors were taken as 5×10^6. The resonance frequency is 900 Hz and T=0.1K. The coil of the inductive circuit with inductance L_t=5μH is a pancake coil and the distance to the facing superconducting mass m_3 is 25 μm. A persistent current of 10 A flows in the circuit. The electrical quality factor of the circuit equals 5×10^6. The input coil on top of the SQUID has an inductance L_i=2μH. The SQUID used in the calculations is one of the previously designed, with $\varepsilon=80\hbar$ at 4.2K, giving $2\hbar$ at 0.1K. It is found that X_{noise} depends on the value of m_3. In all cases the bandwidth is considerably larger than 100 Hz around the resonant frequency of 900 Hz at a level of 10^{-21} m/√Hz.

In order to obtain L_i=2μH, a large number of turns of the input coil is necessary. In general this will lead to unwanted resonances that disturb the proper operation of the SQUID. It is therefore worthwhile to consider an input coil with less turns leading to a smaller value of L_i. However, then an impedance mismatch between L_t an L_i appears. We have varied L_i in our calculations down to 50 nH. The desired

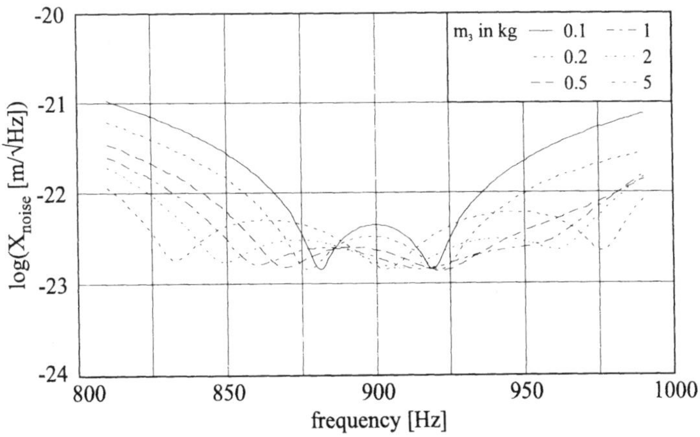

Fig. 4 - Displacement noise spectra for six different values of m_3. Other parameters: L_i = 2 μH, $\varepsilon = 2\hbar$, T = 0.1 K.

displacement sensitivity of 10^{-21} m/√Hz is still reached in a very reasonable bandwidth of 80 Hz.

Due to the high mechanical and electrical quality factors, back action effects of the flux locked loop operation can deteriorate the spectral displacement noise features of the system [5]. The feedback current induces not only a flux in the SQUID hole but also in the input coil circuit via the SQUID. In order to avoid this unwanted coupling we propose to mix the internal feedback with an external feedback to the input circuit so that the back action signal via the SQUID is just canceled by the external feedback signal. Also a large bandwidth of the SQUID read out system will contribute to the suppression of the back action effects.

5. Conclusions

We have shown that nearly quantum-limited dc SQUIDs can be obtained with a relatively large hole size in the SQUID washer enabling a good coupling to an inductive circuit that is used for the read out of the small displacements of a resonant mass gravitational wave antenna. With these SQUIDs, a sensitivity level of 10^{-21} m/√Hz can be reached in a bandwidth of over 100 Hz around the central resonant frequency of 900 Hz.

References

1. G. Blair, The detection of gravitational waves, Cambridge University press, (1991).
2. J. Clarke, Ultrasensitive measuring devices, Physica 126B, (1984), 441-448.
3. G. B. Donaldson, A. Cochran and R. M. Bowman, More SQUID applications , Proceedings of the NATO Advanced Study Institute on The New Superconducting Electronics, H. Weinstock and R. W. Ralston eds., vol.251, (1992), 181-220.
4. J. P. Richard and W. M. Folkner, Motion amplifiers and pasive transducers, presented in ref.1, 180.
5. W. M. Folkner *et al.*, Instrumentation of a resonant gravitational radiator detector with a planar thin-film dc SQUID, J. Appl. Phys. 65, (1989),12.

FROM BIOSENSORS TO BIOSENSING SYSTEMS

Elisabeth Verpoorte
Institute of Microtechnology, University of Neuchâtel
Rue Jaquet-Droz 1, CH-2007 Neuchâtel, Switzerland
E-mail: sabeth.verpoorte@imt.unine.ch

0. Abstract

There are several different approaches to bioanalytical measurements, including biosensors, flow injection analysis, and separation techniques such as capillary electrophoresis. Automating the use of all these techniques usually involves incorporation into flow systems, a concept which is known as the total analysis system (TAS). Optimization of analytical performance in most flow systems, particularly those which are based on high resolution separation methods, can be accomplished through miniaturization of the system to a μTAS. The use of microfabricated devices has become a popular approach for realizing miniaturized analysis systems for measurements on a sub-μL scale. This presentation will focus on the use of μTAS-type systems, both biosensor and non-biosensor based, for biosensing applications. A brief introduction to the analytical methods available for adaptation to bioanalytical μTAS is included in this review. Finally, the potential of μTAS to combine the best aspects of both stand-alone biosensors and benchtop biosensing systems will be discussed.

Keywords: biosensor, biosensing, miniaturized total analysis systems (μTAS), flow injection analysis, capillary electrophoresis, microchips

1. Introduction

Selectivity in bioanalytical measurements can be achieved in several different ways. Many methods exploit naturally occurring molecular associations, such as those between enzymes and specific substrates, antibodies with antigens, or DNA hybridization, to detect the presence of the analytes of interest. There are different strategies for how this may be done. Often, the molecules which are to be used to capture analyte molecules from solution are immobilized onto a surface of some kind. The sample is transported over the surface, to allow selective interaction of the analyte with available surface sites. This binding event is then subsequently monitored in some way. In the case of biosensors, the biologically selective moieties are immobilized onto an interface which is in intimate contact with a transducer to allow direct conversion of chemical information into an electrical signal. Ideally, the response mechanism of the biosensor is then dictated solely by the selective interaction of the probe molecules and the analyte, to the exclusion of all other species present. Direct immersion of the biosensor into a sample should then yield the desired information in a matter of seconds, independent of sample composition.

1.1 Flow systems based on biosensors

In reality, it can be assumed that there are several processes occurring at the interface, inducing both selective and non-selective responses. These processes can have a profound influence on biosensor performance characteristics such as sensitivity, response time, response drift, lifetime, and reversibility [1]. Optimization of biosensor interfaces can prove to be extremely difficult and time-consuming, as it requires a fundamental understanding and control of the interface and the processes occurring there. It is thus common practice to optimize biosensor behaviour in conjunction with some form of sample pretreatment, to lessen the stringent selectivity demands on the sensor. This could include removal of substances from the sample matrix which could adversely affect sensor response by fouling the surface or by inducing a false transducer signal. Prolonged sensor lifetime is one major benefit to be gained through matrix dilution and clean-up. Frequent recalibration is also a necessity when using most biosensors, to account for the changes at the sensor/solution interface which lead to drift in response. Thus, a number of fluidic manipulations are usually associated with optimized biosensor use. Analysis times are no longer solely a function of the sensor's response time, but must also include the time required to pretreat the sample and transport it to the sensor. This can mean an increase from seconds to minutes in the time interval between starting a measurement and receiving the desired chemical information. To improve sample throughput using biosensors, one preferred approach nowadays is to incorporate them into flow systems capable of automatically handling, treating, and detecting the components of interest. In fact, flow manifolds like these can be regarded as biosensing systems, since the biosensor no longer is used as a stand-alone device.

1.2 Biosensing using flow injection analysis

Other commonly employed methodologies in bioanalytics focus not on optimization of the selectivity of a modified sensor surface, but rather on improved sample treatment and separation in flowing systems, to facilitate detection by optical or other means. Flow injection analysis (FIA) is one well-established approach. In its basic form, an FIA system sees the injection of a sample plug into a flowing reagent stream, with subsequent detection of the reaction zone further downstream. Other functions such as dilution and mixing can be built into the system, depending on the requirements of the analysis. Numerous bioanalytical FIA applications have been reported since its introduction at the beginning of the 70's. One early example used capillary FIA for enzyme assays [2]. Small amounts of enzyme-containing sample, in this case glucose oxidase, were injected into a reagent stream. Production of a fluorescent dye whose concentration was directly indicative of the initial enzyme concentration ensured optical detection of the reaction. This type of homogeneous enzymatic assay has many applications in clinical and biotechnological analysis, for the determination of either substrate content or of enzyme activity [3]. Other FIA methods using packed reactors containing immobilized enzymes have also been frequently reported (see, for example, [3,4]). The advantages of these systems include the multiple use of often costly enzymes as reagents, as well as the possibility of combining sample pretreatment by FIA with the specificity imparted by the enzymatic reaction [3,4].

1.3 Biosensing using separation methods

The strategy of separating a mixture into its constituents allows both qualitative and quantitative information about a sample to be obtained. Again, the sample may undergo pretreatment beforehand to simplify the matrix, if necessary. There are a number of powerful separation techniques available to the analyst today, which use differences in the physical and chemical properties of compounds to effect their separation. If done in capillaries having diameters less than 1mm, separations on a time scale of minutes rather than hours are achievable. In high performance liquid chromatography (HPLC), for instance, components in an injected sample plug partition to differing extents between the mobile phase and a stationary phase found on column surfaces, and thus migrate at different rates down the column. A newer and increasingly popular technique is capillary electrophoresis (CE), usually done in fused silica columns of 100 μm diameter or less. Application of an electric field along the length of the capillary results in the generation of electroosmotic flow (EOF), a bulk flow of the buffer originating at the walls of the capillary. The magnitude of the EOF is dependent on the degree of ionization of the silanol groups on the internal wall of the capillary. Charged species can be separated on the basis of their different charge-to-mass ratios, which translate to differing electrophoretic mobilities in the applied electric field. Neutral species, on the other hand, cannot be resolved with CE. An approach for enhanced resolution of neutral solutes in electrically driven separations involves the introduction of a stationary phase to the column. This is done in capillary electrochromatography (CEC), a new micro separation technique in which the capillary tubes are coated with stationary phase or packed with octadecyl silanized silica gels. The eluent is moved through the column by the EOF induced by the applied electric field at the capillary walls and the solid packing, due to the presence of additional free silanol groups at the surface of the micro spheres [5]. Micellar electrokinetic capillary chromatography (MECC) is another type of CEC often used in open tubes. It is an electrically driven separation technique in which surfactants are included in the running buffer, adding in effect a pseudo-stationary phase to the system. Uncharged analytes can be then be separated through differential partitioning between buffer and micelles.

These separation techniques find many applications in bioanalysis - a survey of any separation science, clinical analysis, or protein analysis conference program will attest to this. HPLC, the most well-established of the capillary separation methods, is applied in a wide variety of biomolecular separations. Although the number of applications for CE are still somewhat limited, this younger method has found increasing favour for protein separations, nucleic acid analysis, and DNA sequencing, to name but a few [6]. One new and developing field is the use of CE for homogeneous immunoassays. The immunological reactions are carried out in solution using appropriate reagents, and then reaction products and leftover reactants are separated by CE [7-12]. Advantages offered by CE over other common assay formats such as radioimmunoassay (RIA) or enzyme-linked immunosorbent assay (ELISA), usually done in microtiter plates, are very small scale analysis and significantly shorter analysis times, as well as simplified fluid management. Another very interesting application of CE systems for ultramicroreaction-based chemical analysis using electrophoretically mediated microanalysis (EMMA) was introduced by Regnier et al. in 1992 [13,14]. The application of an electrical potential over a capillary allows electrophoretic mixing of a narrow band of enzyme-containing sample with substrate-containing solution, due to the different electrophoretic mobilities of the various species involved. Product is monitored optically as it passes the detector. Though employing biomolecular recognition mechanisms to ensure analytical selectivity, these systems are in great contrast to biosensors, where the observed signal relies on the interaction of the analyte with an immobilized surface. In fact, the immobilized biosensor surface could be regarded as serving a dual purpose, in that it not only selectively reacts with the analyte of interest, but also simultaneously separates that analyte from other sample components. Often, however, the "separation power" of the biosensor surface leaves something to be desired, as alluded to above. The use of a technique like CE eliminates the need for the immobilized surface, as the separation of enzymatic or immunological reaction products and reactants can now be carried out very effectively in solution. Thus, bioanalytical systems based on high resolution separation techniques, whether they involve the use of biomolecular recognition elements or not, can certainly be considered biosensing systems in their own right.

1.4 Total analysis systems (TAS)

As discussed above, many analyses, whether sensor- or separation-based, are carried out in flowing systems. A standard approach to increased automation and the related achievement of higher throughput is to incorporate all the analytical steps, from the very initial sampling right through to the final detection step, into a flowing system. This has come to be known as the total analysis system, or TAS [15]. This concept is inherently flexible: systems could be configured in many different ways, depending on the requirements of the analysis. The use of capillary-based separation techniques could impart to a TAS the power of detecting many molecular constituents in a single sample. However, most analytical flow systems tend to take the form of some kind of benchtop instrument and have a relatively large footprint. Achievable improvements in analysis times are limited by the overall distances fluids need to be transported between entry and detector points of the system. Achievable separation times are also dictated by capillary geometries, and are thus restricted in most cases by the dimensions of commercially available tubing to several minutes or more. Other possible disadvantages of TAS could include the often significant consumption of reagents and carrier solutions. Should better, faster analytical performance be required from the TAS, one needs to miniaturize it.

1.5 Miniaturized total analysis systems (μTAS)

A concept for the miniaturized total analysis system, or μTAS, first appeared in the literature in 1990 [16]. Essentially, the μTAS can be envisioned as a complete analysis system shrunk down into a single, small module. This leads to substantially reduced transport distances and decreased sample/reagent consumption. In addition, separation power of all the separation methods improves as capillary diameters shrink, so that faster separations having adequate resolution can be accomplished in smaller capillaries. However, the realization of complex flow manifolds having the sub-microlitre volume elements necessary to carry out analyses in times amounting to seconds rather than minutes is not trivial with capillary tubing. In particular, the connecting pieces used to couple capillaries to one another and to peripheral system elements generally have significant dead volumes associated with them. These can prove detrimental to the quality of the analytical result, as they contribute to unwanted sample dispersion. An alternative format which has been increasingly explored in recent years for miniaturizing flow systems sees the use of photolithographically fabricated channel structures incorporating injection, separation, and detection elements [17]. These micro flow manifolds are formed in planar substrates such as silicon and glass, using techniques first developed for structuring silicon in the microelectronics industry. Features having dimensions of as little as 2 μm × 2 μm × 2 μm can easily be fabricated, which means volume elements of microliter down to femtoliter proportions are possible. Elimination of bulky connections between system components and reduction of system dead volumes to vanishingly small values are major advantages which micromachining bring to the analyst. It is not surprising that the popularity of this technology has grown tremendously in the analytical community over the past few years.

Small volume systems continue to be a challenge with regard to fluid propulsion and control within microchannel networks. Some examples of micromachined FIA and liquid chromatography systems, in which liquids are pressure-driven, have been reported [18-25]. Of the microchip-based analysis systems, however, those which have proven particularly successful are based on electrically driven separation techniques. The use of glass as a substrate material opens up the possibility of using electric fields to control solution flows, through use of EOF as a pumping mechanism. This bulk electrolyte flow is generated without the aid of mechanical elements in the system, making it an excellent pumping mechanism in capillaries less than 100 μm in diameter, where sub-μL/min flowrates are generally required. It is therefore possible to define and manipulate pL-sized solution plugs electrokinetically [26-28]. Combining this capability with electrophoretic separation has led to examples of planar integrated devices for free flow electrophoresis [29,30], capillary electrochromatography [31-34], capillary electrophoresis [35-41], and synchronized cyclic capillary electrophoresis [42-44]. Integrated CE in particular demonstrates how miniaturization by micromachining affords a route to very fast separation times, down from tens of minutes to several seconds [37,39,40] and even sub-second levels [41]. These analysis times are equivalent to (or even better) than those which may be expected for stand-alone sensors. Though the experimental systems used to test chips to date have usually been anything but portable, requiring large voltage supplies, light sources, and the like, newer, smaller versions of these components are becoming available. Continuing developments along these lines will allow the periphery around the chip to be miniaturized to an extent that the entire system could in future look much like a sensor to the user. Thus, the idea of separation or other flow systems on microchip basis as another form of chemical sensor becomes conceivable [45].

1.6 μTAS for biosensing in a biosensor-like fashion

The use of microfabricated devices for bioanalytical applications has become a research area in its own right, stimulated by programs like the Human Genome Project in the United States. Modern drug discovery is also driving the development of chip-based analysis systems. The combinatorial synthesis of large libraries of related compounds and the subsequent need to screen these libraries for active pharmaceutical substances requires very high throughput analytical techniques. Massively parallel microchannel systems are possible with micromachining, since many, possibly hundreds, of channels can be placed side-by-side on a single chip. These types of systems are now being explored by a number of research groups [46,47]. This presentation will focus on the use of the following μTAS-type systems, both biosensor and non-biosensor based, for biosensing applications. Their potential to combine the best aspects of both stand-alone biosensors (for instance, ease-of-use) and benchtop biosensing systems (flexibility in the sample pretreatment process) will be emphasized.

2. Miniaturized Biosensing Systems based on Biosensors

A recent example of a biosensor-based biosensing system according to the μTAS concept was developed for real-time monitoring of glucose and lactate for biomedical applications [48,49]. The system consisted of a highly efficient microdialysis sampling interface combined with a microflow manifold based on planar silicon elements and an integrated biosensor array for glucose and lactate determination. Computer software for overall process control and signal display was also developed. A prototype of this system was first built in the laboratory, where parameters for the various system components, such as the microdialysis sampling, were optimized. It was then used in the on-line monitoring of glucose and lactate in the core blood of anaesthetized dogs. System response times were less than 5 minutes during this *in-vivo* testing, demonstrating the utility of low dead volume microflow manifolds made by micromachining. A similar system has been reported for bioprocess monitoring [50].

3. Miniaturized Biosensing Systems based on Electrically Driven Microchip Devices

A number of publications describing microchip devices for enzyme [51] and homogeneous immunoassays [52-54] have appeared in the literature recently. In the example to be presented here, MECC was the separation method of choice to carry out a number of different analyses of biological samples on a planar microelectrophoresis device [55]. It was possible using MECC on chips to separate mixtures of six fluorescein isothiocyanate labelled amino acids within a few seconds. The detection limit for fluorescein isothiocyanate labelled arginine was found to be a very respectable 3.3 nM, corresponding to approximately 40 zmol injected. The feasibility of analyzing human urine and serum samples in uncoated microchannels was demonstrated. A chip-based, homogeneous, MECC immunoassay for theophylline in serum was also presented, and with separation times of about 25 seconds, shown to be about 50 times faster than that developed in a fused silica capillary. These types of systems point the way to biosensor-like biosensing systems.

References

1. M. Thompson and U.J. Krull, Biosensors and the transduction of molecular recognition, Anal. Chem. 63 (1991), 393A-405A.
2. T.A. Kelly and G.D. Christian, Capillary flow injection analysis for enzyme assay with fluorescence detection, Anal. Chem. 54 (1982), 1444-1445.
3. J. Ruzicka and E.H. Hansen, Flow Injection Analysis, 2nd Edition, Wiley, New York, 1988.
4. G. Blankenstein, U. Spohn, F. Preuschoff, J. Thömmes, and M.-R. Kula, Multichannel flow-injection-analysis biosensor system for on-line monitoring of glucose, lactate, glutamine, glutamate and ammonia in animal cell culture, Biotechnol. Appl. Biochem. 20 (1994), 291-307.
5. M.M. Dittmann, K. Wienand, F. Bek, and G.P. Rozing, Theory and practice of capillary electrochromatography, LC-GC Intl. 13 (1995), 800-814.
6. G.J.M. Bruin and A. Paulus, Biopolymer separations with capillary electrophoresis, Anal. Methods Instrum. 2 (1995), 3-26.
7. N.M. Schultz and R.T. Kennedy, Rapid immunoassays using capillary electrophoresis with fluorescence detection, Anal. Chem. 65 (1993), 3161-3165.

8. O.-W. Reif, R. Lausch, T. Scheper, and R. Freitag, Fluorescein isothiocyanate-labeled protein G as an affinity ligand in affinity/immunocapillary electrophoresis with fluorescence detection, Anal. Chem. 66 (1994), 4027-4033.
9. K. Shimura and B. Karger, Affinity probe capillary electrophoresis: analysis of recombinant human growth hormone with a fluorescent labeled antibody fragment, Anal. Chem. 66 (1994), 9-15.
10. N.M. Schultz, L. Huang, and R.T. Kennedy, Capillary electrophoresis-based immunoassay to determine insulin content and insulin secretion from single islets of Langerhans, Anal. Chem. 67 (1995), 924-929.
11. D. Schmalzing, W. Nashabeh, X.-W. Yao, R. Mhatre, F.E. Regnier, N.B. Afeyan, and M. Fuchs, Capillary electrophoresis-based immunoassay for cortisol in serum, Anal. Chem. 67 (1995), 606-612.
12. L. Tao and R.T. Kennedy, On-line competitive immunoassay for insulin based on capillary electrophoresis with laser-induced fluorescence detection, Anal. Chem. 68 (1996), 3899-3906.
13. J. Bao and F.E. Regnier, Ultramicro enzyme assays in a capillary electrophoretic system, J. Chromatog. 608 (1992), 217-224.
14. D. Wu and F.E. Regnier, Native protein separations and enzyme microassays by capillary zone and gel electrophoresis, Anal. Chem. 65 (1993), 2029-2035.
15. H.M. Widmer, Trends in industrial analytical chemistry, Trends Anal. Chem. 2 (1983), VIII-X.
16. A. Manz, N. Graber, and H.M. Widmer, Miniaturized total chemical analysis systems: a novel concept for chemical sensing, Sensors Actuators B 1 (1990), 244-248.
17. see, for e.g., (a) Micro Total Analysis Systems (A. van den Berg, P. Bergveld, eds.; Kluwer Academic Publishers, Dordrecht) 1995, 1. (b) Proceedings of the 2nd International Symposium on Miniaturized Total Analysis Systems, μTAS'96 (Basel, November 19-22, 1996) 1996.
18. A. Manz, Y. Miyahara, J. Miura, Y. Watanabe, H. Miyagi, and K. Sato, Design of an open-tubular column liquid chromatograph using silicon chip technology, Sensors Actuators B 1 (1990), 249-255.
19. G. Ocvirk, E. Verpoorte, A. Manz, M. Grasserbauer, and H.M. Widmer, High performance liquid chromatography partially integrated onto a silicon chip, Anal. Methods Instrum. 2 (1995), 74-82.
20. S. Cowen and D.H. Craston, An on-chip miniature liquid chromatography system: design, construction and characterization, Micro Total Analysis Systems (A. van den Berg, P. Bergveld, eds.; Kluwer Academic Publishers, Dordrecht) 1995, 295-298.
21. E.M.J. Verpoorte, B.H. van der Schoot, S. Jeanneret, A. Manz, H.M. Widmer, and N.F. de Rooij, Three-dimensional micro flow manifolds for miniaturized chemical analysis systems, J. Micromech. Microeng. 4 (1994), 246-256.
22. A van den Berg and P. Bergveld, Developments of μTAS concepts at the MESA research institute, Proceedings of the 2nd International Symposium on Miniaturized Total Analysis Systems, μTAS'96 (Basel, November 19-22, 1996) 1996, 9-15.
23. P. Woias, M. Richter, E. Yacoub-George, H. Wolf, and T. Abel, A micromachined open tubular reactor for heterogeneous immunoassays, Proceedings of the 2nd International Symposium on Miniaturized Total Analysis Systems, μTAS'96 (Basel, November 19-22, 1996) 1996, 256.
24. D.J. Strike, P- Thiébaud, A.C. van der Sluis, M. Koudelka-Hep, and N.F. de Rooij, Glucose measurement using a micromachined open tubular heterogeneous enzyme reactor (MOTHER), Microsystem Technologies 1 (1994), 48-50.

25.J. Drott, K. Lindström, L. Rosengren, and T. Laurell, Porous silicon as the carrier matrix in microstructured enzyme reactors yielding high enzyme activities, J. Micromech. Microeng. 7 (1997), 14-23.
26.A. Manz, C.S. Effenhauser, N. Burggraf, D.J. Harrison, K. Seiler, and K. Fluri, Electroosmotic pumping and electrophoretic separations for miniaturized chemical analysis systems, J. Micromech. Microeng. 4 (1994), 257-265.
27.F. von Heeren, E. Verpoorte, A. Manz, and W. Thormann, Characterization of electrophoretic sample injection and separation in a gel-filled cyclic planar microstruture, J. Microcolumn Sep. 8 (1996), 373-381.
28.S.C. Jacobson, R. Hergenröder, L.B. Koutny, R.J. Warmack, and J.M. Ramsey, Effects of injection schemes and column geometry on the performance of microchip electrophoresis devices, Anal. Chem. 66 (1994), 1107-13.
29.D.E. Raymond, A. Manz, and H.M. Widmer, Continuous sample pretreatment using a free-flow electrophoresis device integrated onto a silicon chip, Anal. Chem. 66 (1994), 2858-2865.
30.D.E. Raymond, A. Manz, and H.M. Widmer, Continuous separation of high molecular weight compounds using a microliter volume free-flow electrophoresis microstructure, Anal. Chem. 68 (1996), 2515-2522.
31.S.C. Jacobson, R. Hergenröder, L.B. Koutny, and J.M. Ramsey, Open channel electrochromatography on a microchip, Anal. Chem. 66 (1994), 2369-2373.
32.A.W. Moore Jr., S.C. Jacobson, and J.M. Ramsey, Microchip separations of neutral species via micellar electrokinetic capillary chromatography, Anal. Chem. 67 (1995), 4184-4189.
33.F. von Heeren, E. Verpoorte, A. Manz, and W. Thormann, Micellar electrokinetic chromatography separations and analyses of biological samples on a cyclic planar microstructure, Anal. Chem. 68 (1996), 2044-2053.
34.J.P. Kutter, S.C. Jacobson, and J.M. Ramsey, Integrated microchip device with electrokinetically controlled solvent mixing for isocratic and gradient elution in micellar electrokinetic chromatography, Anal. Chem. 69 (1997), 5165-5171.
35.D.J. Harrison, A. Manz, Z. Fan, H. Lüdi, and H.M. Widmer, Capillary electrophoresis and sample injection systems integrated on a planar glass chip, Anal. Chem. 64 (1992), 1926-1932.
36.A. Manz, D.J. Harrison, E.M.J. Verpoorte, J.C. Fettinger, H. Lüdi, and H.M. Widmer, Planar chips technology for miniaturization and integration of separation techniques into monitoring systems: capillary electrophoresis on a chip, J. Chromatog. 593 (1992), 253-258.
37.K. Seiler, D.J. Harrison, and A. Manz, Planar glass chips for capillary electrophoresis: repetitive sample injection, quantitation, and separation efficiency, Anal. Chem. 65 (1993), 1481-1488.
38.D.J. Harrison, K. Fluri, K. Seiler, Z. Fan, C.S. Effenhauser, and A. Manz, Micromachining a miniaturized capillary electrophoresis-based chemical analysis system on a chip, Science 261 (1993) 895-897.
39.C.S. Effenhauser, A. Manz, and H.M. Widmer, Glass chips for high-speed capillary electrophoresis separations with submicrometer plate heights, Anal.Chem. 65 (1993) 3550-3560.
40.C.S. Effenhauser, A. Paulus, A. Manz, and H.M. Widmer, High speed separation of "anti-sense" oligonucleotides on a micromachined capillary electrophoresis device, Anal.Chem. 66 (1994), 2949-2953.

41.S.C. Jacobson, R. Hergenröder, L.B. Koutny, and J.M. Ramsey, High-speed separations on a microchip, Anal. Chem. 66 (1994), 1114-1118.
42.N. Burggraf, A. Manz, N.F. de Rooij, and H.M. Widmer, Synchronized cyclic capillary electrophoresis: a novel concept for high-performance separations using low voltages Anal. Methods Instrum. 1 (1993), 55-59.
43.N. Burggraf, A. Manz, C.S. Effenhauser, E. Verpoorte, N.F. de Rooij, and H.M. Widmer, Synchronized cyclic capillary electrophoresis - a novel approach to ion separations in solution, J. High Resolut. Chromatog. 16 (1993), 594-596.
44.N. Burggraf, A. Manz, E. Verpoorte, C.S. Effenhauser, H.M. Widmer, and N.F. de Rooij, A novel approach to ion separations in solution: synchronized cyclic capillary electrophoresis, Sensors Actuators B 20 (1994), 103-110.
45.D.J. Harrison, P.G. Glavina, and A. Manz, Towards miniaturized electrophoresis and chemical analysis systems on silicon: an alternative to chemical sensors, Sensors Actuators B 10 (1993), 107-116.
46.A.T. Woolley and R.A. Mathies, Ultra-high-speed DNA fragment separations using microfabricated capillary array electrophoresis chips, Proc. Natl. Acad. Sci. USA 91 (1994) 11348-11352.
47.A. Manz, Miniaturized chemical analysis systems based on electroosmotic flow, Proceedings of the Tenth Annual International Workshop on Micro Electro Mechanical Systems (Nagoya, Japan, January 26-30, 1997), 1997, 14-18.
48.R. Freaney, A. McShane, T.V. Keaveny, M. McKenna, K. Rabenstein, F.W. Scheller, D. Pfeiffer, G. Urban, I. Moser, G. Jobst, A, Manz, E. Verpoorte, H.M. Widmer, D. Diamond, E. Dempsey, F.J.S. de Viteri, and M. Smyth, Novel instrumentation for real-time monitoring using miniaturized flow systems with integrated biosensors, Ann. Clin. Biochem. 34 (1997), 291-302.
49.E. Dempsey, D. Diamond, M.R. Smyth, G. Urban, G. Jobst, I. Moser, E.M.J. Verpoorte, A. Manz, H.M. Widmer, K. Rabenstein, and R. Freaney, Design and development of a miniaturised total chemical analysis system for on-line lactate and glucose monitoring in biological samples, Anal. Chim. Acta 346 (1997), 341-349.
50.M. Busch, J. Schmidt, S.A. Rothen, C. Leist, B. Sonnleitner, and E. Verpoorte, μTAS meets biotechnology: micromachined flow systems combined with biosensor arrays for bioprocess monitoring, Proceedings of the 2nd International Symposium on Miniaturized Total Analysis Systems, μTAS'96 (Basel, November 19-22, 1996) 1996, 120-121.
51.A.G. Hadd, D.E. Raymond, J.W. Halliwell, S.C. Jacobson, and J.M. Ramsey, Microchip device for performing enzyme assays, Anal. Chem. 69 (1997), 3407-3412.
52.L.B. Koutny, D. Schmalzing, T.A. Taylor, and M. Fuchs, Microchip electrophoretic immunoassay for serum cortisol, Anal. Chem. 68 (1996), 18-22.
53.N. Chiem and D.J. Harrison, Microchip-based capillary electrophoresis for immunoassays: analysis of monoclonal antibodies and theophylline, Anal. Chem. 69 (1997), 373-378.
54.D. Schmalzing, L.B. Koutny, T.A. Taylor, W. Nashabeh, and M. Fuchs, Immunoassay for thyroxine (T4) in serum using capillary electrophoresis and micromachined devices, J. Chromatog. B 697 (1997), 175-180.
55.F. von Heeren, E. Verpoorte, A. Manz, and W. Thormann, Micellar electrokinetic chromatography separations and analyses of biological samples on a cyclic planar microstructure, Anal. Chem. 68 (1996), 2044-2053.

GLASS CHANNELS AND CAPILLARY INJECTORS FOR CAPILLARY ZONE ELECTROPHORESIS

Yolanda Fintschenko[1], Vincent Spiering[2], Johannes van der Moolen[3], Gert-Jan Burger[2], Albert van den Berg[1]

[1] MESA Research Institute, University of Twente
PO Box 217, 7500 AE Enschede, The Netherlands

[2] Twente MicroProducts, PO Box 217, NL-7500 AE Enschede; tmp@eltn.utwente.nl

[3] University of Amsterdam, Nieuwe Achtergracht 166, NL-1018 WV Amsterdam

0. Abstract

Novel glass and silicon microstructures and their application in chemical analysis are presented. The micro technologies comprise (deep) dry etching, thin layer growth and anodic bonding. With this combination it is possible to create high resolution electrically isolating silicon dioxide structures with aspect ratio's similar to those possible in silicon. Main applications are chemical separation methods such as High Performance Liquid Chromatography (HPLC) or Capillary Zone Electrophoresis (CZE). Beside these channel structures, a capillary connector with very low dead and mixing volume has been designed and fabricated for use in correlation CZE, and demonstrated by determining the precision of consecutive single injections.

Keywords: Microchannels, microfabrication, capillary zone electrophoresis (CZE), high performance liquid chromatography (HPLC)

1. Introduction

In the field of chemical analysis, especially in the area of separation techniques, planar microstructures have a large potential, because they offer a number of advantages as compared to fused silica capillary-based systems. Microfabrication techniques allow for the (i) realization of microchannels with variable cross-sections and depths, (ii) easy manipulation of the dimensions and cross-sectional geometry of microchannels, (iii) integration of detection components into the fluid channels, and (iv) minimization of the dead volume associated with interconnections. Also, miniaturization leads to improved separation efficiencies and higher resolution in shorter times than conventional sized methods. Thus, a generic separation method when used on the seconds time scale becomes a generic multianalyte sensor capable of quasi-continuous monitoring. For this reason, the miniaturization of capillary zone electrophoresis (CZE) and high performance liquid chromatography (HPLC) using microfabrication is of interest.

Conventional CZE typically uses fused silica columnar capillaries of varying inner diameters (10 μm - 100 μm) [1]. It is possible to pump a solution through narrow bore capillaries using an electric field rather than moving parts and differential pressure as in high pressure liquid chromatography (HPLC). This is due to the electroosmotic flow or EOF. EOF results from the movement of buffer cations organized at the negatively charged capillary wall as they migrate towards the cathode, dragging the bulk solution in the same direction. The resulting flow has a plug profile instead of the parabolic profile associated with pressure driven flow at similar Reynolds numbers. As a result, ideally with CZE there is no band broadening due to solution movement.

Several groups have been working on the fabrication of planar structures for capillary electrophoresis and liquid chromatography. Recently, Colyer et. al. published a minireview of microchip capillary electrophoresis systems, or CZE "on-a-chip"[2]. Manz *et al.* [3] was the first to use anisotropically etched channels in silicon for liquid chromatography, later followed by work from Ramsey *et al.* [4] and Cowen *et al.* [5]. Simultaneously, work was concentrated on CZE, where Manz and Harrison [6,7] were the first to use planar etched glass structures for this purpose. By etching small bore capillary channels in planar glass substrates, Manz *et al.* were able to integrate the injector with the separation channel on one structure [6,8]. Rapid improvements of instrumentation led to the demonstration of this technology in the separation of amino acids with 100,000 to 160,000 theoretical plates and mostly baseline resolution in 2 minutes [9] to 15 seconds [10]. Other structures described by Jacobson *et al.* are capable of making entire separations in a total of 150 ms [10]. Thus, both the integration of two aspects of the CE system were effected and the analytical figures of merit were improved while decreasing total analysis time. The impact of CZE "on a chip" is the elevation of a separation technique to a quasi-continuous, generic chemical sensor.

Despite these contributions, planar structures have several problems in practice. The microchannels frequently need to be connected to conventional system components such as an injector, sample (micro)vial, detector and waste. Thus, the interconnection techniques and system concept are of crucial importance. Our planar Micro System Fluid (MFS) concept was described earlier [11], and in this paper a capillary connection to microchannels is described. This technique can also be used to connect two capillaries with very low dead volume. This can facilitate the use of capillary interconnects as a standard interconnect for hybrid miniaturized systems.

Since the quality and speed of separations for both LC and CZE applications depends on channel geometry, it is important to employ a technique that enables the fabrication of microchannels with a virtually arbitrary cross-section. It is also interesting to see if separation channels for capillary electrophoresis, where high voltages on the order of several kV's are used, can be fabricated using silicon etching techniques. The conventional micro system technology (MST) way to fabricate glass microchannels is to etch one or two wafers in HF and then bond these on top of each other [7,12]. This method has the disadvantage that in glass only isotropically etched channels are feasible to make. This severely limits the geometry's available for channel design. For this, the Black Silicon Method (BSM) [13,14] is applied. With the BSM technology, deep high aspect ratio channels can be dry etched. The cross-sectional shape can be varied by means of process variable control (type and flow of gasses, pressure). In this way channels shapes from square to elliptical or round can be fabricated. Particular emphasis is laid on a technology that enables the realization of an enclosed, fully insulated microchannel with very thin walls for efficient heat transfer.

Correlation CZE, which uses a repeating pseudo-randomly generated pattern of injections in rapid succession, also requires rapid switching of the injection potential and separation potential [15]. The result is a signal over a wide frequency range, and it is possible to gain the same signal-to-noise increase typically associated with Fourier transform. However, it is important that each injection is reproducible, and that the separation voltage is not interrupted to make each injection. For this reason it is of interest to apply microfabrication to make an electrically isolated connection from the run buffer and the sample to the separation capillary for repetitive injections.

2. Experimental

Transparent glass microchannel fabrication

The main process steps are as follows (Fig. 1.):
1. Channel etching in silicon
2. Channel material deposition on silicon
3. Machining of the Pyrex™ wafer
4. Anodic bonding of the Pyrex™ wafer on top of the silicon wafer
5. Etching the back of the silicon in KOH
6. Addition of a protection/planarization layer over the channels

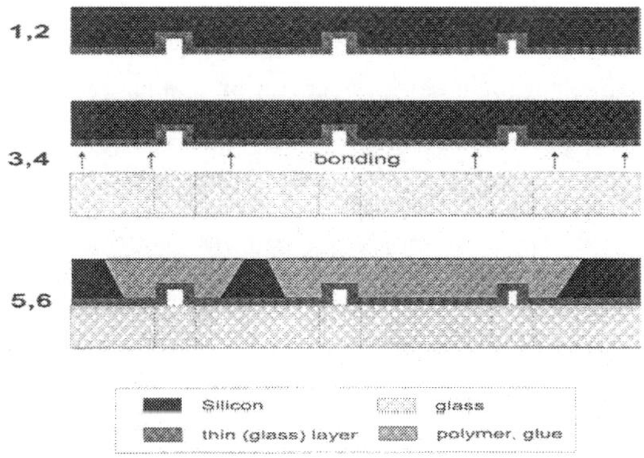

Fig. 1. Process scheme for glass microchannels.

1. Channel etching in silicon
With the BSM technology, deep channels with a variable cross-sectional area can be etched.

2. Channel material deposition on silicon
Materials applied for structures are usually low stress LPCVD silicon-nitride, and/or TEOS oxide, sometimes combined with a thin layer of polysilicon. These layers are required for mechanical strength, electrical isolation, and to stop the etch process for backside KOH-etching. The polysilicon prevents local release at <111> faces of silicon during the KOH etch back process. The total thickness of the layers is restricted by the requirements for successful anodic bonding.

3. Machining of the Pyrex™ wafer
This is accomplished by means of conventional precision machining techniques. The inlet/outlet holes are drilled in the glass wafer.

4. Anodic bonding of the Pyrex™ wafer on top of the silicon wafer
The machined glass wafer is bonded to the silicon wafer with an intermediate layer of oxide and/or nitride.

5. Etching the back of the silicon in KOH
This step can be done without patterns (a complete etch of the back of the silicon wafer) but also with a pattern in silicon to acquire more mechanical strength of the total structure.

Especially in combination with the next step, this will prevent breaking of the tiny glass channels walls.

6. Addition of the protection/planarization layer over the channels

Several layers/materials such as thin films, polymers, and glue can be applied for this purpose. If a transparent material is used, the whole structure remains transparent from the top and the bottom side, which allows for optical detection methods.

Relevant steps are the thin layer deposition followed by an anodic bonding process. Critical to the success of the final structure are the mechanical *and* electrically isolating properties of the thin layers [16] as well as the heat transfer coefficient and absorption spectrum of the various substrates.

Figure 2 shows the outside of a complete glass covered channel on a glass substrate (step 5 carried out without mask). These structures are of great interest in the field of electrophoresis and other applications where electrical isolation of the structure is essential. Currently, specially shaped glass microchannels are to be tested for their fluid resistance and electrical isolation in concentration experiments for pre-separation applications.

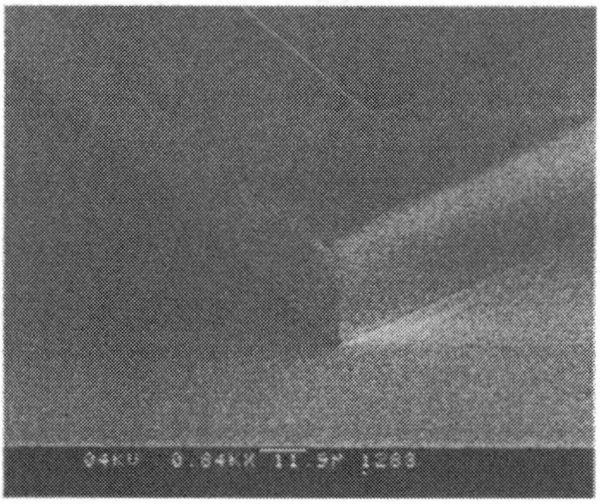

Fig. 2. Complete closed glass microchannel on a glass substrate.

Capillary connector fabrication

Connections between the capillary and the sample injector or between several capillary's are realized via micro channels in silicon. The capillary connection chips were designed and fabricated by Twente MicroProducts [17]. The sample injection holes, channels, and wafer-through holes are realized by a combination of dry and wet etching steps in a silicon substrate. The channel dimensions are matched to the inner diameter of the capillary, thus minimizing the dead and mixing volume of the connection. The channels are electrically isolated with a thin layer of silicon oxide and they are covered by a glass plate. The electrically isolated connectors are fabricated in a very reproducible way. Essential for a suitable connection are the application of high performance dry etch techniques, the design of a pre-aligned connection, and the gluing technique.

3. Results and Discussion

Figures 3 and 4 show a diagram and a SEM of a cross-section of a glued capillary connection with a microchannel. Figure 5 shows a photograph of the microconnector incorporated into the correlation CZE apparatus. The dead volume was less than 0.5 nL, and the alignment was within 5 μm. The structure was able to sustain voltages up to 250V/cm across the channel, and pressures up to 120 bar.

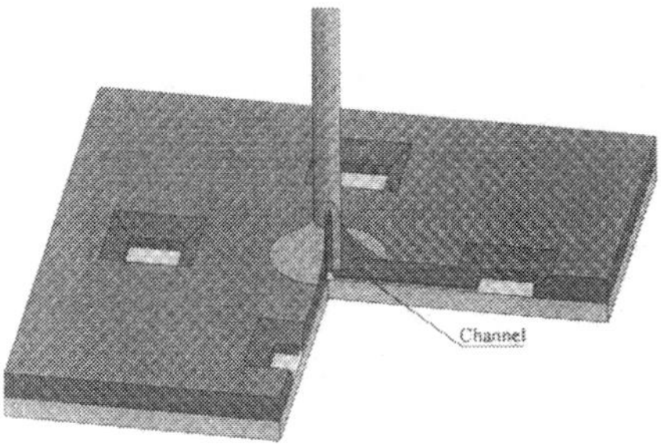

Fig. 3. Capillary connector combined on-chip with sample injector holes.

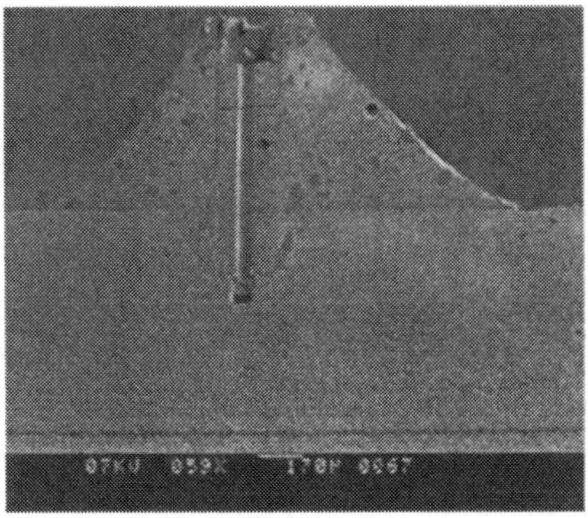

Fig. 4. Cross section of glued 280/75 μm capillary

The connector has been tested by the Laboratory of Analytical Chemistry at the University of Amsterdam with a set-up as given in Figure 5 [15]. A photograph of the injection device using the microchannel connector with a fused silicon capillary is seen in Figure 6.

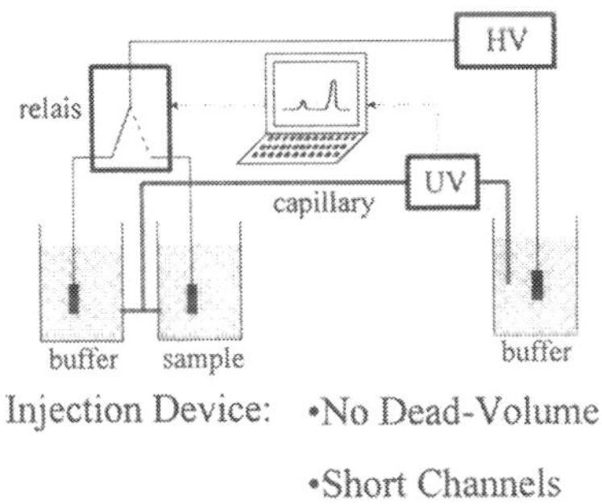

Fig. 5. Correlation CZE system

Fig. 6. Photograph of capillary and two sample injector connected via the chip in a correlation CZE system

After proven precision of eight consecutive injections (see Fig. 7) the so-called correlation capillary zone electrophoresis (CCZE) technique was applied [18]. The capillary connector showed reproducible behavior (RSD's <1 %).

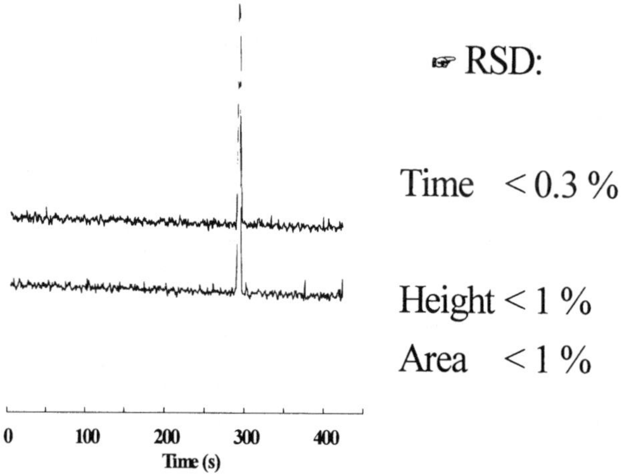

Fig. 7. Precision of eight consecutive injections

The results in Figure 8 show the improvement of the limit of detection as a result of the CCZE technique. A comparison of the typical electropherogram to the correlogram shows the signal to noise improves eight-fold in the latter case.

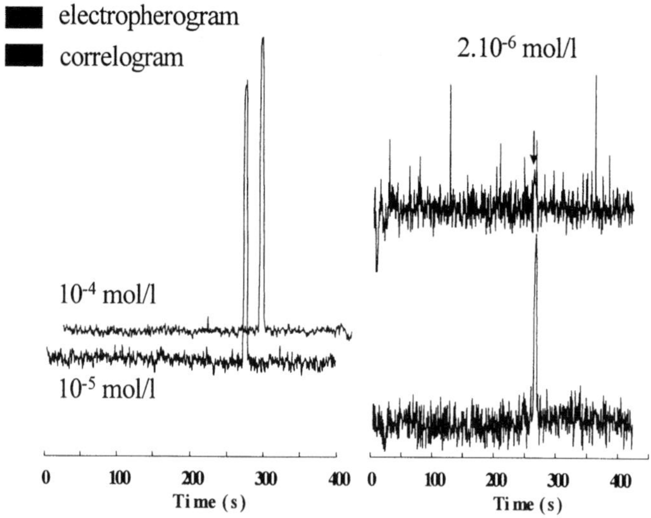

Fig. 8. Electropherogram vs. Correlogram

4. Conclusions

Microchip technology is very well suited for accurate fabrication of complex structures. Beside the fabrication of glass micro channels, it is demonstrated that connecting a microchip injection device to a fused silica capillary can be done quit easily. SEM photographs proved that the connection is excellent, with extremely low dead-volume. The reproducibility in peak height and peak area of the electrokinetic injection was very good, with RSD-values better than 1.1 %. An improved reproducibility with the microchip injection device compared to conventional injections has been demonstrated. With correlation CZE, the injection reproducibility was even better than the reproducibility with conventional CZE with the microchip injection device. Correlation experiments showed a considerable improvement in S/N ratio at high as well as low concentration. An 8-fold improvement in the detection limit has been demonstrated.

Acknowledgements

The authors recognize that (in alphabetical order) Erwin Berenschot, Meint de Boer, and Henri Jansen are behind most of the technological ideas and tricks presented in this paper. We greatly appreciate their useful contributions.

References

1. J.W. Jorgenson, K.D. Lukacs, Zone electrophoresis in open-tubular glass capillaries, Anal. Chem., 53 (1981) 1298.
2. C.L. Colyer, T. Tang, N. Chiem, D.J. Harrison, Clinical potential of microchip capillary electrophoresis systems, Electrophoresis, 18 (1997) 1733-1741.
3. A. Manz, Y. Miyahara, J. Miura, Y. Watanabe, H. Miyagi, and K. Sato, Design of an open-tubular column liquid chromatograph using silicon chip technology, Sensors and Actuators, B1 (1990) 249-255.
4. S.C. Jacobson,. R. Hergenroeder, L.B. Koutny, J.M. Ramsey, Open channel electrochromatography on a microchip, Anal. Chem., 66 (1994) 2369-2373
5. S. Cowen , D.H. Craston, An on-chip miniature liquid chromatography system: design, construction, and characterization, in 'Micro Total Analysis Systems', Kluwer Academic Publishers, Dordrecht NL , 1992, p. 295.
6. A. Manz, D.J. Harrison, E.J.M. Verpoorte, J.G. Fettinger, H. Lüdi, H.M. Widmer, Planar chips technology for miniaturization and integration of separation techniques into monitoring systems. Capillary electrophoresis on a chip. J. Chromatogr., 593 (1992) 253-258.
7. D.J. Harrison, A. Manz, Z.Fan, H. Ludi, J.M. Widmer, Capillary electrophoresis and sample injection systems integrated on a planar glass chip, Anal. Chem., 64 (1992) 1926-1932.
8. K. Seiler, D.J. Harrison, A. Manz, Planar glass chips for capillary electrophoresis: repetitive sample injection, quantitation, and separation efficiency, Anal. Chem., 65 (1993) 1481-1488.
9. C.S. Effenhauser, A. Manz, H.M. Widmer, Glass chips for high-speed capillary electrophoresis separations with submicrometer plate heights, Anal. Chem., 65 (1993) 2637-2642.
10. S.C. Jacobson, R. Hergenroeder, L.B. Koutny, J.M. Ramsey, High-speed separations on a microchip, Anal. Chem., 66 (1994) 1114-1118.
11. T.S.J. Lammerink, V.L. Spiering, M. Elwenspoek, J.H.J. Fluitman, and A. van den Berg, Modular concept for fluid handling systems: a demonstrator micro analysis system, 9th IEEE Workshop on Micro Electro Mechanical Systems (MEMS'96), San Diego, USA, Feb. 11-15,1996, pp. 389-394.
12. M.-A Grétillat, F. Paoletti, P. Thiébaud, S. Roth, M. Koudelka-Hep, N.F. de Rooij, A new fabrication method of borosilicate glass capillary tubes with lateral inlets and outlets,

Proc. Eurosensors X The 10th European Conference on Solid-State Transducers, 1996, 1 (1996) 259-262.

13. H.V. Jansen, Plasma etching in microtechnology, Thesis, The University of Twente, MESA Research Institute, Enschede, The Netherlands, 1996.
14. R.W. Tjerkstra, M. de Boer, E. Berenschot, J.G.E. Gardeniers, A. van den Berg, M. Elwenspoek, Etching technology for microchannels, 10th IEEE Workshop on Micro Electro Mechanical Systems (MEMS 97), Nagoya, Japan., Jan. 26-30 1997, pp. 147-152.
15. J.N. van der Moolen, D.J. Louwerse, H. Poppe, H.C. Smit, Correlation capillary zone electrophoresis: a novel technique to decrease detection limits, Chromatographia, 40 (1995) 368.
16. V.L. Spiering, T.S.J. Lammerink, H. Jansen, J.H.J. Fluitman, A. van den Berg, Technologies and microstructures for separation techniques in chemical analysis, Proc. SPIE - The International Society for Optical Engineering, 2882 (1996) 91-100.
17. Twente MicroProducts, PO Box 217, 7500 AE Enschede, The Netherlands, tel +3153 4894264 fax +3153 4358259, tmp@eltn.utwente.nl.
18. J.N. van der Moolen, H.F.M. Boelens, H. Poppe, H.C. Smit, Origin and correction of bias caused by sample injection and detection in capillary zone electrophoresis, J. Chromatogr. A, 744 (1996) 103-113.

A NEW FABRICATION METHOD FOR IN-PLANE MICRO FLUID HANDLING COMPONENTS AND ANALYSIS SYSTEMS

R.E. Oosterbroek, J.W. Berenschot, S. Schlautmann, T.S.J Lammerink, A. van den Berg and M.C. Elwenspoek
MESA Research Institute, University of Twente,
P.O. Box 217, 7500 AE Enschede, The Netherlands
R.E.Oosterbroek@el.utwente.nl

0. Abstract

This paper reports a new method to fabricate in-plane oriented micro structures with use of an-isotropic wet chemical etching techniques. The method is very engaging due to the simplicity, low demands on cleanroom resources, the high accurate controllability and wafer-thickness independence. Integration in micro systems is facilitated by the unique possibility to obtain in-plane working devices. The functionality is demonstrated by performed analytical and numerical simulations as well as measurements of a duckbill valve.

Keywords: μ-TAS, valves, etching, simulation

1. Introduction

In the field of chemical analysis, there is need on small, portable and cheap (disposable) systems of sensors and actuators. Examples of these actuators are valves to control the fluid streams through the system [1,2]. A way to obtain these small systems is to utilize wet chemical etching techniques, such as with use of potassium hydroxide (KOH). This bulk micro-machining process is fast, cheap and puts only small demands on available resources and is therefore a very interesting alternative to for example more expensive dry etching methods. During KOH etching of silicon, material will be attacked along the {111} planes which causes anisotropy. For [001] wafers this results in a 54.7° etch pit, relatively to the wafer surface. This anisotropy can be exploited to etch thin {111} plates, used as building blocks of micro devices and systems.

2. The etching process

For obtaining {111} plates, figure 1C. shows a known principle. From the top side of the wafer a hole is etched until it stops on the mask at the bottom side. At the same time, a self-stopping cavity is etched from the bottom side, positioned aside of the top hole, such that the {111} wall, which separates both cavities, is not etched through. After removing the etch mask, a thin {111} plate is left.

The dimensions of the plate are defined by, among others, the wafer thickness. This relation means that the design is strongly limited to the wafer thickness variations. Because these variations can be in the order of ±20μ for a 380μm thick wafer, the plate thickness can be controlled with a maximum precision of about 12μm. This precision is far too poor for producing micro devices. Another disadvantage is the necessary double-sided precision aligning of the top and bottom mask. Any method to reduce these drawbacks will be very valuable.

Basically the mentioned drawbacks result from using two masks which both are necessary to define the structure. For avoiding this, we used the effect of switching over of {111} planes when KOH etch cavities on both sides of a wafer meet, such that a convex hole is obtained (figure 1A-2). The sharp *convex* corners are not stable, so that they will be blunted until straight walls are obtained (figure 1A-3). Further etching will result in a *concave* hole, again stopped on {111} planes (figure 1A-4). A thin {111} plate is obtained if a KOH cavity is etched on one side, closely positioned to the concave hole [3]. The function of the top mask can be reduced to switching-over the convex wall to concave. Therefore the wafer thickness dependencies are eliminated and the aligning of the top mask has become non-critical. Expensive high precision double-sided mask alignment is no longer needed.

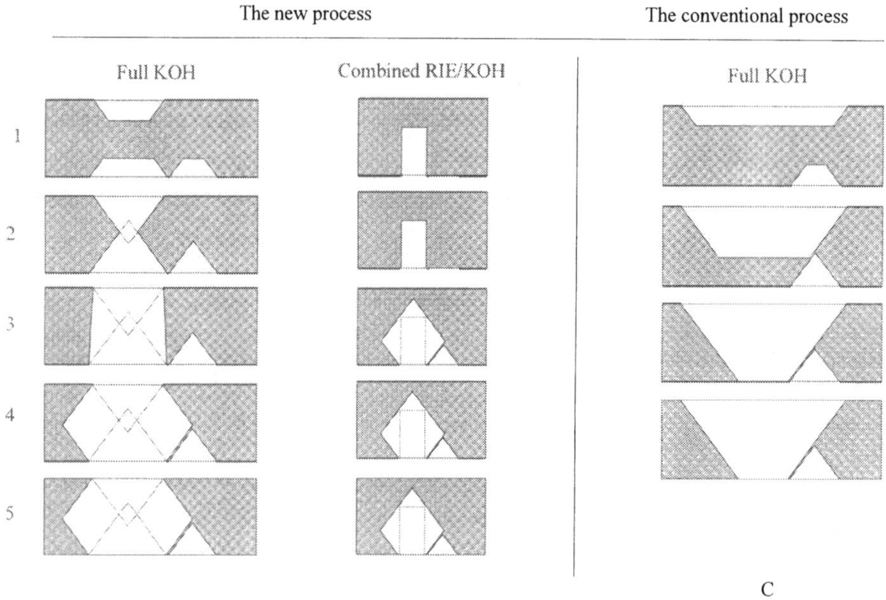

Figure 1: *Process steps of the two new etching methods and the conventional method.*

A variant of method A is shown in figure 1-B. Here, first a vertical trench is etched with an-isotropic reactive ion etching (RIE). During the following KOH etch step, a concave hole will be created like the process before. A structure, closed at the top side is obtained.

3. Designing with the new etching technique

The information of 3D structures are partially defined by the mask and by the crystallographic lay-out of the wafer. Therefore the masks look very elementary as is shown in figure 2. They consist of well-dimensioned and positioned lines. When using the full wet etching method, design parameters of the {111} plate are: the size of the top mask opening, the length and width of the both bottom mask openings, the relative position between the two bottom mask openings and of less influence the aligning between the top and bottom mask. For the combined RIE/KOH method, only the dimensions and relative aligning of the bottom mask openings are of importance, in combination with the obtained etch depth.

For the full wet etching technique, figure 3 shows the effects of changing the mentioned parameters. The height of the plate depends on the position (t_h) of the concave angle of the switched {111} planes relative to the v groove. If the top mask is moved to the left, this angle, and thus the top mask is determining the plate height (figure 3a-1). In this situation, the wafer thickness is controlling the geometry. This is not wished. So moving the top mask to the right makes it thickness independent (figure 3a-3). By a change of the relative aligning of the top and bottom mask, the concave profile can be changed from symmetric to strong a-symmetric to both sides.

Changing the dimensions of the v-grooves, t_m, offers the possibility to obtain plates with different sizes as is shown in figures 3B. This is very interesting for the design of valves with different characteristics in one wafer. For making a passive check valve, we selectively bonded a glass wafer to the bottom side. A duckbill type of valve is obtained because one of the four edges of the plate is left free such that plate will bend and fluid will flow under the plate rim when pressure is applied. A SEM picture of a cross section of the duckbill is shown in figure 2. The hydraulic resistance is strongly direction sensitive because the free plate edge will be pressed against the glass plate when the flow direction is reversed. Using different dimensions of the plate will result in differences in stiffness and thus differences in resistance characteristic of the designed valve.

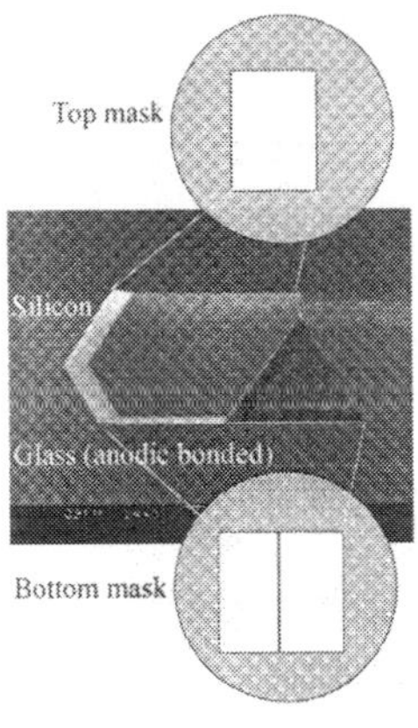

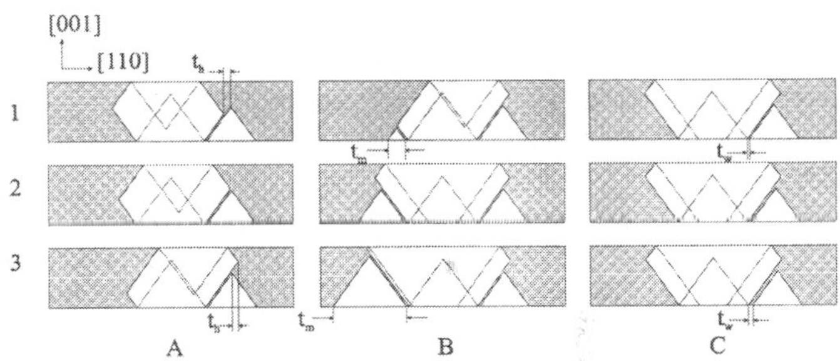

Figure 2: *Mask layouts and cross-section of the "duckbill" valve*

Figure 3: *Variation of the top mask position* t_h, *bottom mask opening* t_m *and distance* t_w

The thickness of the {111} plate can be varied by changing the distance between the openings in the lower mask (figure 3c). Because the crystallographic [111]/[001] selectivity during our etching tests were in the order of 1/100, the plate is etched thinner as well. This offers possibilities to reduce the plate thickness independent of the resolution of the mask pattern generator.

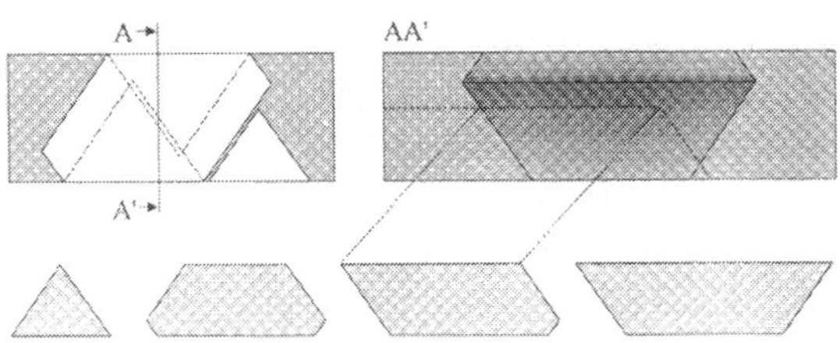

Figure 4: *Influence of the positioning of the v-groove on the {111} plate shape*

In the length direction, the cross-view section is shown in figure 4. The shapes of the plate can be varied from triangular to trapezium by changing the position and the size of the v-groove, relative to the etched-through hole. If triangular membranes are etched, it is possible to obtain high density valve arrays like it is shown in figure 5. Because v-grooves are used for the entrance channels, the design is ideally suitable in combination with capillary tubes. Large, compact arrays can be build, consisting of valves, each with a different resistance characteristic.

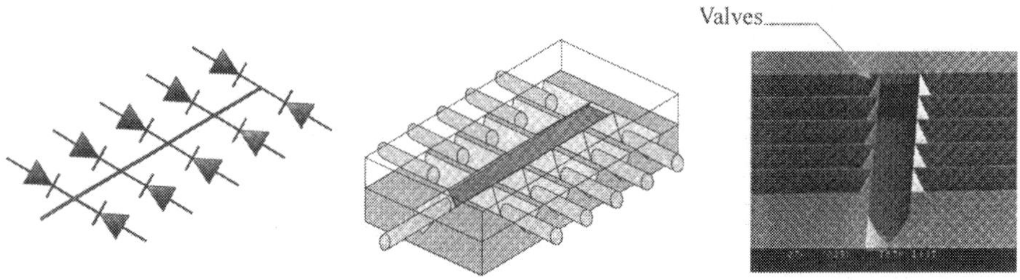

A B C

Figure 5: *Functional (A), art (B) and SEM (C) impression of the 2x5 check valve array*

Passive check valves are often used in micro-pumps [4]. The duckbill valve would be of interest in these applications because of the nice feature of in-plane supply and outlet channels. What is needed is a method to build two valves in one channel, each with an identical open direction (two diodes in series). A possible way to obtain this is shown in figure 6a. Due to etching in the {111} direction, the thinner plate will be etched away, leaving only the identical directed valves.

B C

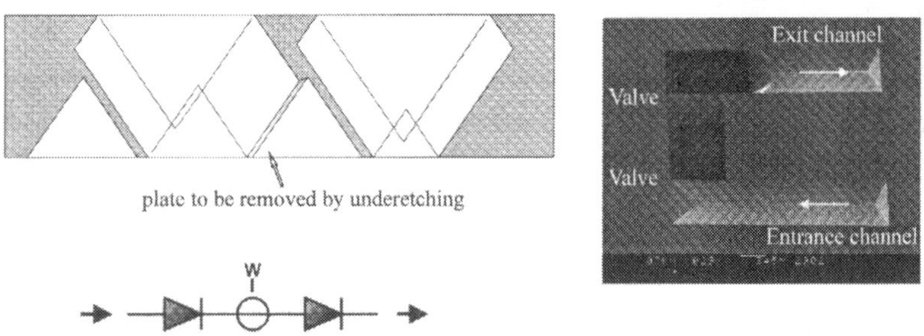

A

Figure 6: *Function diagram (A) and two designs (B,C) for creating duckbill valves in series for micro-pumps*

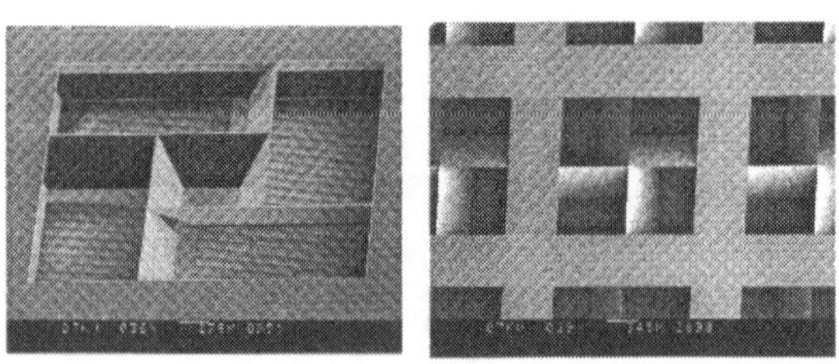

A B

Figure 7: *Examples of 3D spatial designs, e.g. structures for directing flow streams*

A more elegant way is shown in figure 6b. The {111} walls of the exit channel are only partially switched-over due to locally etching-through. The space between the two valves can be used as actuation chamber of the membrane pump.

Till now we concentrated on fluid control applications of the etching technique. Exploring the new etching possibilities resulted in a range of new three dimensional structures. Examples of such experiments are shown in figure 7a and b. Here the {111} plates are oriented in perpendicular directions, creating blades to direct for example flows in a rotating way.

4. Valve computation and characterization

To check whether the presented valve designs meet the wishes for fluid control and to obtain design formula's, analytical as well as numerical computations are performed. The number of geometric shapes of the plate give rise to the need of simplifications. In the used model, we imagined the duckbill valve to consist of a three-sided clamped square plate which is bending-out at the fourth, free edge, such that a flow opening arises.

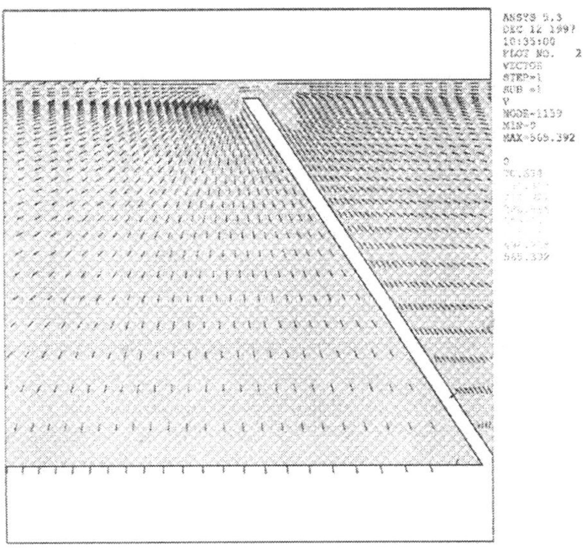

Figure 8: *Cross-section of the flow-pattern (vector plot) through the duckbill valve, computed with a de-coupled mechanical-flow finite element method (ANSYS 5.3). The liquid flows from the right side under the thin {111} plate to the left.*

To verify the analytical results and to extend the model, domain de-coupled finite element computations are performed. The pressure drop over the valve is computed for different deflection heights of the {111} plate. A second series of computations is performed to obtain the mechanical relation between the constant pressure acting on the membrane and the deflection of the free edge. Because both relations are expressed in terms of the deflection, a coupling can be made between both domains to obtain the flow-resistance relation. Results of these computations are shown in figure 8. From the simulations, it can be concluded that the hydraulic resistance of the valve in open position can be designed such that acceptable values are obtained. The leakage of the valve however cannot be modeled and must be obtained from characterization experiments.

First experimental characterizations are done by applying a defined flow on the valve in open as well as closed direction and measuring the pressure drop over the membrane. For a valve with dimensions 1mm length, 5μm thickness and 300μm height, a flow-pressure drop curve is given in figure 9. It shows that there is a high direction sensitivity when the flow-rate is increased. With the derived formulas we aim at optimizing this sensitivity to further reduce the leakage at small flow-rates, and decrease the resistance in the open-direction.

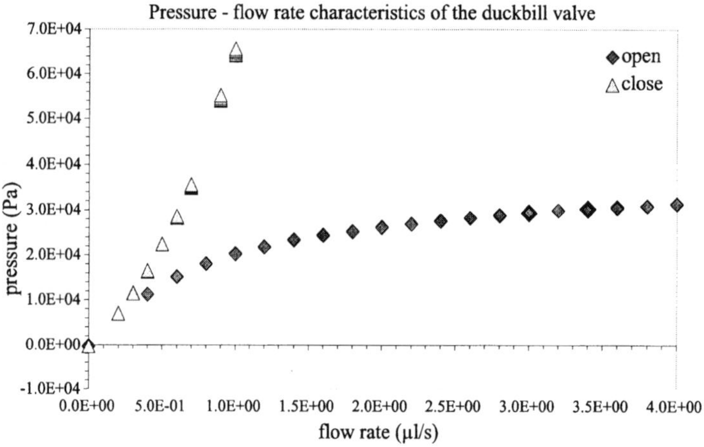

Figure 9: *Measured flow-pressure relation of the duckbill valve in open and close direction*

5. Conclusions

A new, high precision design technique has been developed, using the principle of switching-over of {111} planes. This technique enables to fabricate thin <111> oriented plates which can be used as building blocks in for example fluid control devices and systems. Devices such as passive check valves, valve arrays and valves for micro pumps as well as more sophisticated three dimensional structures have been fabricated to demonstrate the potential of the method. The functionality of the duckbill valve is simulated with analytical and numerical computations and verified by measurements. Theoretical models as well as experiments showed that a good direction-sensitive hydraulic resistance can be obtained with a good open resistance.

References

1. S. Shoji, M. Esashi, Microflow Devices and Systems, Journal of Micromechanics and Microengineering (4) , pp. 157-171, 1994
2. M.C. Elwenspoek, T.S.J. Lammerink, R. Miyake, J.H.J. Fluitman, Towards integrated microliquid handling systems, Journal of Micromechanics and Microengineering (4), pp. 227-245, 1994
3. J.W. Berenschot, R.E. Oosterbroek, T.S.J. Lammerink and M.C. Elwenspoek, Micromachining of {111} plates in <001> oriented silicon, Journal of Micromechanics and Microengineering, March? 1998
4. H.T.G. van Lintel, F.C.M van de Pol, S.Bouwstra, A piezoelectric micropump based on micro-machining of silicon, Sensors and Actuators, 15, 1988, pp. 153-167

A SYSTEM FOR FLUID DOSING IN THE NANOLITER RANGE

Sebastian Böhm, Dorota Pijanowska, Wouter Olthuis and Piet Bergveld
MESA Research Institute, University of Twente,
P.O. Box 217, 7500 AE Enschede, The Netherlands
e-mail: s.bohm@el.utwente.nl

0. Abstract

A system for the dosing of nanoliter amounts of fluids is presented. It is based on the displacement of fluid from a reservoir by the electrochemical production of gas bubbles at noble metal electrodes. The total volume of fluid that is dispensed can be controlled by the current that is sent through the electrodes. This simple principle was applied to calibrate a pH ISFET mounted in a flow through system by dispensing two different pH buffers in the carrier stream leading to the ISFET. The offset and sensitivity resulting from the two point calibration was in close agreement with the manual calibration procedure carried out before. First experiments were performed using a system machined in plastic using conventional techniques but it is shown that by adopting appropriate micromachining techniques a structure results that can easily be integrated in a microsystem for (bio)chemical analyses.

Keywords: electrochemical pump, liquid dosing

1. Introduction

Micro Total Analysis Systems (μTAS) enable the performance of (bio)chemical analyses on the nanoliter scale. Such (bio)chemical microsystems can be divided in modules like microreactors, microfluidic systems and detectors. This modular approach enables the construction of chemical microsystems by assembling the required building blocks on a fluidic board [1]. Because of the small scale there is a need for a system component that enables dosing of nanoliter amounts of liquids such as reagents, indicators or calibration fluids. Such a system consisting of a micropump integrated with a liquid flow sensor produced in silicon, has been described earlier in [2] but the main disadvantage was the complicated construction and the use of different technologies. To enable easy monolithic integration with other system components for optimizing the performance of the μTAS, preferably similar micromachining technologies should be adopted. The dosing system described here can be implemented in silicon by adopting standard etching techniques to form channels and has no moving parts.

The proposed system consists of a reservoir containing the fluid to be dispensed connected at one side to the dosing target, for instance a microchannel (see figure 1). At the other side the reservoir is connected to a separate container in which gas bubbles can be generated by the reservoir is connected to a separate container in which gas bubbles can be generated by the electrolysis of water at noble metal electrodes. As these bubbles evolve and expand they push the fluid out of the fluid reservoir towards the target. If appropriate conditions are chosen (electrode/cell configuration) then the total volume of the gas bubbles and thus the amount of fluid that is dispensed, can accurately be controlled by means of an appropriate combination of on-time and current.

As a model application two dispensing systems were built and filled with different pH buffers for on line calibration of a pH sensitive ISFET in a flow-through cell. Using two different calibration fluids, the offset of the sensor as well as its sensitivity can be measured.

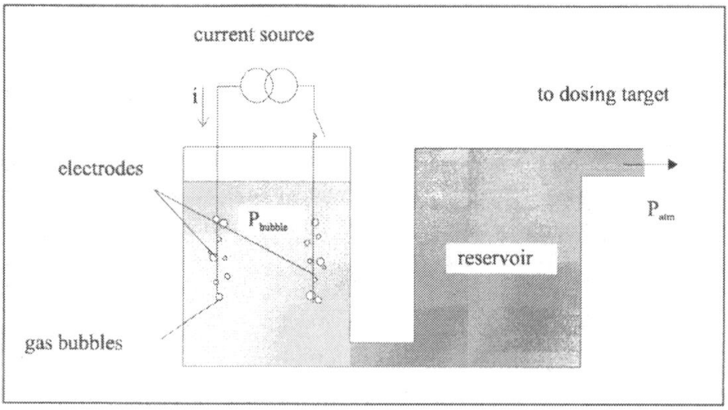

***Figure 1**: Principle of electrochemical controlled dosing system.*

2. Theory

Gas bubbles (oxygen and hydrogen) can be generated at noble metal electrodes by the electrolysis of water:

$$\begin{aligned} &anode: && 2H_2O \rightarrow 4H^+ + 4e^- + O_2 \\ &cathode: && 2H_2O + 2e^- \rightarrow 2OH^- + H_2 \end{aligned} \qquad (1)$$

If the concentration of these gasses exceeds their solubility in water, bubbles are formed with internal pressure P_{bubble} which expand to atmospheric pressure. This pressure difference P_{bubble}-P_{atm} drives the fluid out of the reservoir.

In order to exceed the gas solubility in water, an appropriate large current should be applied. The volume of liquid that is displaced equals the final gas volume of the bubbles when they are fully expanded to atmospheric pressure. Neglecting diffusion of gas in the liquid phase, this volume V is given by:

$$V = \frac{3}{4}\frac{i\,\Delta t}{F} V_m \qquad (2)$$

where i is the amplitude of the current pulse, Δt the duration, V_m the molar gas volume (2,24 10^{-2} m^3 mol^{-1} at standard pressure and temperature) and F is Faraday's number (9,6 10^4 C mol^{-1}). The factor ¾ is due to the fact that for every mole of electrons that is displaced, in total ¾ moles of gasses are produced according to (1).

3. Experimental

Experiments were performed using the flow through system depicted in figure 2. A 75µm thick plastic spacer was applied to define a channel of 750µm x 25mm containing a pH sensitive ISFET. A reference electrode was integrated in the Perspex® cover by inserting a chlorinated silver wire into a cavity filled with saturated KCl. This cavity made contact via a frit with the flow channel containing the ISFET. The channel was connected to a precision syringe pump (CMA/102, CMA Microdialysis, Stockholm Sweden) filled with the carrier fluid (5mM HEPES, pH 7.00). Reservoirs of 0.5ml volume were made in Perspex® and T-connected via PEEK tubing (ID 120µm, 10mm length) to the flow channel. The point of intersection was 10mm upstream of the ISFET and the space between two inlets was also 10mm. After filling the reservoirs with different pH buffers (50mM HAc pH 4.01, 50mM TRIS pH 9.45) a cap containing two platinum wire electrodes of 120µm diameter, 5mm in length and 2mm apart, was used for closure. In this experimental set-up the electrodes were placed directly in the fluid to be dispensed in contrast to figure 1. This simplifies the construction but has the disadvantage that the calibration fluid can be contaminated by the reaction products given in (1).

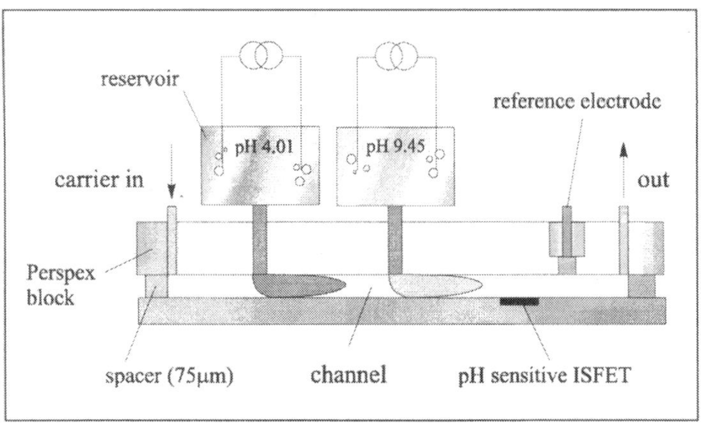

***Figure 2**: Construction of the flow channel with pH ISFET and two reservoirs containing the pH calibration buffers to be dispensed.*

4. Results

To evaluate the amount of calibration fluid needed for a calibration in this particular system, different amounts of pH buffer was injected via the T-connector using another syringe pump. First the channel was flushed with carrier fluid and then a certain volume of calibration was injected. Then the carrier flow was turned on again and the output of the ISFET amplifier was recorded. The peak of the response as function of the volume of the calibration plug is given in figure 3. It can be concluded that above a plug volume of 60 nl there is no increase in peak height. For larger volumes the detected peaks have a flat plateau indicating that the ISFET is measuring in the undiluted calibration buffer.

Then the relation between the current and the final injected volume (equation 2) was investigated. For this purpose the reservoirs were connected to transparent fused silica tubing (ID 50μm) to measure the displacement of the water front after the generation of gas bubbles. Figure 4 gives the relation between the amplitude of the current and the total dispensed volume after a time interval of 10 s. It can be concluded that this total volume is in close agreement with equation (2).

Finally the setup of figure 2 was used to dispense two plugs of different pH buffer in the stagnant carrier solution by applying a current of 80 μA with a duration of 10 s. After dispensing the carrier flow was turned on again with a flow rate of 3μl/min. The output of the ISFET amplifier is plotted in figure 5. The peaks in the response were applied for a two point calibration giving the sensor offset as well as the sensitivity. These parameters were both in close agreement with the manual calibration procedure carried out before.

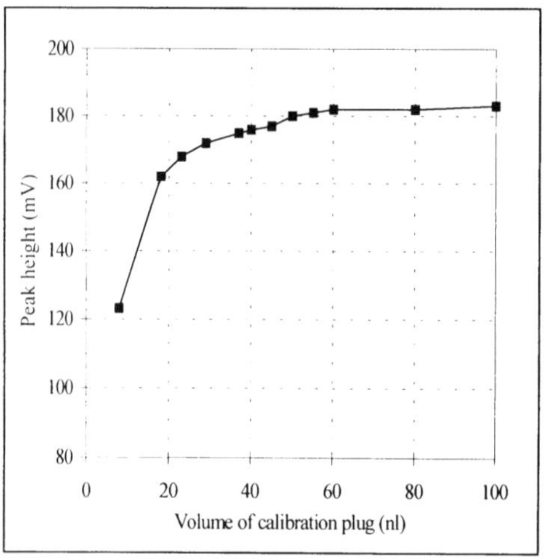

Figure 3: Peak height of the ISFET amplifier output after injecting a volume of calibration buffer and flushing with the carrier fluid.

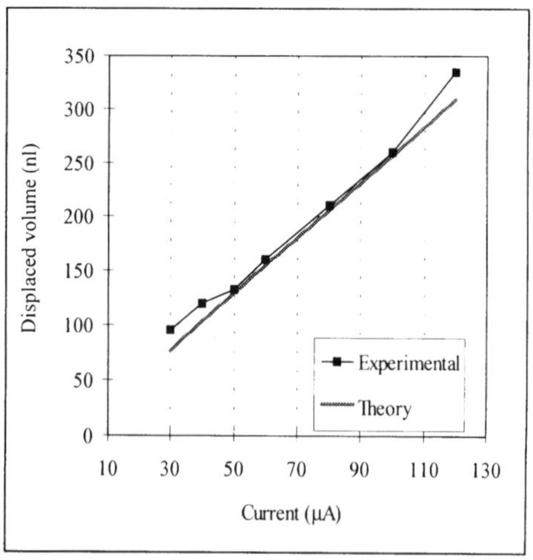

Figure 4: Displaced volume after 10s bubble generation as function of the current. Theoretical displacement is given by equation (2).

5. Conclusions

A simple electrochemical actuated pump for dispensing nanoliter amounts of fluid is described. Gas bubbles generated by the electrolysis of water at noble metal electrodes were applied to displace fluid. To accurately dose a certain volume the relation between the charge applied through the electrodes should generate at atmospheric pressure an equivalent volume of gas bubbles. This relation was evaluated experimentally and was in close agreement with theory. A two point calibration of a pH sensitive ISFET mounted in a flow through cell was performed by using two identical pump systems filled with different pH calibration buffer. Results were identical to a manual calibration performed before.

The above indicates that an accurate controlled fluid dosing system without moving parts can be constructed by applying simple principles and techniques. The system can be implemented in silicon microsystems by the application of standard micromachining techniques.

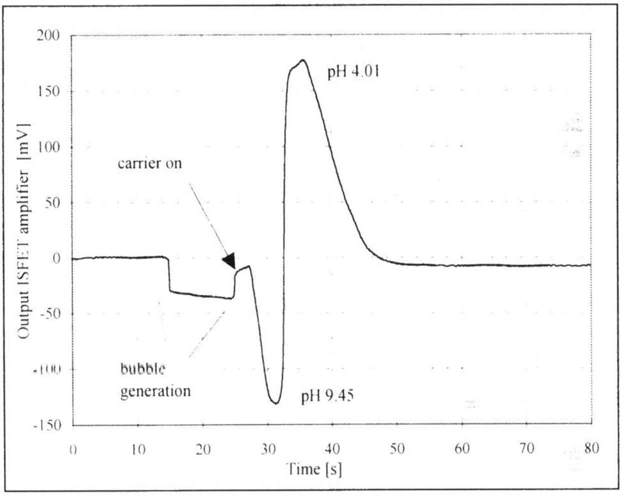

***Figure 5**: Output of the ISFET amplifier after the injection and transportation of two calibration plugs (note the electronic artifact during bubble generation).*

Acknowledgments

This project is financially supported by the Dutch foundation for Fundamental Research on Matter (F.O.M.).

References

1 Fluitman, J. H., A. van den Berg and T. Lammerink, Micromechanical components for µTAS, Micro Total Analysis Systems, (1994) 73-83

2 Elwenspoek, M, T. S. J. Lammerink, R. Miyake and J. H. J. Fluitman, Towards integrated microliquid handling systems, J. Micromech. Microeng 4 (1994) 227-245.

CHEMICALLY MODIFIED FIELD-EFFECT TRANSISTORS FOR MEASUREMENT OF ION ACTIVITIES IN AQUEOUS SOLUTION

Martijn M.G. Antonisse, Ronny J.W. Lugtenberg, Alina Ion, Richard J.M. Egberink, Bianca H.M. Snellink-Ruël, Johan F.J. Engbersen and David N. Reinhoudt

Department of Supramolecular Chemistry and Technology
MESA Research Institute, University of Twente
P.O. Box 217, 7500 AE Enschede, The Netherlands

0. Abstract

Chemically modified field effect transistors for the selective detection of several cation and anion activities in aqueous solution are described. For obtaining sensors of high durability, novel polysiloxane membranes have been developed which contain different side groups to tune their intrinsic properties. These polysiloxane membranes show good performance in life time experiments. The ion selectivity has been tuned by incorporation of various novel ion receptor molecules, yielding sensors with high selectivities for sodium, potassium, lead, cupper, cadmium, silver, nitrate, nitrite, fluoride, and dihydrogen phosphate.

Keywords: CHEMFET, Calixarene, Uranyl Salophene, Polysiloxane

1. Introduction

Miniaturized sensors for the (on line) detection of low concentrations of ions in aqueous solution can find useful application in many areas, e.g. for water quality monitoring, localisation of environmental pollution and biomedical analysis. Until now such miniaturized sensor systems of sufficient stability have not been fully developed for the majority of analytes and analysis is carried out with conventional techniques like chromatography, atomic absorption and emission spectrometry and various wet analytical techniques. However, the innovative technology in semiconductor materials, micro-engineering, and molecular engineering shows great promise for the fabrication of microsensors by IC technology. Solid state pH-sensors based on ion-sensitive field effect transistors are already commercially available and it is expected that these soon will be followed by chemically modified field effect transistors capable to detect activities of other ionic species in solution.

2. Chemically modified field effect transistors

Chemically modified field effect transistors (CHEMFETs) are designed for the selective detection of ions in aqueous solution. The measurement principle of these sensors is, like for the ion selective electrode and glass electrode, based on potentiometric determination of ion activities in aqueous solution. This detection technique is particularly useful in cases where a large dynamic range of ion activities have to be measured. CHEMFETs have some important advantages compared to ion-selective membrane electrodes as fast response time, low noise level, small size and possibilities for mass production. Basically, the CHEMFET is a special modification of the ion sensitive field effect transistor (ISFET), which shows sensitivity for the activity of hydrogen ions in solution. A field effect transistor measures the

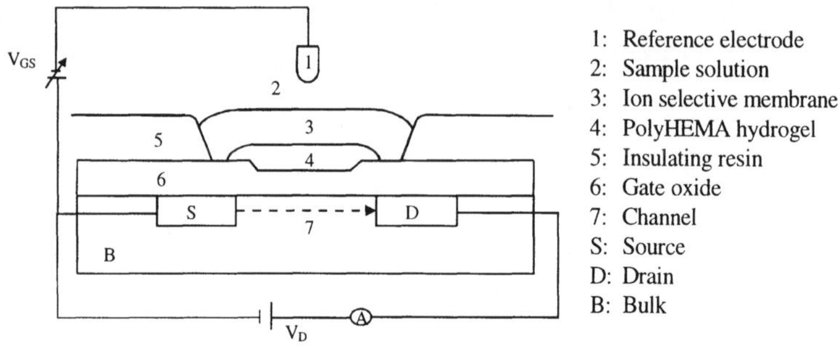

Figure 1. *Schematic drawing of chemically modified field effect transistor (CHEMFET).*

conductance between source and drain (the gate) of a semiconductor as a function of an electrical field perpendicular to the gate oxide surface. In contact with an aqueous solution, the state of ionisation of the surface SiOH groups of a SiO_2 gate oxide in an aqueous solution determines the surface potential and therefore the source-drain current in the transistor. A change in pH of the solution will produce a change in surface potential which can be measured by the change in conductance of the transistor. Typical pH responses thus obtained are 37-40 mV/pH unit.

For the detection of other ionic species than protons, it is necessary to introduce ion receptors which can selectively bind these species. Moreover, the intrinsic pH sensitivity caused by dissociation of the silanol groups at the gate oxide surface, has to be suppressed. Covering of the gate oxide of a silicon dioxide ISFET with a plasticized PVC membrane containing the potassium-selective natural ionophore valinomycin already gives a good potassium response upon variation of the potassium concentration. However, this construction has the severe drawback that the physical adhesion of the membrane to the semiconductor surface is not stable and that plasticizer and ionophore leach slowly from the membrane upon prolonged contact of the sensor with the aqueous solution. Moreover, such a sensor suffers from interfering pH-sensitivity (and carbon dioxide sensitivity) due to the uncontrolled ionization of the surface silanol groups of the gate oxide. The latter problem can be overcome by placing a buffered hydrogel between gate oxide and sensing membrane. We have attached a layer of polyhydroxyethylmethacrylate (polyHEMA) covalently to the gate oxide surface by photopolymerization the monomer on the gate oxide which was pretreated with methacryloylpropyltrimethoxysilane. The methacrylated siloxane groups on the surface ensure the covalent binding of the hydrogel.

The next step in the sensor preparation was the covering of the polyHEMA layer with a sensing membrane. Although plasticized PVC membranes with ionophores as sensing membranes can give quite satisfactory results -no CO_2 and pH interference and a low noise signal- for a limited life time, a durable sensor requires the chemical attachment of the sensing membrane to the semiconductor and a membrane material which does not need a plasticizer. Also the electroactive components in the sensing membrane must be prevented from leaching out from the membrane which can be achieved by covalent attachment. To a sensing membrane suitable for these purposes a number of requirements have to be made: (i) The sensing membrane must be an elastomer (glass transition temperature below zero; (ii) The membrane must be moderately hydrophobic, i.e. it must possess a low partition of ions from the aqueous phase but must have sufficient conductivity in order to obtain fast response times. For the permselectivity of the analyte ions, ionic sites of opposite charge must be

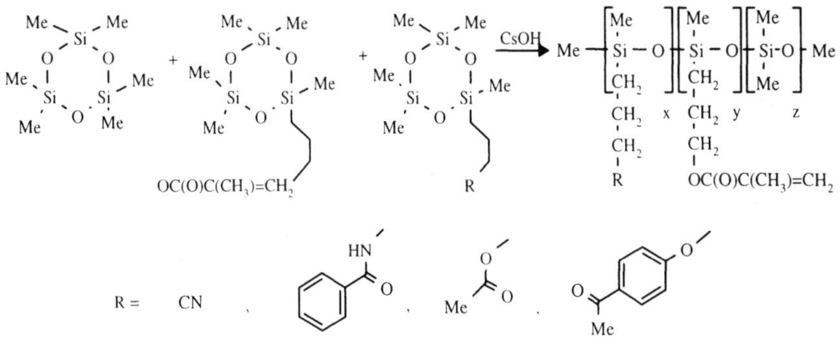

Scheme 1 *Anionic copolymerization of different cyclotrisiloxanes*

present in the membrane or allowed to be added. The presence of charged species in the membrane also lowers the electrical resistance of the membrane. (iii) The electroactive components in the sensing membrane, i.e. receptor molecules and ionic sites, have to be covalently bound to the membrane matrix in order to prevent leaching out. (iv) In order to make sensor fabrication compatible with IC-technology linkage of all components in the membrane matrix and attachment of the membranes to the gate oxide surface must occur by photopolymerisable processes.

3. Polysiloxane membranes for durable sensors

We have recently developed a reproducible synthesis for well-structured siloxane terpolymers, based on anionic copolymerisation of cyclotrisiloxanes (Scheme 1) [1]. The covalent attachment of the membrane to the transducing part of the sensor can be enabled by the photopolymerization of the methacrylate groups. The crosslinking of the methacrylate groups also enhances the mechanical strength of membrane.

The new polymers offer the possibility to tune the polarity and the character of the membrane by changing the amount and type of polar substituents. These changes can have a profound effect on the partition of various ions between sample and membrane phase, resulting in changes in the selectivity of the sensor. This is clearly illustrated for nitrate

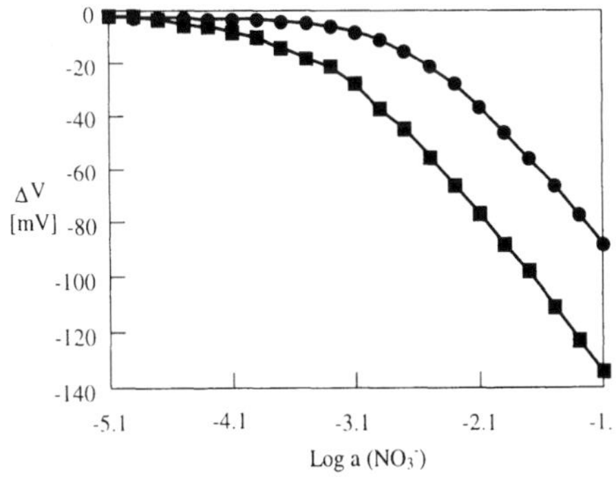

Figure 2. *Effect of the polar substituent on the nitrate over chloride selectivity. Nitrate response in the presence of 0.1 M Cl⁻ for CHEMFETs with polysiloxane membranes containing cyano substituents (■) or amide substituents (●).*

selective CHEMFETs based on a polysiloxane ion-exchange membrane [2]. As these membranes contain only lipophilic cationic groups (tetraalkylammonium) and no anion selective receptor molecules, the selectivity is determined by the differences in energy of transfer from water to the polysiloxane membrane (Hofmeister selectivity). Sensors with polysiloxane membranes with cyanopropyl substituents show a much lower interference of halogenide ions than sensors having membranes of polysiloxane with amide substituents ($\log K^{Pot}_{NO_3,Cl}$ = -2.3 for cyanopropyl, $\log K^{Pot}_{NO_3,Cl}$ = -1.6 for amide respectively, Figure 2). This difference reflects the more favorable interaction of the halogenide ions with the amide moieties, lowering the solvation energy of these ions in the membrane phase.

4. Membranes with cation selective receptor molecules

Receptors applied in sensor membranes should posses high selectivity, especially over interfering ions that will possibly also be present in high concentrations in the sample. Furthermore, the receptor molecule must be lipophilic in order to make it soluble in the membrane matrix and to prevent fast leaching to the contacting aqueous solution.

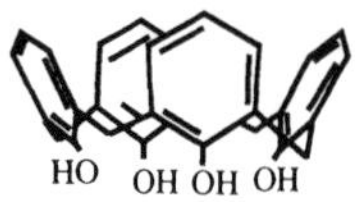

1

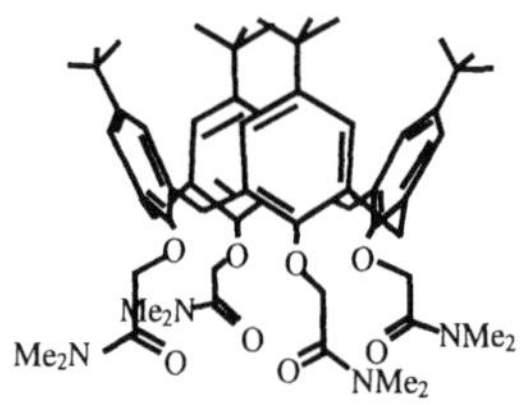

2

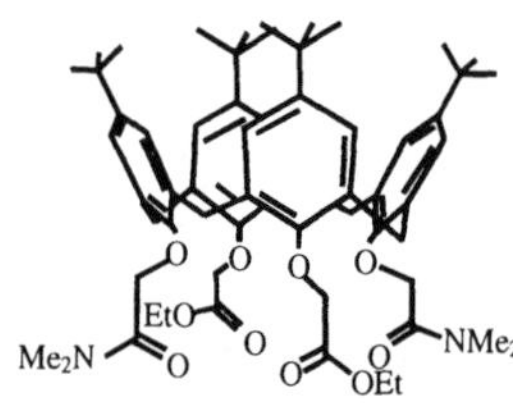

3

Calix[4]arene **1** is a versatile lipophilic three-dimensional building block that can be functionalized with various ligating substituents for selective ion binding and can be provided with a functional group which enables covalent anchoring to the membrane matrix.

The selectivity of the receptor can be influenced by the type and number of substituents and the conformation of the calix[4]arene. For example, functionalization of the calixarene at the *lower* rim with amide moieties (**2**) results in the right positioning of the coordinating carbonyl and phenol oxygens atoms for complexation of Na^+ with selectivity over other alkali cations. The selectivity of Na^+ over K^+ can even be improved by the use of two diametrally positioned amide ligands instead of four, in combination with two ester

Table 1 *Selectivity ($\log K_{Na,j}$) and sensitivity (slope, mV.decade^{-1}) of durable Na^+selective polysiloxane CHEMFETs.*[a]

interfering ion	log $K_{Na,j}$ and slope (mV.decade^{-1})[b] 0 weeks	15 weeks	40 weeks
0.5 M Mg^{2+}	-3.7 [59]	-3.7 [59]	-3.0 [59]
1 M K^+	-2.6 [60]	-2.2 [59]	-1.4 [56]
0.5 M Ca^{2+}	-3.7 [59]	-3.5 [59]	-2.9 [56]

[a] 1Wt% ionophore and 50 mol% (with respect to the ionophore) KTTFPB. [b] Log $K_{Na,j}$ +/- 0.1; slope +/- 2 mV.decade^{-1} [Na^+].

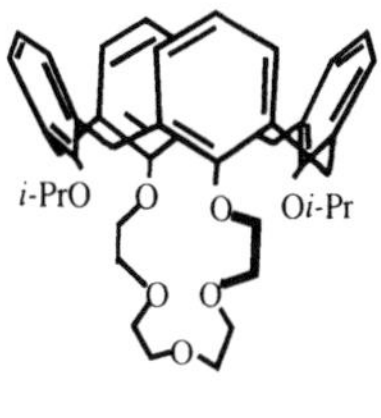

4

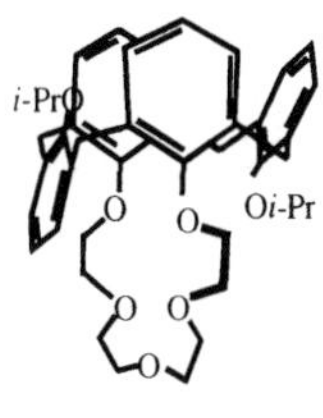

5

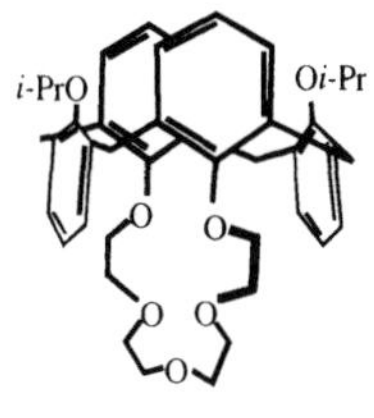

6

ligands (**3**). For application in the sensor membranes of CHEMFETs, 1 wt% of the Na^+ selective receptor calix[4]arene tetraethyl ester and 50 mol% (with respect to receptor) of potassium tetrakis[3,5-bis(trifluoromethyl)phenyl]borate (KTTFPB) was incorporated into cyanopropyl-functionalized siloxane terpolymers. The Na^+ selectivities and sensitivities of the obtained CHEMFETs measured in the presence of different interfering ions are shown in columns 1 and 2 of Table 1. These polysiloxane CHEMFETs exhibit good Na^+ selectivities (log $K_{Na,j}$ = -3.7, -2.6, and -3.7) and Nerstian responses (59 - 69 mV.decade^{-1}) for Na^+ in the presence of Mg^{2+}, K^+ and Ca^{2+} ions, respectively [3].

A crown ether moiety can easily be combined with the calix[4]arene building block. The preorganization of the binding cavity of the crown ether by the calixarene skeleton enhances the binding selectivity and the molecule becomes more lipophilic. For K^+ selective sensors, a number of calix[4]arene crown ether ionophores were developed with five oxygen atoms in the crown ether bridge. These receptors clearly show the influence of the conformation of the calix[4]arene on the selectivity. The *cone* conformer **4** has the lowest selectivity of K^+ over Na^+ (log $K^{Pot}_{K,Na}$ = -2.5). The selectivity of the *partial cone* conformer **5** is much better (log $K^{Pot}_{K,Na}$ = -3.9), but the highest K^+ selectivity was obtained for *1,3-alternate* calix[4]arene **6** (log $K^{Pot}_{K,Na}$ = -4.2). The high selectivity of the latter can be attributed to favourable cation-π interactions between K^+ and the aromatic rings. In Figure 3 the response curve in the presence of 1 M Na^+ of a CHEMFET with this ionophore is given. The obtained selectivity of **6** is even higher under identical conditions than that of the natural occuring ionophore valinomycin.

Changes in the nature of the donating atoms influences the selectivity. Soft heavy metal ions can be selectively bound by providing the calix[4]arene with soft thioether, thioamide, or thiocarbamate groups [4]. Again small changes in the ligand structure can alter the selectivity of the ionophore. Thioamide **7** is highly selective for Pb^{2+}, thiocarbamate **8** yields well functioning Cu^{2+} selective CHEMFETS and the dimethylthiocarbamoyl-methoxyethoxy substituents of **9** introduce Cd^{2+} selectivity. With dithioether **10** Nernstian Ag^+ response in the presence of K^+, Ca^{2+}, Cd^{2+}, Hg^{2+}, and Cu^{2+} has been obtained.

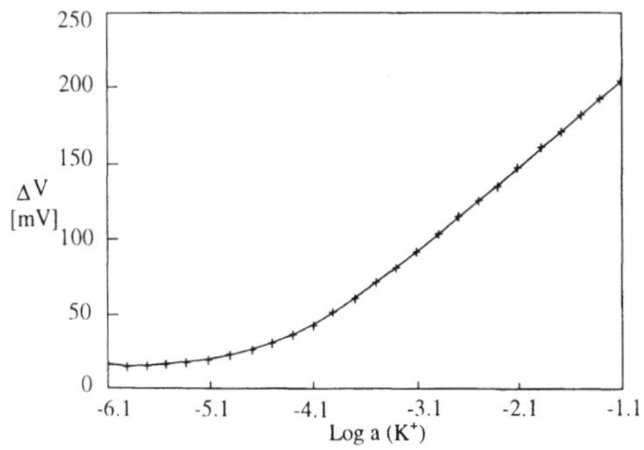

Figure 3. *K^+ response of CHEMFET containing ionophore **6** in the presence of 1 M Na^+.*

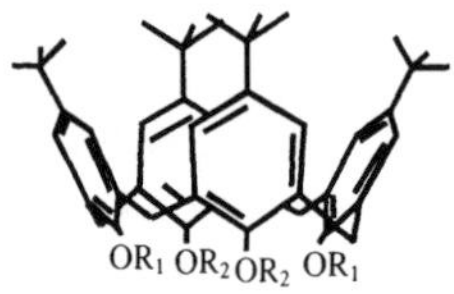

7 R_1=R_2= $CH_2C(S)N(CH_3)_2$
8 R_1=R_2= $CH_2CH_2SC(S)N(CH_2CH_3)_2$
9 R_1=R_2= $CH_2CH_2OCH_2C(S)N(CH_3)_2$
10 R_1= $CH_2CH_2SCH_3$, R_2= $CH_2CH_2CH_3$

CHEMFETs with Pb^{2+} selective calix[4]arene derivative **7** in membranes of cyanopropyl or acetoxypropyl functionalized polysiloxane did not show good sensitivities (sub-Nernstian slopes). However, CHEMFETs with membranes of polysiloxanes with benzoylaminopropyl substituents gave linear Pb^{2+} responses, although a high interference of Ca^{2+} was observed. This can probably be attributed to the interaction between this hard ion with the hard donating carbonyl oxygen atoms of the amide substituents of the polysiloxane favouring partition of the calcium ion. As can be seen in Figure 4, CHEMFETs with polysiloxane membranes with acetylphenoxypropyl substituents having relatively softer ester groups, show selective Pb^{2+} responses even in the presence of Ca^{2+}. Another example of the importance of the nature and polarity of the substituents present at the polysiloxane backbone is seen for CHEMFETs with Cd^{2+} selective ionophore **9**. CHEMFETs incorporating this ionophore in 3-cyanopropyl functionalized siloxane membranes are Cd^{2+} selective in the presence of Ca^{2+} and Pb^{2+} ions (log $K^{Pot}_{Cd,j}$ = -2.2 and -1.4, respectively). These selectivities can be impoved to -4.1 and -2.0, respectively, by incorporating **9** in polysiloxane membranes with acetoxypropyl substituents. Compared with **9** in DOP-plasticized PVC membranes, these results are even more striking because with the latter membranes no selectivity of Cd^{2+} over Pb^{2+} could be obtained.

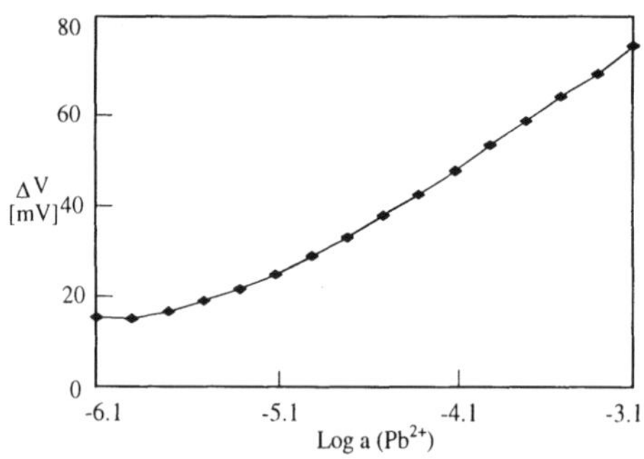

Figure 4. *Pb^{2+} response in the presence of 0.1 M Ca^{2+} of CHEMFET with acetylphenoxypropyl functionalized polysiloxane membrane containing calix[4]arene ionophore 7.*

5. Durability of Na^+ selective CHEMFETs

The durability of the sensors have been tested by the exposure of these sensors to a continuously refreshed stream of tap water (25 $mL.min^{-1}$). The CHEMFETs were conditioned in a 0.1 M NaCl solution for one night before measuring their characteristics (*i.e.* selectivity and sensitivity). The polysiloxane based CHEMFETs retained their Na^+ selectivity in the presence of different alkali and alkaline earth metal ions for a period of more than 40 weeks (Table 1, colums 3 and 4). Despite of the decrease in Na^+/K^+ selectivity from -2.6 to -1.4 after 40 weeks, the CHEMFETs still have (near) Nernstian responses. Even after 40 weeks of continuously exposure to streaming tap water the CHEMFETs are still very selective (log $K_{Na,j}$ = -3.0 and -2.9) for Na^+ in the presence of Mg^{2+} and Ca^{2+} ions, respectively. Similar results have been obtained with nitrate- selective CHEMFETs, incorporating a quaternary ammonium groups covalently bound polysiloxane membranes, and with potassium-selective CHEMFETs, incorporating covalently bound crown ether analogs.

6. Anion selective CHEMFETs

The number of anion receptors known in literature that are applied in ion selective sensors is very limited. The receptors for selective anion binding in the membrane of sensors have to meet high requirements because of the relatively large differences found in the energies of transfer from water to an organic medium like the (sensor)membrane between hydrophilic anions (e.g. phosphate or fluoride) and more lipophilic anions (e.g. nitrate and perchlorate). These differences strongly favor the partition of lipophilic ions like nitrate into the membrane even in the presence of hundred to thousand-fold excess of more hydrophilic phosphate or fluoride ions giving rise to sensor sensitivity for these lipophilic ions (*i.e.* the well-known Hofmeister selectivity). In the reversed case, when a hydrophilic ion has to be measured in the presence of lipophilic ions, the natural partition of the lipophilic ion in the membrane phase has to be overcompensated by strong and highly selective binding of the hydrophilic ion to (preferentially neutral) receptor molecules incorporated in the membrane. One of the best described receptors is a cobalt pophyrin which yields nitrite selectivty when applied in membranes of ion selective electrodes. Also applied in PVC and polysiloxane membranes on CHEMFETs good nitrite sensitivities and selectivities are obtained (table 2) [5]. Common ions like halides or nitrate do not interfere the nitrite respons even when the activity of these ions is hundred to thousand fold in excess.

Recently we have developed a novel class of uranyl salophene based anion receptors which introduce selectivity for certain hydrophilic anions. In the salophene unit the Lewis acidic UO_2 cation is bound via four of its equatorial coordination sites. The fifth site in the same plane is available for anion binding (Chart 1). Lipophilic substituents are introduced to make the receptor well soluble in the membrane matrix of the ion selective membrane. Receptor **11** was used as ionphore in plasticized PVC membranes cast on top of CHEMFETs [6]. This uranyl salophene receptor introduces a high sensitivity for dihydrogen phosphate with almost Nernstian slope of the response curve, and the activity of this ion can be determined in the sub-millimolar level. The high selectivity of this receptor is reflected by the low interference of lipophilic ions in fixed interference measurements (Table 2). Phosphate can be measured with a twenty-fold selectivity over nitrate. This is a more than 10^5 fold increase in phosphate selectivity compared with the Hofmeister selectivity of ion exchange membranes in which nitrate is favored 10^4 times over phosphate.

Table 2. *Sensor characteristics of anion selective CHEMFETs with nitrite, phosphate or fluoride selectivity and PVC/NPOE or polysiloxane (PS) polymer membranes.*

Receptor	Polymer	Primary ion (i)	Log $K^{Pot}_{i,j}$			
			NO_3^-	Br^-	Cl^-	SO_4^{2-}
Co-Porph.	PVC	NO_2^- [a]	-2.6	-2.6	-2.6	-2.9
Co-Porph.	Ketone PS	NO_2^- [a]	-2.0	-2.2	-2.4	-2.7
11	PVC	$H_2PO_4^-$ [b]	-1.3	-1.7	-1.8	-2.3
12	PVC	F^- [a]	-2.1	-2.1	-2.0	-2.5
12	Sulfone PS	F^- [a]	-2.6	-2.5	-2.6	-3.0
13	PVC	OAc^-	-0.3	-1.2	-1.2	-2.5

[a] [j] = 0.1 M in 0.01 M MES pH = 6.0; [b] [j] = 0.1 M in 0.01 M MES pH = 4.5

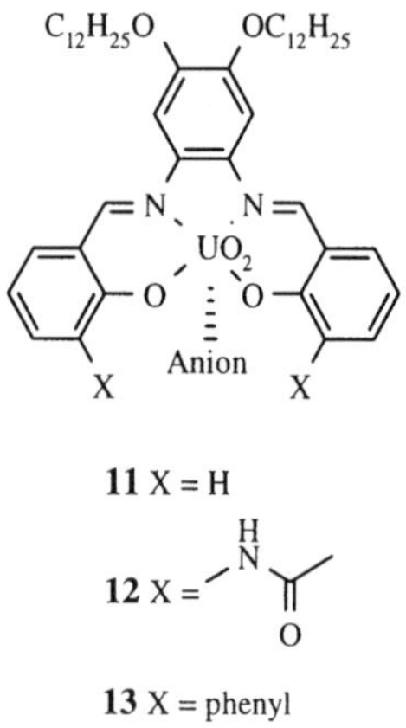

Chart 1

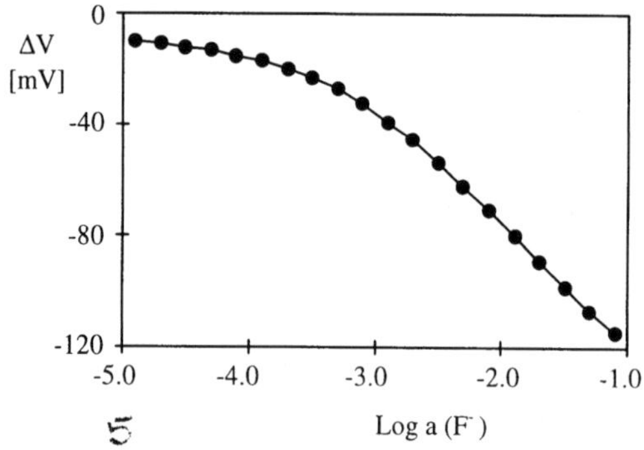

***Figure 5** Fluoride response of CHEMFET with novel lipophilic uranylsalophene with amide substituents, in the presence of 0.1 M $NaClO_4$.*

The selectivity of the anion binding of uranyl salophenes can be influenced by introduction of additional binding sites in close proximity to the anion coordination site of the uranyl cation. The amide substituents of salophene **12** can donate hydrogen bonds to the bound anion and reduce the size of the binding cleft. This introduces very high binding selectivity for the small, strongly hydrogen bond accepting, fluoride anion. Well functioning fluoride selective CHEMFETs were obtained that show excellent selectivity over more lipophilic ions, like other halides and nitrate (Table 2). As presented in Figure 5, also in the presence of high concentrations of the highly lipophilic perchlorate ion (0.1 M), detection of fluoride is possible at millimolar level. The phenyl substituents of uranyl salophene derivative **13** result in a apolar anion binding cleft and selectivity for the organic acetate anion is obtained. CHEMFETs with **13** as anion receptor respond to acetate at concentrations $\geq 3 \times 10^{-4}$ M with an almost Nernstian slope of -56 mV/decade. The CHEMFETs are acetate selective over more lipophilic anions like Cl^-, Br^-, and NO_3^- (Log $K^{Pot}_{OAc,j}$ = -1.2, -1.2, and -0.3, respectively). Furthermore, over $H_2PO_4^-$ a 250 fold selectivity is obtained. The presence of the phenyl substituents reduces the binding strength between the $H_2PO_4^-$ anion and the uranyl salophene and consequently the selectivity of acetate and $H_2PO_4^-$ is reversed compared with the $H_2PO_4^-$ selective CHEMFETS based on salophene **11**.

7. Conclusions

In summary, we have successfully developed potentiometric cation and anion sensors based on CHEMFETs. High durability is obtained with ion-selective membranes based on modified siloxane polymers. The incorporation of calixarene cation receptors or uranyl salophene anion receptors results in a high selectivity of the detection of several analytes.

Acknowledgment

This study is financially supported by the Technology Foundation (STW), Technical Science Branch of the Netherlands Organization for Scientific Research (NWO).

References

1 R.J.W. Lugtenberg, M.M.G. Antonisse, R.J.M. Egberink, J.F.J. Engbersen, D.N. Reinhoudt *Polysiloxane based CHEMFETs for the detection of heavy metal ions*, J. Chem. Soc., Perkin Trans. 2, (1996), 1937-1941.

2 M.M.G. Antonisse, R.J.W. Lugtenberg, R.J.M. Egberink, J.F.J. Engbersen, D.N. Reinhoudt *Durable nitrate-selective CHEMFETs based on new polysiloxane membranes*, Anal. Chim. Acta, 332 (1996), 123-129.

3 H. Gankema, R.J.W. Lugtenberg, J.F.J. Engbersen, D.N. Reinhoudt, M. Möller *Crosslinkable polar siloxane copolymers for ion detection devices*, Adv. Mater., 6 (1994), 944-947.

4 R.J.W. Lugtenberg, R.J.M. Egberink, J.F.J. Engbersen, D.N. Reinhoudt *Pb^{2+} and Cd^{2+} selective chemically modified field effect transistors based on thioamide functionalized 1,3-alternate calix[4]arenes*, J. Chem. Soc., Perkin Trans. 2, (1997), 1353-1357.

5 M.M.G. Antonisse, B.H.M Snellink-Ruël, J.F.J. Engbersen, D.N. Reinhoudt *Chemically modified field effect transistors with nitrite or fluoride selectivity*, submitted for publication.

6 M.M.G. Antonisse, B.H.M Snellink-Ruël, I. Yigit, J.F.J. Engbersen, D.N. Reinhoudt *Neutral anion receptors; synthesis and evaluation as sensing molecules in chemically modified field effect transistors (CHEMFETs)*, J. Org. Chem., 62 (1997), in press.

SELECTIVE OPTODE MEMBRANES FOR HEAVY METAL ION DETECTION.

Niels J. van der Veen, Richard J.M. Egberink, Johan F.J. Engbersen and David N. Reinhoudt

Department of Supramolecular Chemistry and Technology, MESA Research Institute, University of Twente, P.O. Box 217, 7500AE, Enschede, The Netherlands

0. Abstract

Optode membranes for Lead(II) based on an ion-exchange mechanism and neutral ionophores show that the calix[4]arene ionophores developed for CHEMFETS can be used in optical sensing schemes. To achieve pH independent sensor response, chromoionophores have been developed based on these neutral ionophores, which show a optical response to the complexation of metal ions.

Keywords: optode membrane, chromoionophores, heavy metal ions

1. Introduction

The monitoring of heavy metal concentrations in surface water, harbor sludge, ground or waste water will be crucial in solving environmental problems related to these pollutants. Ionophores selective for several heavy metal ions have been used in chemically modified field effect transistors (CHEMFETs)[1], which involve the transduction of the supramolecular recognition of the ions by the selector molecules into an electrical signal. In principle, these ionophores can also be applied in optode membranes. The use of optical transduction has a number of advantages, such as the absence of a reference electrode, the insensitivity to electric or magnetic fields, the possible variation of transduction schemes and the possibility of remote sensing using fiber optics.

A well studied optical transduction scheme was developed by Simon *et al.* It involves the use of pH dyes by an ion exchange process, which changes the absorption spectrum of the optode membrane. The method has been used to develop optode membranes for detection of metal ions such as lead(II)[2] and silver(I).[3] The principle disadvantages of these optode membranes is that the response is intrinsically pH dependent and that plasticized PVC is used, which limits the durability because of leaching of the plasticizer. We therefore work on the development of optode membranes based on polysiloxanes, which are intrinsically elastomeric and therefore do not need plasticizers. The membrane processing with these polymers is compatible with IC technology.[4] This would make it possible to photochemically bind the optode membrane to a silica waveguide and so integrate membrane and optics in a device. For pH independent optode membranes a different transduction scheme should be employed. We present the synthetic strategy for preparation of selective heavy metal ion chromoionophores which could be used in such a scheme.

2. Application of calix[4]arene receptors in ion exchange optode membranes

The metal ion receptors developed for use in CHEMFETs can be used in ion exchange optode membranes involving pH dyes. These liquid polymer membranes contain ionophores, a pH dye and lipophilized anionic sites. When the membrane is exposed to an analyte solution, the complexation of ions by the receptor causes an influx of positive charge into the membrane. This triggers deprotonation of the pH dye, which is accompanied by a change in absorption spectrum. The two equilibria involved, ion complexation-decomplexation by the receptor and pH dye protonation-deprotonation, compete and result in a reversible color response of the membrane. This mechanism is illustrated in figure 1.

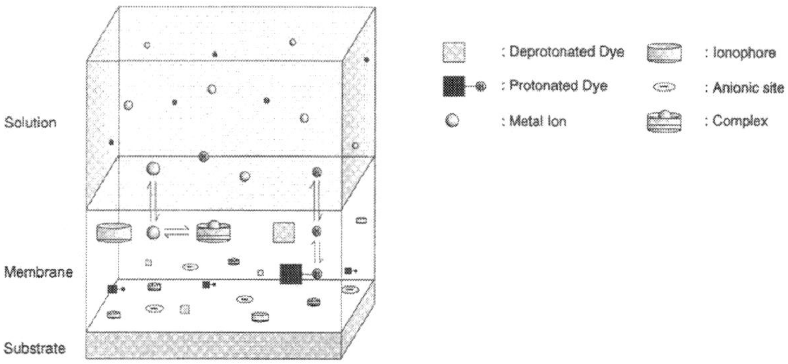

Figure 1: *Mechanism of response of an optode membrane containing a pH dye*

In figure 2, the response of a DOS-plasticized PVC membrane containing a calix[4]arene based Pb(II) receptor is demonstrated. As the concentrations of Pb(II) get higher, the absorption spectrum of the membrane shifts from the spectrum dominated by the protonated pH dye to the spectrum dominated by the deprotonated dye. The response is reversible.

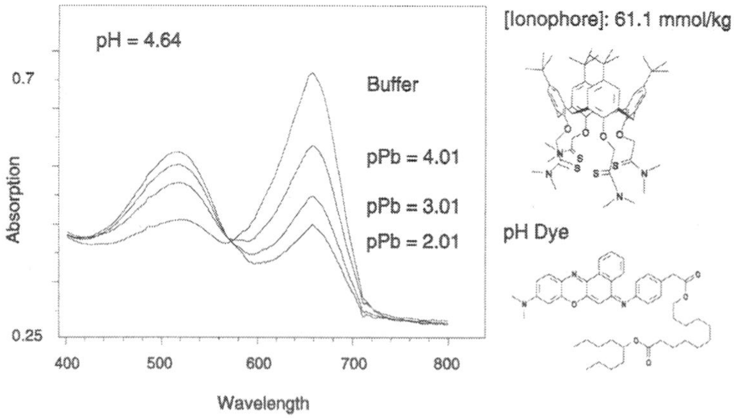

Figure 2: *Response of a Lead selective plasticized PVC optode membrane based on ion-exchange.*

3. Chromoionophores derived from selective heavy metal ion receptors

Optode membrane response should preferably be pH insensitive. This cannot be achieved when the transduction involves pH dyes. An improved transduction scheme requires chromoionophores which couple the recognition event directly to a color change. The synthetic strategy for making selective chromoionophores is to couple chromophores to known receptors. Therefore, chromoionophores **2** and **4** in figure 3 are currently being synthesized as derivatives of the Pb(II) and Cd(II) ionophores **1** and **3**, respectively. Recent results will be reported.

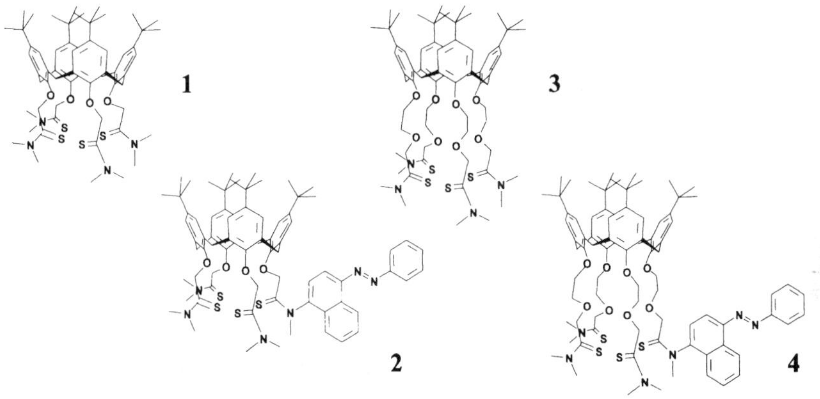

Figure 3: *Chromoionophores for selective recognition of lead and cadmium ions*

The molecule depicted in figure 4 is a compound *en route* to a chromoionophore selective for Lead(II) ions. The dye functionalities are conjugated with the amide carbonyl groups, which are responsible for the complexation of metal ions. These amide groups will later be transformed to thio-amide groups to obtain selectivity for heavy metal ions. The molecule has no acidic or basic sites, which should make its response to changes in pH small. In figure 5 the response is shown of this chromoionophore to complexation of Sodium ions. It can be seen that by coupling of chromophores to the metal complexing amide-moieties transduction of the complexation in an optical signal is indeed possible. The spectral changes upon complexation of the metal ion are clear.

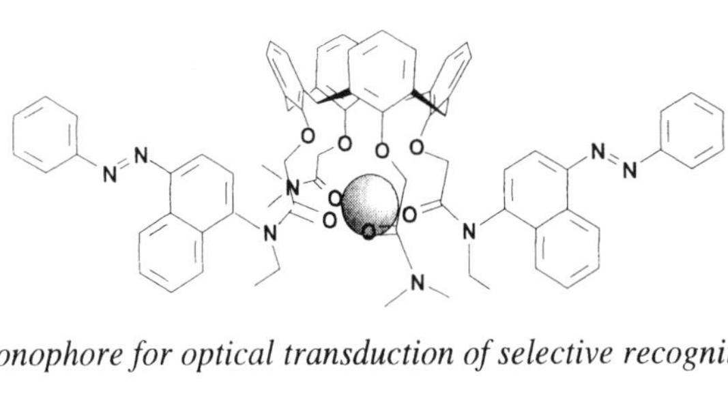

Figure 4: *Chromoionophore for optical transduction of selective recognition of metal ions.*

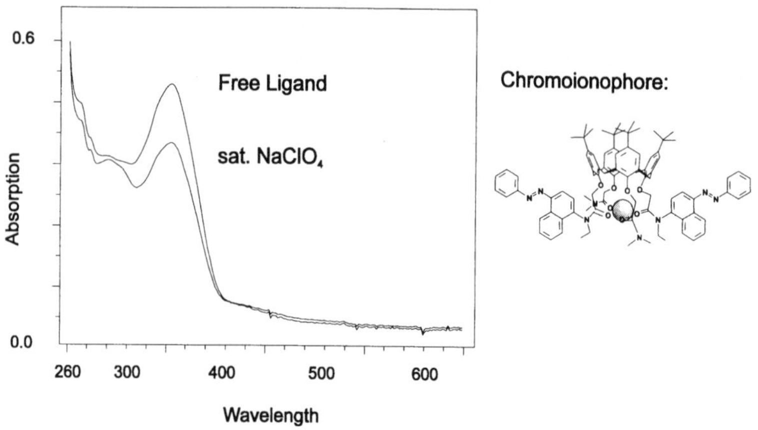

Figure 5: *Optical transduction of the complexation of Sodium ions by a chromoionophore.*

4. Conclusions

It has been shown that calix[4]arene neutral ionophores can be used in optical transduction schemes. Also, chromoionophores based on integration of a dye functionality into the chemical structure of these ionophores respond to complexation of metal ions by changing their spectral characteristics.

5. Acknowledgment

We thank the Technology Foundation (STW) and the Netherlands Organization for Scientific Research (NWO) for financial support.

6. References

1 P.L.H.M. Cobben, R.J.M. Egberink, J.G. Bomer, P. Bergveld, W. Verboom and D.N. Reinhoudt, *J. Am. Chem. Soc.* **114** (1992) 10573.

2 M. Lerchi, E. Bakker, B. Rusterholz and W. Simon, *Anal. Chem.* **64** (1992) 1534-1540.

3 M. Lerchi, F. Orsini, Z. Cimerman, E. Pretsch, D.A. Chowdhury and S. Kamata, *Anal. Chem.* **68** (1996) 3210-3214.

4 R.J.W. Lugtenberg, M.M.G. Antonisse, R.J.M. Egberink, J.F.J. Engbersen, D.N. Reinhoudt, *J. Chem. Soc., Perkin Trans. 2*, (1996) 1937-1941.

COVALENT COUPLING OF POLYAMINES TO AMINOBEAD-COVERED ISFETs

Geert Besselink, Gijs Dirks, Wouter Olthuis and Piet Bergveld
MESA Research Institute, University of Twente
P.O. Box 217, 7500 AE Enschede, The Netherlands

0. Abstract

The so-called ion-step method is a novel potentiometric approach for determination of protein concentration, in which modified ISFETs are used [1, 2]. Aim of our current research is to develop a generic technology for immobilization of proteins (and other biomolecules) to the ISFET gate in order to confer specific binding properties to the ISFET. In this report, a new modification procedure is introduced, namely the covering of the ISFET surface with a *monolayer* of latex beads followed by the *covalent* coupling of (poly)amines to these beads. The resulting ISFET devices could be used for detection of heparin in the concentration range of 0.025-0.75 U/ml. Considerable differences were found concerning the sensitivity of heparin detection when employing different types of polyamine ligands.

Keywords: ISFET, modification, latex, beads, polyamine, heparin

1. Introduction

Former research in our group has resulted in a laboratory set-up for an ISFET-based heparin sensor, in which the so-called ion-step method is used [3,4]. Current research is aimed at the development of a biosensor chipcard system that is based on the former principle and which can be applied for the specific detection not only of heparin but for any other relevant biomolecule. One important research goal therefore is the development of a generic ligand immobilization technology for conferring specific binding properties to the ISFET surface.
In the past, ISFET modification was realized by passive adsorption of solutes to the bare ISFET surface (e.g. of protamine [3]) or to a thick microporous polystyrene membrane (e.g. of protamine [4] and lysozyme [2]). We have developed a method for *covalent* immobilization of polyamines to the ISFET device. First step of the corresponding procedure is to cover the ISFET surface with a *monolayer* of latex beads (which contains surface-bound amino groups). The thickness of the resulting membrane layer equals the diameter of the constituent beads (0.9 μm). In the second modification step, the latex-bound amino groups are activated by treatment with glutaraldehyde (GA), and polyamine solutes are coupled to the GA-activated bead surface. The polyamine-modified aminobead-ISFETs were evaluated by examination of the binding of heparin to the respective devices. For comparison, three different types of immobilized ligands were used: protamine, polylysine and (home-made) aminodextran.

2. Experimental

2.1. Materials

Aminobeads (Polybead™, Diameter 0.913±0.024 μm) were obtained from Polysciences, Inc. (Warrington, PA). L-lysine (free acid), protamine sulfate (from Salmon, grade X) and poly-L-lysine hydrobromides (with molecular weights (based on viscosity measurements) of 2 kDa (denoted as PLL 2 kDa); 4 kDa; 22,4 kDa (PLL 22 kDa), and 128 kDa) were all obtained from Sigma Chemical Co. (St. Louis, MO). Glutaraldehyde (50% in water), $NaBH_3CN$ and Tween® 20 (zur Synthese) were purchased from Merck-Schuchardt (Hohenbrunn, Germany). Dextran T70 was supplied by Pharmacia Biotech AB (Uppsala, Sweden). Heparin (Thromboliquine®; 5,000 U/ml) was obtained from Organon Teknika BV (Boxtel, The Netherlands).

2.2. Methods

Ta_2O_5-ISFETs were fabricated in the MESA cleanroom laboratory and encapsulated in Hysol epoxy as described elsewhere [5]. Before each experiment, ISFETs were cleaned by rubbing them with detergent solution (1% Alconox™) by using a cotton bud. About 0.5 μL Aminobead suspension (10% in demineralized water) was put on the ISFET gate and the bead suspension was allowed to dry up. Within about 20 sec after completeness of the drying process, non-attached beads were rinsed away with demineralized water, after which ISFETs were obtained that were covered with a monolayer of beads. The aminobead-covered ISFETs were treated for 30 min. with glutaraldehyde (5% in 0.2 M sodium phosphate, pH 7.5) after which the devices were washed extensively with demineralized water (minimal 2 h total washing time, during which the washing fluid was renewed thrice). Glutaraldehyde-treated aminobead-devices were incubated with lysine or polyamine (10 mg/ml in PBS; incubation time 30 min. (or shorter when indicated)), followed by washing in demineralized water. Finally, the ISFETs were treated with $NaBH_3CN$ (20 mg/ml in PBS) for 10 min. in order to stabilize the newly formed bonds between polyamine and GA-activated substrate. Success and magnitude of ISFET-modification with (poly)amines was assessed by using the ion-step method itself. Ion stepping was performed as described elsewhere [5]. Modified ISFETs were incubated for 1 min. with different concentrations of heparin (in PBS). After this, ion-step responses of the resulting ISFETs were determined and the effect of heparin incubation on the ion-step response amplitude was investigated. After each measurement, ISFETs were regenerated by washing for 5 min. with 4 M NaCl and thereafter a new incubation with heparin was conducted. Synthesis of aminodextran was carried out by a modification of the procedure of Gregorius et al. [6], which will be described in a forthcoming paper.

3. Results and discussion

Glutaraldehyde-activated aminobead-ISFETs that were incubated with polylysine, showed a markedly changed ion-step response (figure 1).

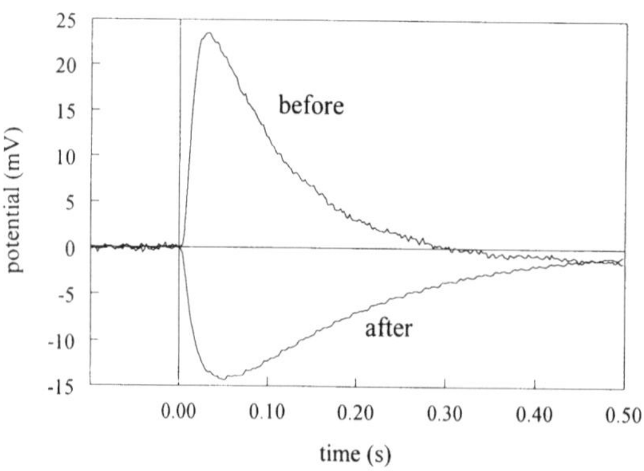

FIGURE 1. Ion-step responses of one and the same aminobead-ISFET before and after incubation with polylysine (Mw 22 kDa).

The reversal of the sign of the ion-step response can be explained to be the result of immobilization of a substantial amount of positive charge, originating from the protonated side chains of ISFET-bound polylysine. Incubation of GA-treated aminobead-ISFETs with different types of polyamines in all cases resulted in ISFET devices exhibiting a negative ion-step response (figure 2). The tendency within the polylysine series was that the ion-step response amplitude became more negative with an increasing molecular size of the polyamine. This tendency is likely to be explained by a higher maximal surface concentration (μg/cm^2) to be expected when immobilizing larger polymers. A higher absolute surface concentration of polylysine will be associated with a larger amount of positive charge on the ISFET gate, and consequently with a more negative ion-step response.

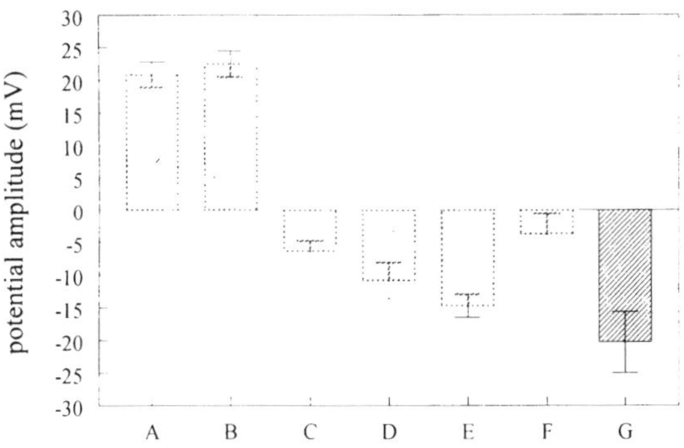

FIGURE 2. Ion-step response amplitudes determined on differently treated aminobead-ISFETs (mean ± s.d., n=5-11). Code meanings: A=GA-treated; B=+lysine; C=+PLL 2 kDa; D=+PLL 22 kDa; E=+PLL 128 kDa; F=+protamine; G=+aminodextran (home-made).

In figures 3 and 4, ion-step responses are shown, obtained from (non-)activated aminobead-ISFETs, before and after incubation with polylysine and subsequent washing with Tween 20. The magnitude of the effect of PLL incubation on the ion-step response was comparable for activated and non-activated aminobead-ISFETs which indicates that the extent of binding of

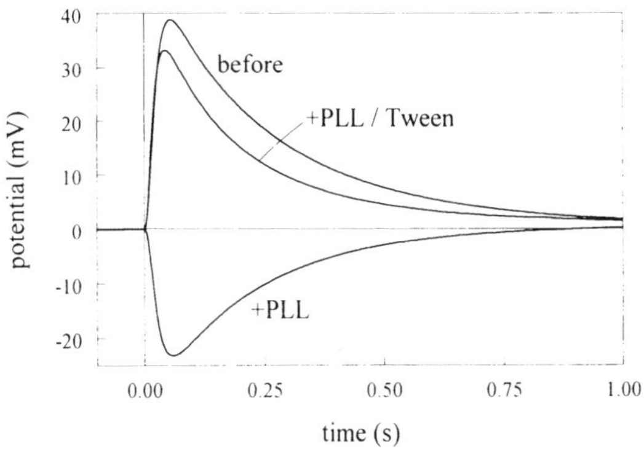

FIGURE 3. Ion-step responses of a ***non-activated*** *aminobead-ISFET before ("before") and after ("+PLL") incubation (for 10 min.) with PLL 4 kDa, and after washing of the PLL-treated ISFET with 5% Tween 20 ("+PLL / Tween") (washing time 2 min.).*

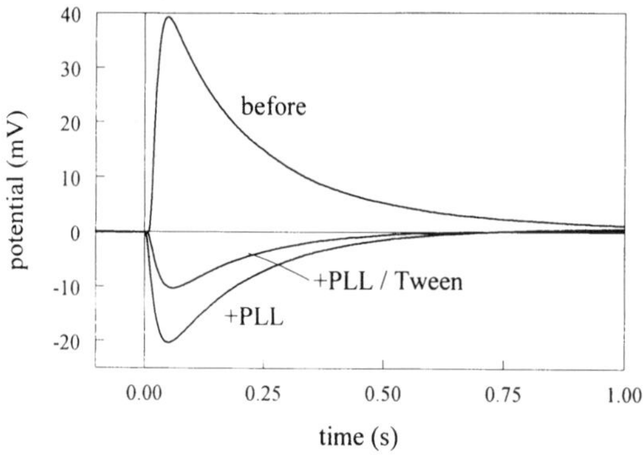

FIGURE 4. Ion-step responses of an ***activated*** *aminobead-ISFET before ("before") and after ("+PLL") incubation (for 10 min.) with PLL 4 kDa, and after washing of the PLL-treated ISFET with 5% Tween 20 ("+PLL / Tween") (washing time 2 min.).*

PLL to these ISFETs is similar. However, washing with Tween 20 removed PLL initially bound to non-activated ISFETs while PLL, bound to (previously) activated aminobead-ISFETs was largely retained, despite the washing with Tween (compare figures 3 and 4). This observation forms a strong indication for the covalent nature of the attachment of PLL to GA-treated aminobead-ISFETs. Binding of PLL to aminobead-ISFETs proceeded very rapid as 1 min. incubation time was already sufficient for obtaining a near-maximal immobilization yield (results not shown).

The different polyamine-modified aminobead-ISFETs could be used for detection of heparin, as shown in figure 5. The effective range of detection was about 0.025-0.75 U/ml whereas aminodextran appeared to be the most effective among the three ligands tested. These results partly agree with the observations of Van Kerkhof et al. [4] who found a detection range of heparin of 0.1-2 U/ml when using ISFETs covered with a thick membrane (which is a polystyrene-based solid phase). However, several hours of incubation with heparin are needed for attaining substantial amplitudes when using thick membranes. In the case of a latex bead monolayer, incubation times of not longer than 1 min are needed.

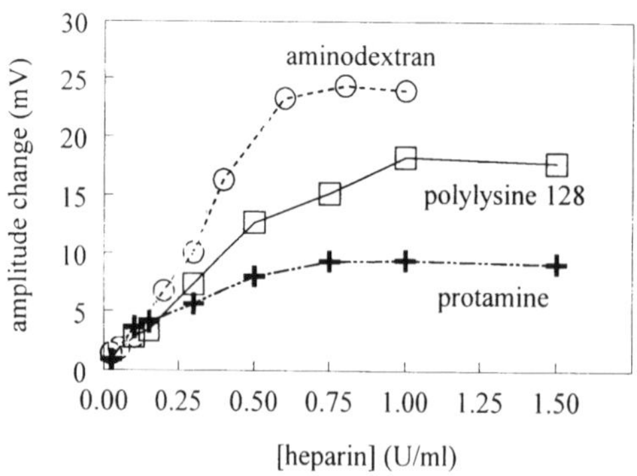

FIGURE 5. Change in ion-step response amplitude of aminobead-ISFETs modified with different polyamines, as a function of heparin concentration used during incubation (time of incubation 1 min.).

From follow-up experiments it appeared that not only polylysine but also other proteins, such as antibodies, could be immobilized to (GA-activated) aminobead-ISFETs. Future work is aimed at the preparation of ImmunoFets by immobilization of antigens or antibodies to aminobead-ISFETs.

4. Conclusions

- ISFETs easily can be functionalized with primary amino groups by covering the ISFET surface with a monolayer of aminobeads.
- Synthetic polyamines can be covalently coupled to GA-treated aminobead-ISFETs in substantial amounts as demonstrated by a markedly changed ion-step response of the modified ISFET.
- Polyamine-modified ISFETs can be used for detection of heparin in the concentration range of 0.025-0.75 U/ml.

Acknowledgements
This work was financially supported by the Dutch Technology Foundation (STW).

References

1. R.B.M. Schasfoort, R.P.H. Kooyman, P. Bergveld and J. Greve, A new approach to ImmunoFet operation, Biosensors and Bioelectronics, 5 (1990) 103-124.
2. J.C.T. Eijkel, Potentiometric detection and characterization of adsorbed protein using stimulus-response measurement techniques, Ph.D. thesis (1995), University of Twente, Enschede, The Netherlands, ISBN 90-9008615-3.
3. J.C. van Kerkhof, P. Bergveld and R.B.M. Schasfoort, The ISFET-based heparin sensor with a monolayer of protamine as affinity ligand, Biosensors and Bioelectronics 10 (1995) 269-282.
4. J.C. van Kerkhof, P. Bergveld and R.B.M. Schasfoort, Development of an ISFET-based heparin sensor using the ion-step measuring method, Biosensors and Bioelectronics 8 (1993) 463-472.
5. G.A.J. Besselink and P. Bergveld, Chapter 12: "ISFET affinity sensor" in: vol. 7 of the series "Methods in Biotechnology", eds. A. Mulchandani and K. Rogers, Humana Press Inc., Totowa, NJ.; book in press.
6. K. Gregorius, S. Mouritsen and H.I. Elsner, Hydrocoating: a new method for coupling biomolecules to solid phases, Journal of Immunological Methods 181 (1995) 65-73.

DNA BINDING KINETICS OF Oct-1 LINKER MUTANT PROTEINS STUDIED WITH THE INSTRUMENT FOR BIOMOLECULAR INTERACTION SENSING (IBIS)

Hans C. van Leeuwen and Peter C. van der Vliet

Laboratory for Physiological Chemistry, Utrecht University,

Stratenum, Universiteitsweg 100,

3508 TA Utrecht, The Netherlands

Richard B.M. Schasfoort

Intersens Instruments B.V.

Scheltussingel 156,

3814 BH Amersfoort, The Netherlands

Email: IBIS@pi.net

0. Abstract

The binding constants of two proteins (a wild type and a mutant of Oct-1) that interact with a viral DNA are measured with a gel electrophoresis technique and with the Instrument for Biomolecular Interaction Sensing (IBIS). IBIS enables scientists to monitor biomolecular interactions in real-time and label-free for kinetic analysis of obtained interaction plots. The detection principle of IBIS is Surface Plasmon Resonance (SPR). IBIS detects changes in refractive index at the sensor surface. These changes are directly related to the amount of sensor surface-bound biomolecules. The SPR-measurements with the protein Oct-1 confirm biochemical band shifts measurements showing a difference in binding affinity between the wild-type and mutant Oct-1 protein (2.9 fold measured from IBIS; 2.5 fold from gel-electrophoreses experiments).

Keywords: Adenovirus, DNA, IBIS, Biosensor, Gel-electrophoresis

1. Introduction

Adenovirus requires several proteins for efficient duplication of its dsDNA genome. These proteins have all been cloned and their function, structure and characteristics are currently under study [2]. An essential question of these studies is which part of the protein is involved in specific binding of the viral DNA.

DNA binding can be reasonably easy measured using gel-electrophoreses techniques that requires radioactivity labeled DNA fragments. Disadvantage of this technique, besides working with radioactivity, is that for quantitative analysis of the binding constants at least three time consuming assays are needed (one for on - and off-rate and one for equilibrium constant). Inaccuracies in these assays are the exact protein concentration, which depends on chemical measurement (Bradford), purity of the protein and the exact DNA concentration, determined by Cherenkow counting or deduced from labeling efficiency.

This paper describes the efforts taken to evaluate DNA binding kinetics with the IBIS biosensor and the comparison of the equilibrium constant with gel-electrophoresis values.

2. Experimental

The cellular protein Oct-1 binds to an 8 basepair sequence in the origin of the adenovirus and enhances replication ~5-10 times. Biochemical experiments suggested that a 'flexible' part of the protein (dashed line, see figure 1) could influence recognition of the DNA [1].

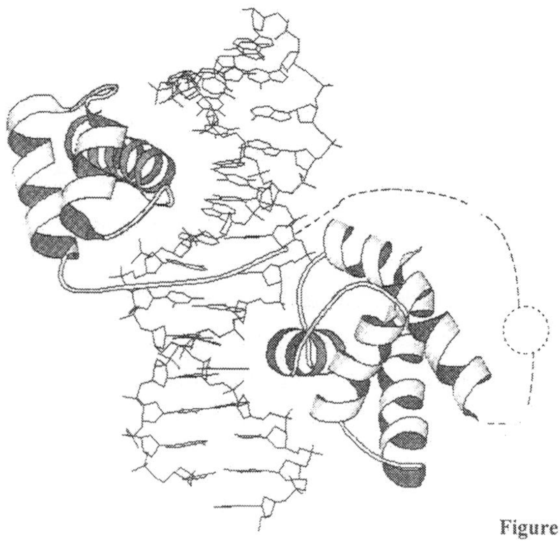

Figure 1. Structure of the Oct-1 DNA binding domain. The dashed line is the flexible linker. The approximate position of the mutated glutamic acid is indicated as a circle.

A conserved glutamic acid residue (dashed circle) in this region was anticipated to be able to influence the conformation of the linker (Fig.1). By replacing this negatively charged glutamic acid residue to a positively charged lysine (Glu>Lys) and subsequent testing of this mutant protein for DNA binding, we hoped to confirm the role of the linker region in DNA binding.

Gel-electrophoresis experiments showed that the equilibrium constant (K_D) of the mutant protein was 2 to 3 fold higher compared to wild-type. In such an experiment the K_D is determined by measuring the amount of protein (Pt [M]) required to bind 50% of a given amount of DNA (Db $_{50}$ [M]) (K_D = Pt -Db $_{50}$, see [2). Because of the possible inaccuracy of several variables (see Introduction) it was difficult to interpretative from the determined 2.5 fold difference in affinity. With the SPR technique the results of the gel-electrophoresis experiments were tried to be confirmed.

3. Material and Methods

- **DNA fragments** were obtained by hybridization of two complementary DNA oligonucleotides (42 bp). One of the oligo's was synthesized with a 5' biotin group (Pharmacia) which allowed us to immobilize the DNA on a streptavidine chip (sensor chip SA, BIAcore).
- The chip was loaded with small amounts at a time (15 ngr) up to ~ 100 mdegrees. A rule of thumb for DNA is that 120 mdegree shift corresponds to 1 ng/mm^2. For the 42 bp DNA fragment this equals ~ 0.1 pmol/mm^2.
- Measurements in the cuvet were performed in a phosphate buffered saline (PBS).
- IBIS conditions are: cuvet volume = 50μl, mix volume = 16.7, mix speed = 15μl/s, and pipette tip distance to surface = 2 mm

4. Kinetic measurements

Association-curves were measured with a protein concentration range between 0.1 μM and 0.5μM. The association phase of a one-to-one interaction in SPR biosensors is described in the literature [4] by monophasic kinetics. Biphasic models are required if two parallel interactions occur simultaneously at the sensor surface.

The kinetic analysis of the interaction plots was best performed by applying a biphasic interaction model. From other measurements it was observed that higher concentrations could not be fitted with a biphasic model. The biphasic character of the studied interaction is probably caused by a quick specific binding of Oct-1 to its binding site and a second slower a-specific binding to the DNA backbone. It is also possible that interactions between Oct-1 and streptavidine or dextran occur. The determined k_s values were plotted against the protein concentration both for the wild-type and the mutant protein.

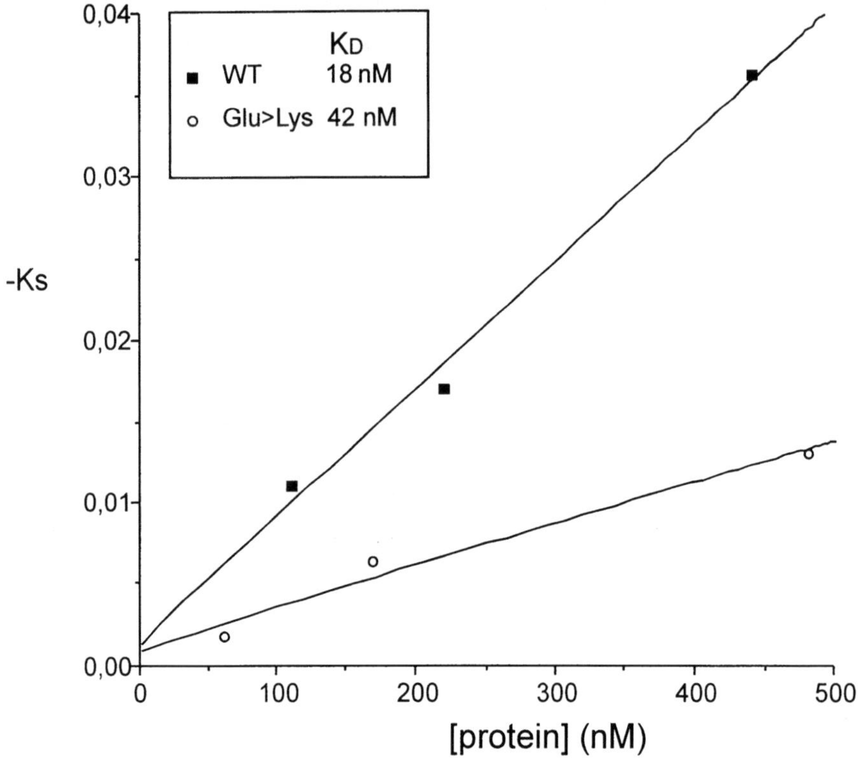

Fig. 2. Oct-1 Linker Glu>Lys mutant has a lower on-rate.
The slope of the plotted line ($-k_S = k_aC + k_d$) is equal to the association rate constant and the y-intercept is equal to the dissociation rate constant. Equilibrium dissociation constants ($K_D=k_d/k_a$) are indicated in the box.

5. Discussion and conclusions

Directly visible in figure 1 is the difference in slopes (k_a) between the wild-type and mutant protein. Since the y-intercepts (k_d) are almost identical, we conclude that the difference in affinity of both proteins as previously detected with gel-electrophoreses experiments, is caused by a lower association constant of the mutant protein. Apparently the newly introduced positive charge influences the conformation of the protein before DNA binding but has hardly any effect once bound. The calculated equilibrium constant (k_a/ k_d) is 2.9 fold higher for the mutant protein.

Comparison with gel-electrophoreses

SPR measurements with the Oct-1 protein confirm biochemical band shift measurements showing a difference in binding affinity between the wild-type and mutant

Oct-1 protein (2.9 fold measured from SPR; 2.5 fold from gel-electrophoreses experiments, see Table I).

	K_D electrophoresis (K_D = Pt -Db_{50})	**K_D SPR** ($K_D = k_d/k_a$)
Wild type	2.2 nM	18nM
Mutant E>K	5.6 nM	42nM
Ratio	2.5	2.9

Table I. Comparison of equilibrium dissociation constants of the Oct-1 protein and Oct-1 Glu>Lys (E>K) mutant protein, assayed with gelelectrophoresis and with surface plasmon resonance.

However the dissociation equilibrium constants vary considerably (see Table). These differences are most likely caused by the fact that DNA binding for gel-electrophoresis are performed in a 4% Ficoll containing buffer. Because Ficoll is very viscous it is not appropriate for SPR measurements. Another difference is that SPR measurements were performed at 20° C while gel-electrophoresis experiments are performed at 4°C.

The relative binding affinities between the two proteins correlate well showing that IBIS-measurements can be used for comparative studies.

Acknowledgments

The authors acknowledge the fruitful cooperation with members of the Department of Physiological Chemistry, Utrecht University especially Joost van Oosterhout.

References

1. van Leeuwen, H.C., Strating, M., Rensen, M., de Laat, W. and van der Vliet, P.C. Linker length and composition influence flexibility of Oct-1 DNA binding. *EMBO J.* 16: (1997). 2043-53.
2. J.J. and van der Vliet, P.C. The DNA-binding specificity of the bipartite POU domain and its subdomains. *EMBO J.* (1992a). 11: 4993-5002.
3. Morton, T.A., D.G. Myszka, and Chaiken,I. Interpreting complex binding kinetics from optical biosensors: a comparison of analysis by linearization, the integrated rate equation, and numerical integration. *Anal. Biochem.* 227: (1995) 176-185.
4. O'Shannessy, D.J., M. Brigham-Burke, K.K. Soneson, P. Hensley, and Brooks, I. Determination of rate and equilibrium binding constants for macromolecular interactions using surface plasmon resonance: use of nonlinear least square analysis methods. *Anal. Biochem.* 212: (1993). 457-468.

ENTREPRENEURSHIP IN SENSOR-LAND

Simon Middelhoek
DIMES, Delft University of Technology,
P.O. Box 5031, 2600 GA Delft, The Netherlands,
email: s.middelhoek@et.tudelft.nl

0. Abstract

Researchers in the Netherlands were among the first to realize that the lack of suitable sensors and actuators would be certain to hamper the further development of information processing systems and their introduction into new markets. As a result MESA and DIMES started sensor programs which acquired a worldwide reputation. It was expected that the academic research would lead to an increased activity in the relevant Dutch industries. Unfortunately, management in most of these industries lacked the appropriate vision and entrepreneurial skills to make use of the available information. Consequently, the academic sensor programs did not receive the necessary industrial follow up and have therefore failed to contribute to the GNP so far. The only possible way out of this impasse is to encourage and financially support MESA and DIMES graduates to start-up their own small sensor companies. In this paper some relevant data supporting this view is presented.

Keywords: entrepreneur, sensors, micro instruments

1. Introduction

It is not an easy task to find out who in the world was the first to discover that silicon sensors were going to be of great importance to the information processing industry. In each "major sensor" country, one can find scientists who did some early work that is related to the now burgeoning field of sensors and actuators. This field is also called MEMS, microsystems or micro instruments (my favorite name, which I will use quite frequently in the rest of this paper). In The Netherlands, Gieles [1] of Philips Research known for his work on miniature pressure sensors and Bergveld [2] of the University Twente known for his work on ISFETs, undoubtedly belong to the pioneers in the sensor field. Delft was an early player in the field as well with the first silicon flowmeter [3]. The first major TRANSDUCERS conference was held in Delft in 1983 and scientists in The Netherlands also launched the annual EUROSENSORS conferences. Moreover, the first scientific journal "Sensors and Actuators," solely devoted to the field of solid state sensors and actuators, started in The Netherlands. The important and early role of Dutch scientists in shaping the micro instrument technology field is generally not disputed.

Around 1970 a number of groups in the USA, notably the group of Prof. J. Angell at Stanford University and the group of Prof. W. Ko at Case Western University did innovative work in the sensor field as well. Other industrialized countries like Germany, France, the United Kingdom and Japan started much later, whereas a small country like Sweden, notably the group around Prof. Lundstrom [4] at the University of Linköping also realized the importance of the chemical sensor field quite early.

Today, practically all countries with some activity in the instrumentation industry are also active in the micro instrument technology field. The reason for this is that micro instruments are considered to be a key-technology, which allows one to apply the wonders of IC design and IC batch-processing (a 25 4" wafer batch gives roughly half a million pressure sensors) to many more application areas like, for instance, the airbag. In 1997 about 25 million crash sensors for triggering airbag deployment have been sold worldwide. The technology of micro instruments can improve existing information processing systems and make them more cost-effective. It also enables the manufacturing of products like the already mentioned crash sensor, which could not have been done with previously existing technologies. Market analysts expect a micro instruments market size for the year 2000 that lies between 1 and 20 billion US $. Their predictions vary considerably, as quite different definitions of the field are used in these market surveys. Anyhow, all surveys agree that a very large market lies ahead.

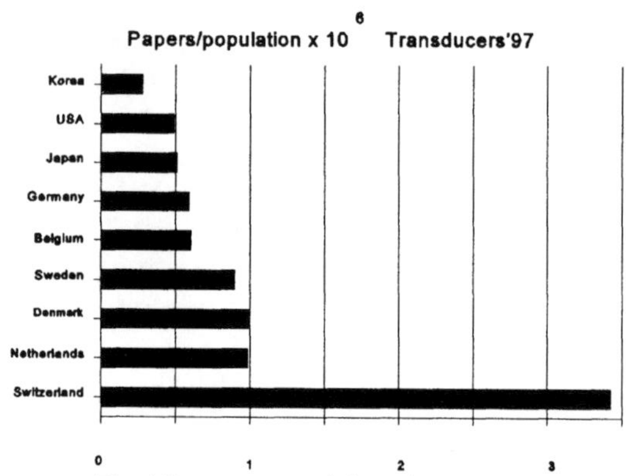

Fig. 1 Sensor research in various countries

In order, to obtain some insight into how much a country participates in the micro instruments field, the contributions of the major industrialized countries to the last international TRANSDUCERS'97 conference in Chicago were evaluated. In Fig. 1 the number of presentations (oral and poster) per country has been divided by the number of inhabitants of that country. From the figure it appears that Switzerland has obtained the position of most important country in this field by far. Switzerland has three internationally known academic centers i.e. the University of Neuchâtel, the ETH Zürich and the EPFL Lausanne. Next are the countries of The Netherlands, Sweden and newcomer Denmark. Not surprisingly we also find Belgium, Germany, Japan, and the USA on the list of the 9 most important sensor countries. Korea only recently started to support the sensor field and advanced quite rapidly to No 9 on the list. In practically all 9 important sensor countries, sensor research is financially supported by the respective governments. Especially in Germany, the USA and Switzerland large budgets are reserved for the sensor field. In The Netherlands NWO/STW was and is very diligent in supporting the field [5].

2. Sensor enterprises in various countries

Most governments are willing to spend some money on special priority programs, with the expectation that this money will be used to bring novel devices to the prototype stage, so that industry can take over without too much risk involved. Further they hope that in the long run the technical field will contribute to the Gross National Product (GNP) and will strengthen the industrial base of the country. It is of interest to see if, with respect to micro instruments, the expectations of these governments have been fulfilled so far. Next, we will briefly study the situation in the USA, Germany, Switzerland and also The Netherlands.

2.1 The USA.

Everybody knows about Silicon Valley and how in only 40 years apricot and prune orchards were replaced by 7000 information technology companies, with 11 new companies being created every week [6]. Also several sensor companies were started in the course of the years, with Stanford University and Berkeley providing the well educated entrepreneurs, scientists and engineers. Also former employees of the larger electronics companies became entrepreneurs and together with the academics created many small start-ups. About 80 companies are currently involved in micro instrumentation, of which 60 are very small entrepreneurial sensor start-up companies [7]. The founders of former NOVASENSORS gave an interesting genealogy of the sensor companies in the USA between 1960 and 1990 in their book [8]. In Fig. 2 a part of this genealogy is displayed, supplemented with more recent developments. The diagram does not show all the companies active in the field. Many companies that have already existed for many years, like Honeywell, Rosemount, Kulite and Endevco are not

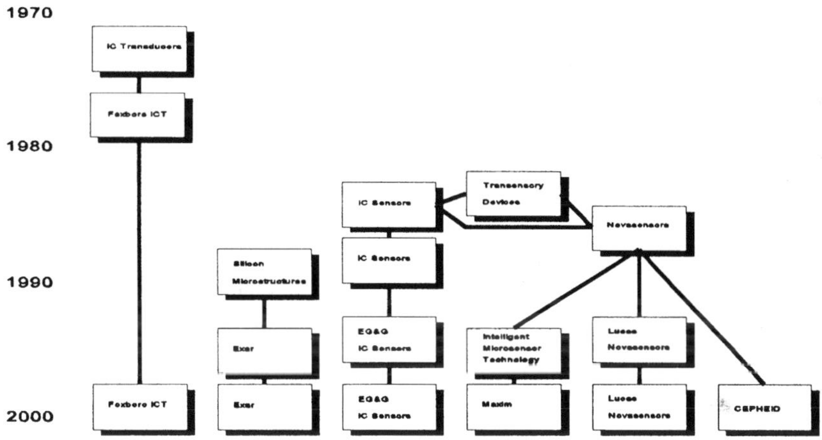

Fig.2 Genealogy of sensor start-ups in California [8].

included in the display either.

Many of the sensor companies will experience difficult times from time to time, and they will be sold and resold. However, at present they manage rather well as the demand for mass-produced and cost-effective sensors is rapidly growing. These companies provide employment to thousands of people. They sell millions of pressure sensors, accelerometers and microstructures like valves, micro-fluidic devices and ink-jet printheads. It is the automotive market especially that shows a great demand for all kinds of innovative sensors. The size of the automotive market induced the big American electronics companies to try to enter the sensor field as well. However, their attempts have not been very successful so far. Apparently the manufacturing and packaging of sensors requires specialized knowledge not yet readily available at the big companies.

Much of the research at the universities in the USA is supported by governmental programs like DARPA, NSF, NHI or by industry. Though the sensor market still has to be developed further, one can say, that in the USA the university research programs receive quite a good follow-up in the relevant industries. Many vacancies in this field still exist sometimes advertised on the Internet [9]. It is interesting to note that at practically all companies in the USA involved in sensor products, one can find graduates from the Dutch universities. Of course, one hopes, that these people will return to Europe and at some time in the future will use their knowledge and drive acquired in California to start up their own companies in The Netherlands.

2.2 Germany

Though Germany was not amongst the first countries to realize that sensors are of importance to their industries, the BMBF (Bundes Ministerium für Bildung und Forschung) currently spends huge sums of money on sensor and sensor technology programs to further this field and to make up for the leeway. Practically all technical universities and colleges have already introduced the micro instrument field in their curricula. Several start-ups are trying their luck and many of the laboratories of the Fraunhofer-Gesellschaft have initiated important and successful sensor programs. Much information can be found on the VDI/VDE-IT web site [7].

Thanks to the ample financial support of the BMBF Germany has progressed vary rapidly and, at present, has the largest program in micro instrumentation in Europe. Also international companies like Siemens and Bosch have become active in the field as the German automotive industry promises to be an important customer for mass-produced micro instruments. It is still too early to see whether all activities in Germany will continue if the BMBF ceases to support the field in the future.

2.3 Switzerland

Because of the watch industry Switzerland has a long tradition in the field of micro-technology. It is no surprise, that the Swiss industry got interested in the upcoming field of micro instruments rather early. The first program was started at the small University of Neuchâtel by a graduate from the University of Twente. Soon after that the ETH Zürich initiated a sensor program, focussing on using as much IC standard processes as possible. Quite recently the EPFL in Lausanne realized the importance of the sensor field for Switzerland as well. At present, a fully equipped processing laboratory is being built and the EPFL is enlarging its faculty by adding 7 new members specialized in micro instruments.

In addition, academics and industry managed to convince the Swiss government of the importance of the field. This resulted in several priority programs of which MINAST is the most influential. Since most Swiss industries are as conservative as industries in the rest of Europe, it took a little bit longer to convince them that micro instruments suit the high quality and miniaturization image of Swiss industry very well. Currently, the MINAST program stimulates intensive cooperation of academics with industry. University groups can only obtain financial support; if they find an industrial partner, that is willing to spend the same budget within its company. Over 80 companies currently participate in the MINAST program, showing that Swiss industry is expecting to improve their product palette with many innovative products. Besides industry and universities, the CSEM [10] in Neuchâtel is also very active in promoting the field. It can be expected that when governmental subsidy will be reduced in the near future, the activities and the cooperation between academics and industry will continue, giving Switzerland an important and lasting position in this field. It is striking again to note how many graduates from Dutch universities are active in the sensor field in Switzerland.

24 The Netherlands

As already mentioned, Philips Research, the University of Twente (MESA) and Delft University of Technology (DIMES) recognized quite early the importance of the sensor field for the information processing industry. The early work of the universities attracted the attention of several American companies, which soon sought contact with the groups in Twente and Delft. Though enthusiasm and expectations were great in those early days, none of the collaborations led to something useful or marketable. Making money with silicon sensors apparently was more difficult than anticipated both by industry and the universities in those early years.

Also the pioneering activity at Philips did not get much follow-up and even today, Philips does very little in this field. Philips easily could have established a very strong position here, using their own data and that of the universities. The consumer product palette of Philips needs many sensors and since Philips possesses very sophisticated IC facilities, they could have become the main supplier of sensors and sensor systems for the European consumer and automotive industry. Rather trustworthy market studies predict that in the year 2000, 20% of the price of a car will cover the cost of automotive electronics. Unfortunately, the prevailing risk avoiding behavior of Philips middle management makes it very unlikely that Philips will become successful in these markets.

Traditionally, The Netherlands is a trade and transport country and much less a manufacturing country. This explains that practically all high-tech measurement and control equipment used in The Netherlands is imported and that only in minor fields, Dutch companies have an international market. Up to now these small and medium industries have shown very little interest in the new field of sensors. The only industrial activity in this field can be found at small start-up companies. An incomplete list of these companies with their special products is shown in Table 1 [11]. In view of the indifferent attitude of the Dutch large and medium-sized industries with respect to innovation, it is not surprising that many graduates of the Dutch universities moved to other countries and, nowadays, many of them can be found in California, Switzerland, Denmark, Germany, Belgium, the United Kingdom and even Singapore.

3. The Dutch polder disease

3.1 Industry in the Netherlands

For a stable economy it makes sense to strive for a balanced spectrum of contributions (industry, traffic, trade, building, services) to the GNP. In times of a declining economy, the effects on employment might be less severe as in the case of a one-sided palette of enterprises. Nowadays, the Dutch economy depends very much on trade and services and much less on the manufacturing industry (10%). Though we find a few very large international corporations in The Netherlands, the contribution of small (<10 employees) and medium sized (10-100 employees) companies is less than that in the rest of Europe. In Europe (49) and in Belgium (49) for instance, one finds twice as many businesses per 1000 inhabitants than in The Netherlands (28).

With respect to the manufacturing industry, The Netherlands has always been weaker than its neighbouring countries and this situation is still worsening. The Netherlands used to have important industries for producing railway material, street cars, ships, trucks, aeroplanes, communication, measurement and control equipment, personal and large computers. Except for trucks (narrow escape) most of these industries disappeared. In most cases it was not the world economy that forced these industries out of business, nor was it the quality of the engineers. Frequently these industries could not survive because they were managed by career hungry managers that did possess social and communication skills, in addition to a large repertoire of risk-avoiding behaviour, but lacked vital technical insight and intuition.

Table 1. Start-up sensor companies and consultancies

Company name	Contact person	Products
3T	Drs. D.J. Sprangers	Sensors and microstructure
Delft Measurement Systems	Ir. J.K. Zelisse	Thickness sensors
ESBE Services	Ing. S. Bootsma	Consultancy
Mierij Meteo	Ing. A. Hagedoorn	Meteorological sensors
Middelhoek Consultancy	Prof.dr.ir. S. Middelhoek	Consultancy
OKO technologies	Dr. G. Vdovin	Adaptive mirrors
SensArt	Dr.ir. G.C.M. Meijer	Smart sensors
Sentron	-	ISFET's, pressure sensors
Smartec	Ir. R. de Boer	Temperature sensors
Smart Sensor Devices	Dr.ir. F.R. Riedijk	Design of smart sensors
Sprenkels Consultancy	Dr.ir. A. Sprenkels	Sensors and interfaces
Twente MicroProducts	Dr. J. Elders	Microsystems
Xensor Integration	Dr.ir. A.W. Herwaarden	Sensors and microstructures

3.2 Management

Why is it that in Europe and, in particular, in The Netherlands, the wrong people are at the head of medium and large organizations? The main reason is, in my opinion, that managers are not sufficiently judged by their achievements. Without idealizing the situation in the USA, an important difference between The Netherlands and the USA is that high ranking managers can only maintain themselves as long as their company or division is doing well. When IBM started to decline, the president was politely, but firmly asked to step aside for his successor. In The Netherlands this seldomly occurs and when it occurs a handsome "golden handshake" is available.

In The Netherlands, it seems that an invisible net separates the common employees from the "business class" people like managers of medium and large companies, university professors, judges, notaries, members of the parliament, etc. When you are above the net, you are above the law as well and immune to criticism. People above the net use their position to further increase their power and immunity. Such people even enrich themselves with clever financial arrangements, even when employees have to be laid off, because the company is loosing markets or is going through the umpteenth reorganization.

In my opinion, managers can be roughly divided into two groups:

1. those that are power hungry and like to command as many people as possible, even when they do not have the talents to do so and
2. those people that have enough talent to manage, but have to be persuaded into taking responsibility for a group of (former) colleagues.

It seems that in The Netherlands the first group is so efficient in communicating through the "old boys network", often based on student fraternities, that the second group seldomly comes into the limelight. Unfortunately the same picture is not only found in industry, but also representative for people in politics. For instance, only 8 members of the Lower House, and 6 Ministers and Under Secretaries have a Dr. degree and it seems, that being a drop-out of school is a good recommendation to become minister or president of a large company.

Even when an industry gets in serious trouble or when a ministry or other governmental organization shows many signs of weak management, these failing managers easily survive at the expense of Dutch society. In a recent book, one can learn how the situation at Philips detoriated through the years [12]. Career managers are not able to judge the strengths of new projects from a technical or commercial point of view and consequently make all the presidential blunders possible.

The Dutch polder disease has many lurking dangers. The most important of which are:

1. the industrial base of The Netherlands will further decline and science based industries will leave the country or will go bankrupt. In a few decades Philips might be a mere light bulb and philishave factory (but hopefully it will still be in the IC business).
2. foreign companies will not put out to contract Dutch companies and might only be willing to set up distribution centres in The Netherlands.
3. consultancies will thrive as never before. The Netherlands already have the largest number of consultancies per business in Europe. Career managers need their services to stay in power and to be provided with alibis.
4. companies will show much risk-avoiding behaviour. R&D expenditure of Dutch companies has already declined significantly and is currently only half (1% of GNP) of that of the USA (1.95% of GNP) and Japan (1.93% of GNP) and smaller than most of the European countries [13].
5. most companies have a very short range outlook and no plans for the future. For instance, even a prominent company like Philips recently announced that the new strategy plan has been postponed to the end of 1998.
6. the position of technical talent will become weaker. Personnel managers of big companies talk with exaggeration about the "nerd" behaviour of technical people and stress that they only are on the look out for people with "social" and "communicative" skills and above all a "broad" technical knowledge and good "employability". New employees should know nothing about everything. The result, to be expected, is that fewer and fewer VWO graduates will choose for technical studies. In The Netherlands only 1 out of 16 students perform a technical study, whereas this number is 1 out of 5 in Germany [14]. The chances for technically talented people to earn a good salary are not good and the chances to be chased around by career managers very large. The number of technical university students decreases every year. The number of EE students in Delft declined from an all time high of 400 to only 144 currently, and next year a further decline is expected.

7. the emphasis on "shareholder value" will further discourage technical people to work for such companies, as shareholders apparently come before employees.

To offset above developments, government and industry should start to glamourize the technical profession again. Practically everything around us came into being thanks to the skills of technical inventors and this should not be taken for granted. We would not have the Internet without PC's, modems and high capacity telephone connections. Salaries of technical people should be increased and they should always be higher than those of the people that only have social and communication skills. At the same time, technical talent should be persuaded to take more management responsibilities and the position and promotion of managers of all levels should much more depend on the measurable results of their units.

3.3 Government should support technical start-ups

In The Netherlands a lot of technical talent is available and also many Dutch emigrated specialists are waiting for a proper moment to return to The Netherlands. Dutch society could make much better use of this technical talent. As has been shown by several studies [15], most large corporations in the USA do no longer influence the labor-market by creating new positions. Yet, the unemployment ratio in the USA is remarkably low. The reason for this is, that the previous large layoffs from big corporations have been overcompensated by a large number of start-up companies of which many fail, but also of which many become world market players (see fig. 2). In The Netherlands, the number (4000) of technical people laid off per year by large industries is barely compensated by the creation of new jobs by small and medium companies [16]. The start-up companies in the USA are usually launched by technical people, but soon also commercial people are hired to explore the market. The start-up companies are financially supported by venture capitalists and often by governmental programs. People in these companies tend to work very hard and the spirit in these companies is contagious. They really are entrepreneurs. And it is this entrepreneurial spirit that keeps unemployment in the USA low and the work force rather flexible. If The Netherlands in the coming century would do more than merely transporting goods made by country A to country B with transport material made by country C, some of this entrepreneurial spirit would have to come back to The Netherlands. Government, in my opinion, should be much more active in developing programs that make it very easy and attractive for groups of technical people to become entrepreneurs. The University Twente already initiated in 1984 the so-called TOP plan, which offers facilities to former students and which is already used by hundreds of entrepreneurs.

The government could further increase the budget of STW to support university programs related to technical applications and at the same time could ask STW to start a new program to support, in a very simple bureaucratic way, small start-up companies that want to develop new products. Like the STW program, the projects should be reviewed by specialists and when a positive decision is reached, 100% of the budget should be provided to the start-up without any further restrictions. Certainly many of the start-ups will fail, but the net return, without doubt, will be much larger than when the money is poured in the bottomless pits of many large Dutch industries. If many university graduates would find it very attractive to start their own companies with governmental help, large industries might start to change their personnel policies, as they would have to offer positions to technical talent, which are equally attractive as those in the small start-up companies.

4. How to survive as a sensor start-up company in the Netherlands

The basic material for micro instruments is silicon and to fabricate such structures, silicon processing steps supplemented by bulk or surface micro machining have to be used. This creates a severe problem for small businesses, as they can not afford to make an investment in the order of US $ 100 000 000 for a production facility necessary to support such production processes. The first Californian companies active in this field started with much simpler production facilities, but are now faced with new large investments to rebuild their facilities. Therefore, in the micro instruments field a reorientation with respect to manufacturing is taking place. For many years, the production by these Californian companies consisted of rather simple devices like pressure sensors or Hall plates. Today, everybody is becoming more interested in so-called smart devices, as the instrumentation field is heading for bus-oriented instrumentation systems. This requires that electronic circuits are added to the sensor device and when possible integrated on the same chip. These electronic circuits should have the same quality as standard circuits that are on the market now. For amplification, autocalibration and bus-compatible output mixed processes like BICMOS might often be required. Such manufacturing facilities go far beyond the possibilities of small sensor companies.

The approach followed by many small start-up companies is to focus more on micro instrument consultancy and design, which requires less investment. Also these companies seek further cooperation with commercial foundries or research institutes like MESA, DIMES or IMEC. For example, in the USA a small and successful start-up like Silicon Microstructures (Fig.2) joined EXAR recently for this reason. Big companies like Motorola, Analog Devices, Siemens and Philips already possess excellent manufacturing capabilities. Nevertheless, up to now, they have not been able to reap the benefits of this situation and have not yet built a strong position in the smart sensor field.

Only a few sensors exist that can be completely made with a standard bipolar or CMOS process. A (smart) Hall device is such an example. In most cases etching steps are needed to shape non-standard membranes, beams or mirrors. A processing sequence for making smart devices should be chosen such, that these special processes come at the end of the processing sequence. These processes could be executed in the laboratory of the start-up company. Also special packaging steps could be performed there. Then again, specialized packaging companies could be of great help. In the sensor field there is a tendency to go to very complex devices from a processing point of view, leading to a very large number of masking steps. It is my opinion, that a much more successful path would be to keep the processing steps as simple as possible and to compensate the deficiencies of the devices by adding electronic circuits that can be made in the same technology. This requires designers who are knowledgeable about both sensor physics and electronic circuitry. Such designers are very rare and more of them should be educated by the universities.

Summarizing, it seems that start-up companies should concentrate on design and after-processing, testing and packaging and they should seek cooperation with silicon foundries for the purpose of standard silicon processing. In the case that a sensor device becomes very successful, mass production will be necessary. As big companies have a lot experience in mass production, they might want to enter the market as well and probably promise much lower prices. This is happening now in the accelerometer market. Therefore, it is advisable that start-up companies leave these areas in time and concentrate more on small series in niche markets. In these markets financial return can be much higher. If the sensor device proves to be a crucial part of an innovative instrumentation system, the start-up might even consider to produce the complete instrument.

5.Conclusions

The micro instrument field generates a lot of new opportunities for businesses around the world. Though the academic world in The Netherlands started very early to recognize the importance of the field, it unfortunately has not led to a contribution to the Dutch GNP. In the USA and Switzerland the situation seems more favourable, often thanks to graduates from the Dutch universities.

As Philips and the Dutch medium sized industries in the instrumentation branch show very little drive for innovation, and, moreover, The Netherlands have very few small to medium sized companies in comparison to the rest of Europe, the Dutch government should stimulate entrepreneurs to start their own companies. An efficient way would be to ask STW to organize the project evaluations. The presence of a large number of start-up companies might lead to a more innovative and competitive atmosphere in The Netherlands, abating the Dutch Polder disease. It might even challenge big companies like Philips to take the micro instrument field and technical talent more seriously.

References.

1. A.C.M. Gieles, Subminiature silicon pressure transducer, Digest IEEE ISSCC, Philadelphia (1969) 108-109.
2. P. Bergveld, Development of an ion-sensitive solid-state device for neurophysiological measurements, IEEE Trans. Biomed. Eng. BME-19 (1970) 70-71.
3. A.F.P. van Putten and S. Middelhoek, Integrated silicon anemometer, Electron. Letters 10 (1974) 425-426.
4. I. Lundstrom, S. Shivaraman, C. Svensson and L. Lundqvist, Hydrogen sensitive MOS field-effect transistor, Appl. Phys. Letters 26 (1975) 55-57.
5. http://www.stw.nl/projecten/index.html
6. Silicon Valley, Business Week, 25 Augustus 1997
7. Micro System Technology, http://www.vdivde-it.de/IT/mst/mst.html
8. J.Bryzek, K. Petersen, J.R. Mallon, L. Christel and F. Pourahmadi, Silicon Sensors and Microstructures, NovaSensor, Fremont, 1990.
9. Jobs, http://mems.isi.edu/archives/mems_associations/Bay-Area/
10. CSEM, http://www.csem.ch/competen.html
11. Lijst Sensorproducenten, STW-Sensornieuws, September 1997.
12. M. Metze, Let's make things better, Sun, Nijmegen, 1997.
13. Test of Dutch competitiveness 1995, http://info.minez.nl/nota/ctoets/hfdst2.htm.
14. Betas start to edge ahead of arts graduate chiefs, Financial Times, 23 October 1997.
15. Test of Dutch competitiveness 1996, http://www.minez.nl/nota/ctoets96/
16. MKB nota, http://info.minez.nl/nota.mkb/bijl.htm

A NOVEL CONCEPT IN THIN FILM MAGNETIC HEADS: THE SENSOR LAST DESIGN

A.H. van Heeren*, C.W.M.P. Sillen, F.A. Pronk*, L. Postma****

* Philips Magnetic Heads & Modules
Lodewijkstraat 1, 5652 AC Eindhoven, The Netherlands
** Philips Research Laboratories
Prof. Holstlaan 4, 5656 AA Eindhoven, The Netherlands

0. Abstract

Almost all commercially available magnetic heads are constructed according the "sensor first design". In this design at first the MR read-out sensor, based on a highly sensitive and thin permalloy layer, is processed. On top of the sensor the other parts of the head are processed. However, this concept has various drawbacks for the next generation of dataheads. The development of planarization in wafer processing technology opened the way to the sensor-last design. The solutions for these (and other) problem has resulted in a robust production process. The total oxide thickness was decreased from 40 μm down to 17 μm. The sensor last design is now successfully introduced in the manufacturing of the DIGAMAX™ system.

Keywords: thin film magnetic head, sensor-last design, planarization, magnetoresistance, wafer technology, CMP, DIGAMAX

1. Introduction

1.1. History

The idea of fabricating a magnetic recording head using techniques similar to those used in integrated circuit technology, was generated in the early sixties. The first thin film head was proposed by Gregg [1] in 1961, but it was not until 1970 that research results were published on the fabrication of horizontal and vertical thin film magnetic heads [2, 3]. The thin film head is applied in computer disk drives, where it is integrated in the slider that flies over the data recording disk surface, and in tape drive systems. Thin film heads replaced ferrite heads, which have several serious disadvantages.

The first product containing a single channel thin film head was IBM's 3370 disk drive, launched in 1979. Design of thin film heads has evolved from early single-turn to multi-turn designs with up to at least 50 turns. The high number of turns is mainly necessary for the reading process. More recently, magnetoresistive read heads were introduced.

Magnetoresistive (MR) read heads produce a very high signal per unit track width and can provide media noise limited system performance at very high areal storage density. The signal is independent of media speed and the heads are relatively simple to produce.

Multi channel thin film heads with separate read and write elements have been applied to digital audio and data recording. In this application, an inductive write head can be used in conjunction with a magnetoresistive element (MRE) for reading. This combination is a powerful design for advanced recording applications [4].

Within Philips in 1970, work started on "integrated magnetism". The subjects included thin film magnetic heads, inductive heads for writing and MR heads for reading. It was focused on consumer applications such as audio and video recording and on professional recording applications. In 1982 Philips realised many different types of thin film heads such as 32- and 64-track write heads and read heads for professional CLS and VLS (Communication and respectively Voice Logging Systems), magnetic MR sensors, a 4-track auto-reverse hi-fi read-only head and 22-track write and read heads for an S-DAT (Stationary Digital Audio on Tape) recording system [5].

In 1993 the Digital Compact Cassette (DCC) system was launched. The first heads to demonstrate the DCC system were made using known technology (separate write and read heads, ferrite substrates). Philips managed to realise the read / write head needed for the DCC system using also existing Seagate technological capabilities [6]. The cross-section of such a head is shown in Fig. 1a: first the MR read head is processed: the inductive write head is built on top (*sensor-first* design). Polymer (baked photoresist) structures are used to increase the efficiency of the recording and playback yokes and to eliminate step coverage problems.

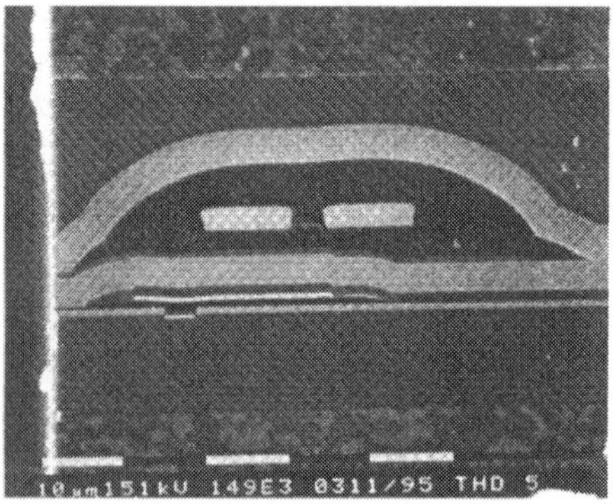

Fig. 1a. Non-planar head, sensor-first design

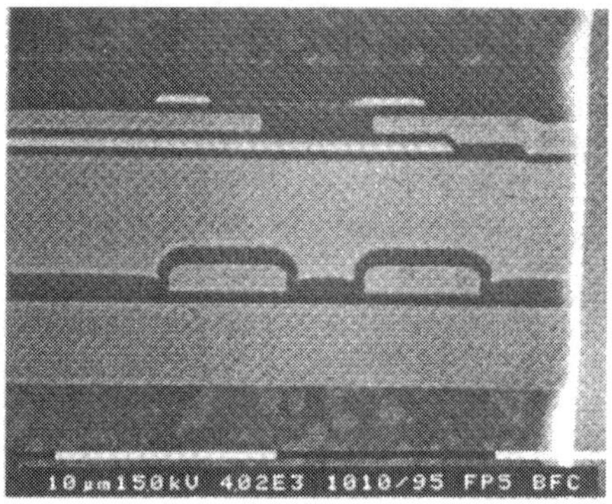

Fig. 1b. Planar head, sensor-last design

1.2. From sensor-first to sensor-last

The first DCC head was applied in home decks, but its power consumption was rather high for portable applications. One of the main reasons was caused by the thickness of the oxide layer separating the MR overlay and the permalloy yoke, influencing the read efficiency in a negative way [7]. This thickness was needed to prevent short circuits.

By reversing the design problems like these are solved. In the *sensor-last* design techniques for local planarization needed to be introduced to solve the step coverage problems and to have a perfect surface for MRE deposition. The applied planarization technique is a combination of chemical and mechanical polishing (CMP, see Sec. 2.). It is executed a number of times during the production of the head. A cross-section of a planar head is shown in Fig. 1b. Polishing processes have been developed for material combinations that exist in thin film heads, such as oxide-magnetic material and oxide-copper [8].

Some advantages of the sensor-last technology are:

-Sensor-last enables freedom in design to obtain high efficiency and a lower power consumption.

-Sensor-last allows high temperatures for optimising the magnetic properties of the flux guide material and the possibility of using exotic materials (NiFe MR and in particular Giant MR layers can potentially be degraded or damaged at high temperatures).
-Large steps no longer occur in the head, implying no step coverage problems and fewer short-circuits, less stress and no problems in wet chemical etching of several materials.
-Sensor-last facilitates the reduction of the smallest dimensions.
-The necessary planar technology offers opportunities for more efficient processing, resulting in smaller technology gaps (i.e. the gap between the substrate and the cover bar) which are favourable with respect to wear of the magnetic tape heads.

2. The planar process

2.1. Introduction

Planarization has long been a part of thin film head processing. It is applied to obtain a flat surface at the start, a substrate covered with a polished Al_2O_3 layer and also at the finished wafer surface on which the cover bar is bonded. The process which is used nowadays is still a more or less conventional mechanical polish of an Al_2O_3 layer. The shift to MR thin film heads triggered the need for more smooth surfaces. By applying a chemical mechanical polishing (CMP) process this requirement could more easily be realised [8]. This explains a tendency in the thin film head industry to shift to CMP planarization processing.

An improvement of the integral process efficiency could be obtained by shifting from the final one-step planarization to "local" planarization. Total oxide thickness decreased from 40 to 50 μm down to 17 to 20 μm (see Fig. 1a and Fig. 1b). Moreover, a 40 μm Cu plating step (including the dry resist process) needed for electrical feed-through could be avoided.

Since well-controlled smooth surface quality is a requirement for this construction, a CMP process has been chosen for the sensor-last process flow. This choice became possible because at the very moment this decision had to be made, the release of CMP in the semiconductor industry started. In this industry the process is used for the realisation of multi-contact layer processes.

2.2. Process description

The principles of the process are very similar to those of the semiconductor industry. However, a few typical differences connected to the application of CMP for the present magnetic heads are (i) the presence of a heterogeneous surface of SiO_2 and NiFe, (ii) the use of a 2.8 mm thick ceramic Al_2O_3-TiC (alsimag) substrate and (iii) the occurrence of step heights as high as 1 to 6 μm.

2.2.1 Surface

At Philips Magnetic Heads & Modules, SiO_2 has been chosen instead of Al_2O_3 as the oxide material. One of the reasons is that substantially higher deposition rates can be obtained in the PE-CVD process used for SiO_2 deposition.

In magnetic heads flux guides are the thickest structures. For this reason after each flux guide manufacturing step in general a planarization step is done. From the design point of view it is better to deal with the tolerance from the planarization process as part of the tolerance of the resulting flux guide thickness after planarization (see Fig.2). The consequence of this choice is to planarise a heterogeneous surface with the need to balance the removal rate of SiO_2 to that of

NiFe or any other used flux guide material. Adequate tuning of and control over polishing parameters will keep the resulting step between SiO_2 and NiFe within specification.

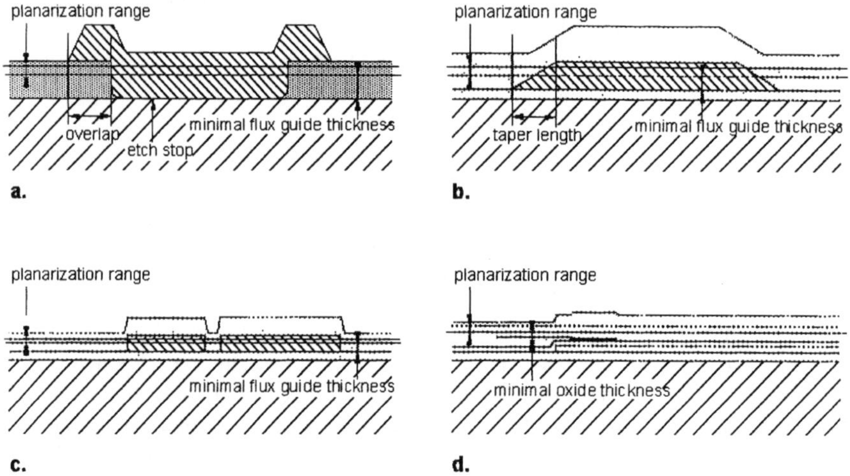

Fig.2. Planarization options: a. buried flux guide, b. taper etched flux guide, c. reader flux guide (plating through mask), d. top planarization

2.2.2. Substrate

The non-uniform removal rate over the wafer diameter is typical for a polishing process. This can be corrected for by introducing a bow to the substrate while loaded in the machine. Special carriers are available for achieving this. The use of a thick ceramic substrate with its relatively high stiffness makes it somewhat tougher to correct. Moreover, other parameters add to this non-uniform process: (i) non-uniformity of the oxide layers and of NiFe layers, (ii) type of planarization step, (iii) initial bow of the substrate, (iv) application of a flat and (v) varying non-uniformity due to wear of the polishing pads. Process reproducibility is obtained by tight off-line control of the removal rate and the uniformity factor (i.e. the ratio between removal outside / removal centre).

2.2.3. Process types of planarization

To create a sensor-last design without polymer and overcoat layers, the thin film layers need to be planarised at four locations in order to have suitable flat surfaces for writer coils, bias conductor, MRE and chip surface for mechanical mounting of the cover bar (after deposition of the writer flux guide, the shared flux guide, the reader flux guides and the protective oxide layer respectively). Various planarization types are described in the following:

"Buried" flux guide (Fig. 2a): This type is chosen for flux guides that need good width control (within +/- 2 μm) in order to define track width in relatively thick layers (3 to 6 μm). This is

obtained by the etch of the SiO_2 layer (4 to 6 μm) in which the flux guide is buried. Special attention has to be given to the step coverage of NiFe over the etched pattern. An additional process parameter for this type is the amount of overlap of the NiFe over the SiO_2. A trade-off has to be made between the robustness of the etching and the planarization

"Taper etched" flux guide (Fig. 2b): This is used if no tight specification on the flux guide width is needed. The smooth slope of the flux guide results in a good step coverage of SiO_2 over the relatively thick NiFe pattern (5 to 8 μm).

Reader flux guide (Fig. 2c): The flux guide is formed by the "box/lid" or "plating through mask" method which gives flux guide width control defined by the lithographic process. This method is more accurate (+/- 0.5 μm) than the wet chemical etching process of Fig. 2a. However, step coverage over the straight edges is the demerit of this process which only makes it usable for relatively thin layers (0.5 to 1.5 μm). Attention also has to be paid to sufficient spacing between the patterns in order to prevent formation of voids. Because of the application for relatively thin layers, this type of planarization is sensitive to non-uniform polish and even to effects of local non-planarity due to underlying planarised structures.

"Top" planarization (Fig. 2d): This type is usually a finishing step of the wafer processing applied to planarise the top structure (in our case the MRE) mainly consisting of leads (total structure height 0.7 μm). It is typically a homogeneous SiO_2 surface planarization, used to obtain a flat surface in order to achieve a good performance of the cover bar glue process.

3. Conclusions

The sensor-first design has various (electrical / magnetic and technological) drawbacks in comparison with the sensor-last design. By introducing the planarization technique in the wafer processing, the sensor-last design has become feasible. Further optimisation of design of processes has resulted in a technology that is used for manufacturing the heads for the DIGAMAX™ system [9]. In combination with the sensor-last design the CMP process and its ability to introduce local planarization has proven to be beneficial in the production of magnetic heads, in performance as well as in process efficiency, and in machining and assembly.

The sensor-last design also offers possibilities for the application of new materials needed for new generations of recording systems.

Acknowledgements

Many colleagues of Philips Research (including the magnetic head workshop) and of the Development and Engineering department of Philips Magnetic Heads & Modules have contributed to the work as described above. For their contributions the authors wish to thank all of them.

References

[1] D.P. Gregg, US Patent 3.344.237 (1967).
[2] L.T. Romankiw, I.M. Croll and M. Hatzakis, IEEE Trans. on Magn. 6, 597 (1970).
[3] J.P. Lazzari and I. Melnick, IEEE Trans. on Magn. 6, 601 (1970).
[4] D. Markham and F. Jeffers, Proc. Magn. Mat. Proc. Devices, ECS 90-8, 185 (1990).
[5] E.L.M. Raemaekers, IEEE Trans. on Magn. 18, 1143 (1982).
[6] V. Zieren, G. Somers, J. Ruigrok, M. de Jongh, A. van Straalen, W. Folkerts, E. Draaisma,

F. Pronk and T. Mitchell, IEEE Trans. on Magn. 29, 3064 (1993).
[7] P.J. van der Zaag, J.J.M. Ruigrok and M.E. Gillies, "New options for thin film recording heads through ferrite layers", Philips Journal of Research, to be published.
[8] L. Postma and J. Haisma, Patent Applications EP-A 0617409, EP-A 0617410, EP-A 0634464, EP-A 0634465 and EP-A 0658881 (1993).
[9] S.B. Luitjens, W. Folkerts, H.W. van Kesteren and J.J.M. Ruigrok, "Trends in digital magnetic recording: the application of thin film heads for tape recording", Philips Journal of Research, to be published.

MEMBRANE FABRICATION WITH GALVANIC ETCH-STOP

C.M.A. Ashruf, E.L. Oemar, P.J. French, P.M. Sarro, P.M.M.C. Bressers⁺, J.J. Kelly*

Delft University of Technology, DIMES, Dept. of Electrical Engineering, Lab. for Electronic Instrumentation, Mekelweg 4, 2628 CD Delft, The Netherlands

⁺ Netherlands Energy Research Foundation ECN, Westerduinweg 3, P.O. Box 1, 1755 ZG Petten, The Netherlands

* Utrecht University, Debye Institute, P.O. Box 80000, 3508 TA Utrecht, The Netherlands

0. Abstract

A novel passivation mechanism for silicon etching in a TMAH solution is presented. This technique is based on a Au/Si/TMAH galvanic cell and does not require external contacts to the wafer or a wafer holder. In this paper the theory is explained. Furthermore, we demonstrate how the etch-stop technique is used for fabrication of n-type epi membranes.

Keywords: Silicon etching, electrochemical etch-stop, galvanic cell

1. Introduction

Micromechanical sensors such as pressure sensors are often based on thin silicon membranes. These membranes are usually fabricated with an electrochemically controlled etch-stop in an alkaline solution [1]. A p^+ etch-stop is not really suitable for this purpose since a high boron concentration introduces stress into the silicon [2]. With double clamped structures this may result in fracture. With electrochemically controlled etching of silicon membranes with a uniform reproducible thickness are obtained. The disadvantage of this fabrication method is the need for an etch holder. Etch holders are used to shield the electrical contacts to the wafer from the etchant. It is difficult to make a holder which does not leak in highly corrosive environment such as an alkaline solution at a temperature of up to 90 °C. Another problem with an etch holder is the mechanical stress which may be introduced into the silicon wafer during etching. With fabrication of fragile thin membranes this may cause fracture. One fractured membrane exposes the electrical contact to the silicon wafer at the front side to the etchant causing a total loss of the wafer.

Recently a new electrochemical etch-stop technique has been reported [3,4]. This technique does not require external electrical contacts to the wafer. An etch holder is not used. This has several advantages which will be discussed in later sections. The mechanism of the etch-stop has not been explained in above mentioned references. In this paper we explain the mechanism of membrane formation.

2. Theory

To understand the principle of the galvanic etch-stop technique we briefly summarize the process of electrochemically controlled etching in alkaline solutions. In Fig.1. a current-potential curve of p-type silicon is shown. In this figure the potential of a silicon working electrode is regulated with respect to a standard calomel electrode (hereafter: SCE). The current is measured between the silicon working electrode and a large area platinum counter electrode. Briefly, the following three potential regions can be defined: (V < OCP), (OCP < V < PP), (V > PP). These are given in Fig.1.

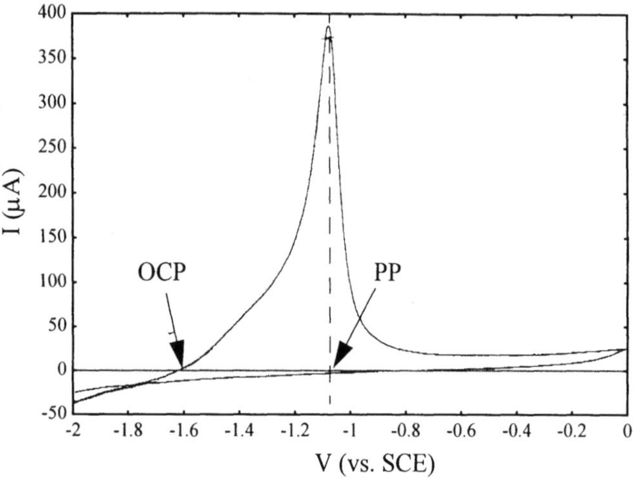

Fig.1. Measured current-potential curve of a p-type Si working electrode in a 25% TMAH solution at 80 °C. The potential is regulated with respect to a standard calomel electrode (SCE).

From the literature [5] we know that at potentials negative with respect to the open circuit potential (hereafter: OCP) a large cathodic current is observed for n-type silicon. In this potential region water is reduced into hydrogen by conduction band electrons. For p-type silicon only a small cathodic current is observed (see Fig.1.). Water is not reduced due to the absence of conduction band electrons. At OCP silicon etches chemically and the etch rate is dependent on the temperature and composition of the solution. At potentials positive with respect to OCP an anodic current is observed for both types of Si. At the passivation potential (hereafter: PP) a peak current of about equal magnitude is measured. In the potential region between OCP and PP silicon is electrochemically oxidized and the reaction products dissolve into the solution. In the potential region positive with respect to PP an oxide layer is formed on the silicon surface. The etchant no longer has access to the Si-Si surface bonds and etching ceases. The small anodic current observed in this passive range is determined by the rate at which oxide is chemically dissolved in the solution. To summarize, silicon is passivated if the potential is positive with respect to PP. In order to assume a potential in this range a peak current must be provided.

Instead of using an external power source the peak current may also be provided by a galvanic element. Fig.2. schematically shows the formation of a galvanic cell by an external contact between a silicon wafer and an inert metal electrode (case (A)) or by evaporation of the metal on the wafer (case (B)).

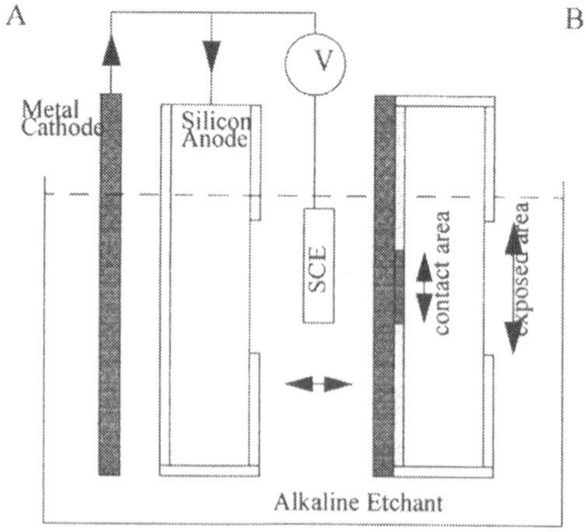

Fig.2. Formation of a galvanic cell by connecting a silicon wafer to a gold electrode (case (A)) or by evaporation of a Au/Cr film on the wafer (case (B)).

In general, a galvanic cell consists of two electrodes, i.e. an anode at which an oxidation reaction occurs and the cathode at which a reduction reaction occurs. In our case silicon is the anode and a gold electrode forms the cathode. Although n-type and p-type Si are oxidized by a different mechanism it is sufficient for our discussion to consider only p-type Si. With p-type silicon valence band holes are required for the formation of an passivating oxide layer. In the case of a galvanic cell, these holes are generated by the reduction of an oxidizing agent at the cathode. The principle of galvanic element formation can be understood with the help of the current-potential curves in Fig.3.

Fig.3. Current-potential curves of; (a) the partial anodic silicon oxidation, (b) the partial cathodic reduction with low and (c) high concentration of oxidizing agent.

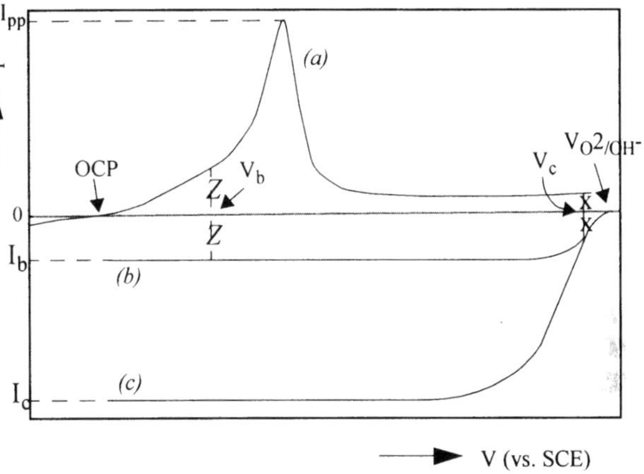

Curve (a) shows the anodic partial current as a function of potential of a silicon electrode in a TMAH solution without oxidizing agent. Curve (b) shows the partial cathodic current from the reduction of an oxidizing agent in the solution. Reduction of the oxidizing agent takes place at the inert gold electrode. Curve (c) shows the partial cathodic current with a higher concentration of oxidizing agent. The rate of the reduction reaction is diffusion controlled and independent of the applied potential for a wide potential range. If the two electrodes are connected the system assumes a rest-potential corresponding to the potential at which the two partial currents have equal magnitude. As mentioned earlier, to passivate silicon galvanically the hole flux needs to be sufficiently large. If the concentration of oxidizing agent is too low for passivation (curve (b)) the rest-potential shifts to a value V_b. If the concentration is increased and the cathodic current is equal or larger than the anodic peak current (curve (c)) the rest-potential shifts to a value V_c; silicon is passivated. A shift of the rest-potential to the passive region (i.e. to a potential positive with respect to PP) is only possible if the redox potential of the oxidizing agent is positive with respect to PP (about -1.1 V with respect to SCE). We used oxygen as oxidizing agent which has a redox potential of about 0.17 V with respect to SCE.

3. Membrane fabrication

The galvanic passivation mechanism has been used with a pn configuration to fabricate n-type Si membranes. A p-type wafer was provided with an n-type epi-layer. A layer of 1000 Å silicon nitride layer was deposited on both sides of the wafer. Next, windows were opened in the nitride to form an etch-mask at the back side and a contact mask at the front side. Finally, a film of 600 Å Cr and 6000 Å Au was evaporated at the front side. The resulting structure is shown in Fig.4a. The wafer was immersed in a 25% TMAH solution at 80 °C. Initially, p-type Si was exposed to the etchant. As discussed above, for the passivation of p-type Si a sufficiently large hole flux is required. However, the required current flows in the reverse direction and in a high quality pn junction this current is limited to a low value. As a consequence, p-type Si was etched until the pn junction was reached. At this point, the pn junction was destroyed enabling the current required for the passivation of the n-type epi to flow. Etching stopped resulting in an n-type epi membrane. Using this technique membranes of 4.8 μm have been fabricated. The result is shown in Fig.4b.

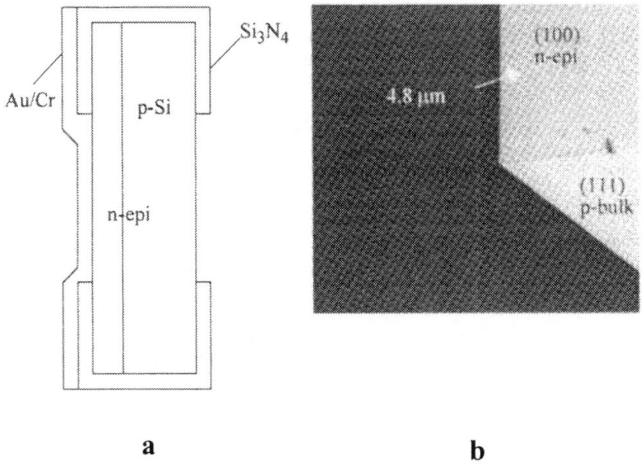

Fig.4. Schematic of the samples which have been etched in a 25% TMAH solution (a). The result after etch-stop is an n-epi membrane of 4.8 μm (b).

This process is similar to the pn electrochemically controlled etch-stop. However, with the galvanic etch-stop external electrodes and contacts are not required. An etch holder is not used and batch fabrication of micromechanical structures may be easier to implement.

4. Discussion and conclusions

Usually thin silicon membranes are fabricated with an electrochemically controlled etch-stop. The etch-stop is accomplished by externally applying a potential positive with respect to the passivation potential. Contacting a silicon sample and at the same time protect the electrical contact from the etchant is possible with an etch holder but this may cause problems. A new galvanic etch-stop technique which does not require a wafer holder has been presented. This etch-stop seems promising for batch fabrication of membranes. Important issues which need to be considered in the future are the exact location of the etch-stop, the dependency of this location on the conditions of the etchant and the reproducibility and uniformity of the thickness of the membranes.

Acknowledgements

The authors wish to thank DIMES for fabricating the samples. This project was sponsored, in part, by the STW project No. DEL55.3780.

References

[1] B. Kloeck, S.D. Collins, N.F. de Rooij, R.L. Smith, Study of electrochemical etch-stop for high-precision thickness control of silicon membranes, *IEEE Electron Dev.* 36 (1989), 663-669.
[2] S.D. Collins, Etch stop techniques for micromachining,*J. Electrochem. Soc.* 144 (1997), 2242-2262.
[3] P.J. French, M. Nagao, M. Esashi, Electrochemical etch-stop in TMAH without externally applied bias, *Sens. Act.* 56 (1996), 279-280.
[4] C.M.A. Ashruf, P.J. French, P.M. Sarro, M. Nagao, M. Esashi, Fabrication of micromechanical structures with a new electrodeless electrochemical etch-stop, *Proc. Transducers'97* (1997), 703-706.
[5] P.M.M.C. Bressers, S.A.S.P. Pagano, J.J. Kelly, Ferricyanide reduction as a probe for the surface chemistry of silicon in aqueous alkaline solutions,*J. Electrochem. Soc.* 391 (1995), 159-168.

FABRICATION TECHNOLOGY OF GEIGER MODE AVALANCHE PHOTODIODES

W.J. Kindt[1] and H.W. van Zeijl[2]

[1]Electronic Instrumentation laboratory, Faculty of Information Technology and Systems, Department of Electrical Engineering, Delft University of Technology
Mekelweg 4, 2628 CD, Delft, The Netherlands

[2]DIMES, Delft University of Technology,
Feldmannweg 17, 2628 CT, Delft, The Netherlands

0. Abstract

As a first assessment for the fabrication of Geiger mode avalanche photodiode arrays, the individual pixels are made. Geiger mode avalanche photodiodes are very sensitive for defects, because these cause dark counts and afterpulsing. To investigate the influence of the processing on these effects, different fabrication methods are used to create experimental devices. The lowest dark count rates and afterpulsing probabilities are obtained for devices with cathodes formed using a polysilicium cathode process. This is due to the gettering of impurities from the junction area by the polysilicon layer.

Keywords: Avalanche Photodiodes, Geiger mode, single photon counting, fabrication technology

1. Introduction

Avalanche Photodiodes (APDs) can be operated in the so called Geiger mode to count single photons. In the Geiger mode, the APD is biased above its breakdown voltage. If a photon is absorbed in the diode, it will generate an electron-hole pair which will initiate avalanche breakdown. After breakdown initiation, the breakdown current is quenched and the diode is re-charged for the detection of another photon.

Because of their high quantum efficiency, Geiger mode APDs (often called SPADs, single photon avalanche diodes), are excellent devices for measuring the photon flux of weak optical signals. They also have a very high time resolution, and therefore they are often used in photon timing measurements.

Little work has been done to create Geiger mode avalanche photodiode arrays (GAPAs) although there would be many applications for these devices [1]. A two-dimensional GAPA could be used for faint light level photon counting imaging, with, for example, applications in astronomy. A one dimensional GAPA could be used for photon timing measurements with spectral sensitivity, for instance for spectral-resolved measurements of fast fluorescent decays.

The main problem involved with the creation of GAPAs is optical cross talk between the pixels of the array. During breakdown a pixel may emit a few photons which can be detected by neighboring pixels. This could be prevented by an optical isolation between neighboring pixels, e.g. trenches filled with an opaque material [2].

The fabrication of Geiger mode APDs is difficult because the devices are extremely sensitive for defects. Generation centers in the depletion layer of the diode cause thermal generation of electron-hole pairs. These will be detected as if they were generated by a photon, resulting in dark counts. Trapping centers in the space charge region may trap some of the charge carriers generated during the avalanche discharge. If these carriers are released after the APD has been recharged, they will also initiate a breakdown pulse. This effect is called afterpulsing.

As an experiment with the fabrication technology of GAPAs, we have fabricated the single pixels. Edge breakdown in the pixel structure is prevented using an anode implantation through a window that is slightly smaller than the cathode area. Different fabrication technologies have been used to fabricate the APDs, to investigate the influence of the processing technology on the dark count rate and the afterpulsing probability of the resulting devices.

2. Prototype fabrication

Prototypes of Geiger mode APDs were fabricated in a 10 Ωcm p-type epilayer on a p^+-type substrate. These wafers are normally used in a standard CMOS process at DIMES.

On some wafers an intrinsic gettering step is used at the beginning of the process. In this step, the oxygen in the bulk of the wafer is precipitated during a 6 hour thermal treatment at 750 °C. This step results in gettering sites in the bulk of the material.

First the p-type high-field area definition is implanted by a boron implantation of 150 keV. The dose of this implantation is used to control the breakdown voltage. With a dose of $6*10^{12}$ cm^{-2}, a breakdown voltage of 33 V results. The anode implantation is followed by a drive in at 1100°C for two hours in an inert environment. The drive in increases the breakdown voltage homogeneity. After the drive in, a boron implantation is performed to create substrate contacts. This implantation is annealed during the following cathode processing.

Five different processes were used to create prototypes. These processes differ with respect to the processing of the cathode of the diodes and with respect to the use of the gettering step, as presented in Table 1. In the first process, the cathode was realized using Arsenic implantation with a dose of $5*10^{15}$ cm^{-2} and an energy of 40 keV. In the second process, Phosphor was used instead of Arsenic. The third process is similar to the first process, only in this process the intrinsic gettering step is used. In the fourth process, the cathode was realized using three implantations with a dose of $1*10^{14}$ cm^{-2} and energy of 40 keV. This dose is below the amorphisation threshold. Between these implantations the wafers were annealed. This process was used to eliminate end of range ion implantation damage. The fifth process uses polysilicon deposition and Arsenic implantation followed by an anneal to create polysilicium cathodes.

On the wafers with the implanted cathodes, a thermal oxide is grown for passivation and isolation after the implantation. On the wafers with the polysilicium cathodes, before the polysilicium cathode processing, also a thermal oxide is grown for passivation. The cathodes are formed in windows etched in this passivation oxide. After patterning the polysilicon an oxide is deposited for isolation. The process is then completed with contact hole etching and metallization.

Table 1: The five different processes

Process	cathode processing	gettering
1	As implantation	no
2	Ph implantation	no
3	As implantation	yes
4	three As implantations below amorphisation threshold	yes
5	polysilicium cathode	no

3. Measurement results

Figure 1 shows the dark count rates measured at the experimental devices at +20°C, -10°C and -40°C, as a function of the excess bias voltage. This voltage is defined as the difference between the bias voltage and the breakdown voltage of the diode. From the results, it is clear that the devices with the polysilicium cathodes have a much lower dark count rate. Because the maximum count rate is limited, only the polysilicium cathode devices have been measured at large bias voltages. The most probable explanation for the low dark count rate observed at devices with polysilicon, is that the polysilicon has gettered impurities that act as Shockley-Read-Hall generation centers from the active region of the APDs. At room temperature, the dark count rates for devices with or without the intrinsic getter step, do not differ much. Apparently the intrinsic getterstep does not remove the defects responsible for SRH-generation.

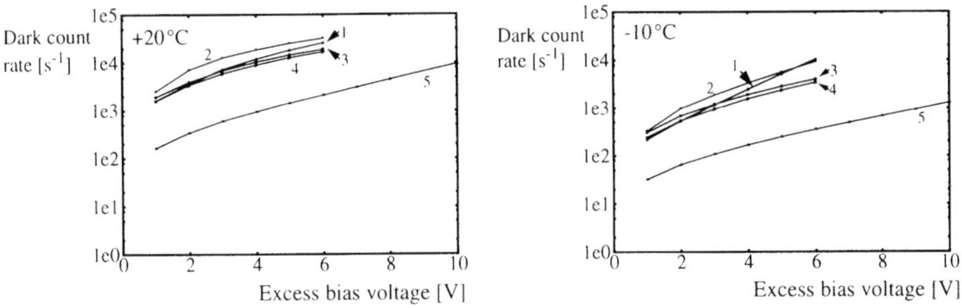

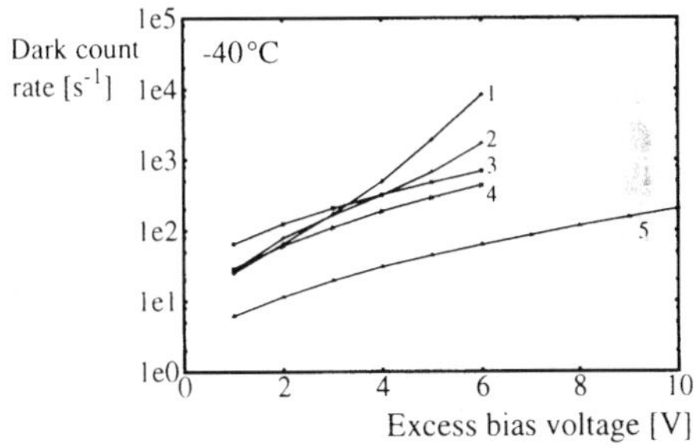

Figure 1 Measured dark count rate for devices fabricated using process 1-5 as a function of the excess bias voltage at different temperatures.

At -40°C however, the intrinsic gettering step does have an effect. Without this step, the dark count rate increases rapidly with the bias voltage for the devices without the intrinsic gettering step or the polysilicon cathode. By measuring the autocorrelation of the output signal of the APD, a strong correlation between the output pulses was determined. From the measured autocorrelation function, the afterpulsing probabilities can be determined [3]. Table 2 gives some initial results of afterpulsing probability measurements for devices biased at 4 volts above breakdown at -40°C. The afterpulsing on the devices without the intrinsic getter step is excessive. The increased dark count rate for these devices at low temperatures is caused by afterpulsing. Because the decay time of the traps increases with decreasing temperature, afterpulsing has a larger effect at low temperatures [4]. Also, because the charge that flows through the diode during breakdown increases with increasing bias voltage, also the probability that a charge carrier will become trapped increases. The afterpulsing probability therefore increases with increasing bias voltage. This explains the increase of the dark count rate with increasing bias voltage.

Table 2: Afterpulsing probabilities measured at -40°C and 4 volts above breakdown

Process	Afterpulsing probability
1	0.93
2	0.81
3	0.10
4	0.15
5	0.03

Intrinsic gettering and especially gettering by polysilicon are very effective in reducing the afterpulsing probability. A more detailed experimental analysis of afterpulsing in the prototypes, as a function of bias voltage and temperature, is required and will be performed in the near future.

4. Conclusions

Geiger mode APDs are very sensitive for defects, because these cause dark counts and afterpulsing. Therefore, careful processing is necessary for the fabrication of APDs with good characteristics. Experime nts with different technologies for the fabrication of the cathode of the diodes and with an intrinsic gettering step were described. The intrinsic gettering step does reduce the afterpulsing probability, but does not reduce the dark count rate. Devices fabricated with a polysilicon cathode show a very low dark count rate and afterpulsing probability. Apparently, polysilicon acts as a very effective getter for the defects that cause dark counts, and it is also more effective than intrinsic gettering for the removal of the defects that cause afterpulsing.

Acknowledgment

This project was supported by the Technology Foundation (STW) under project no. DEL 44.3294

References

[1] F. Zappa et al., "Integrated array of avalanche photodiodes for single-photon counting", Proceedings of the 27th European solid-state device research conference, Stuttgart, Germany, sep. 22-24, 1997, pp. 600-603.

[2] A. Mathewson, J.C. Alderman, J. Ryan, R.M. Redfern and G.T. Wrixon, "Photon counting arrays for spatially varying, low light level signals", Proc. of SPIE, Vol. 2022, Photodetectors and power meters. 1993, pp. 312-143.

[3] H.Z. Cummins and E.R. Pike, Eds, "Photon correlation and light beating spectroscopy", Plenum, London, 1974, pp. 151-224.

[4] A.S. Lightstone and R.J. McIntyre, "Photon counting silicon avalanche photodiodes for photon correlation spectroscopy", Proceedings Photon correlation techniques and applications, OSA, Washington, D.C., may 31- jun 2, 1988, pp.183-191.

DRY RELEASE OF METAL STRUCTURES IN OXYGEN PLASMA

M. Bartek, S.H. Kong and R. F. Wolffenbuttel
Delft University of Technology, Department of Electrical Engineering,
Laboratory for Electronic Instrumentation, Mekelweg 4, 2628 CD Delft,
The Netherlands, Phone: +31-15-2781602, Fax: +31-15-2785755,
E-mail: S.H.Kong@et.tudelft.nl

0. Abstract

A surface micromachining fabrication scheme based on organic sacrificial layers and dry-release in oxygen plasma is presented. Free-standing metal structures are formed using conventional photoresists as a sacrificial layer on which a metal structural layer is evaporated and patterned. Structures are subsequently dry-released in an oxygen plasma. For the HPR504 and HiPR6517 photoresists, the lateral under-etching rate (power 1000 W) of about 1 μm/min was found. The presented dry-release technique is fully compatible with standard silicon IC processing and can be applied as an post processing process module. The technique eliminates any sticking problem and simplifies the processing.

Keywords: surface micromachining, dry release, oxygen plasma

1. Introduction

Surface micromachining processes used for the fabrication of micromechanical systems are mostly based on the 'sacrificial-layer' technology. Here, a structural layer is deposited on a suitable sacrificial under-layer, and subsequently patterned and released using wet etching of the sacrificial material.

Commonly used sacrificial layers (e.g. silicon dioxide, phospho-silicate glass, etc.) have some drawbacks. Often the release step is difficult to combine with already integrated devices due to aggressive etchants used (attack of oxide or Al). The wet etching release processes are also susceptible to the sticking of the released structures.

Recently, new processes have been introduced, which are based on using of polyimide sacrificial layers and dry-release in oxygen plasma [1, 2, 3]. Advantage is that dry-release step eliminates any sticking problem. Conventional photoresists as a sacrificial layer have also been reported, but only in combination with electroplating and wet release in acetone or developer which are susceptible to sticking [4].

In this paper, a fabrication scheme based on photoresist sacrificial layers and dry-release in oxygen plasma is introduced. The process, which is suitable for metallic structures, eliminates any sticking problem and simplifies the processing.

2. Fabrication sequence

The fabrication sequence is schematically shown in Fig. 1. The substrates are (100) silicon wafers with a thermal oxide isolation layer. As a sacrificial layer the HPR504 or HiPR6517 photoresists have been used. After the photoresist patterning and an optional postbake, a 500 nm thick aluminum layer has been e-beam evaporated at pressure of $2x10^{-6}$ Torr. Then second photoresist layer has been deposited to pattern the Al layer using standard wet etching. Final structures have been dry-released in a TEPLA 300 oxygen plasma asher.

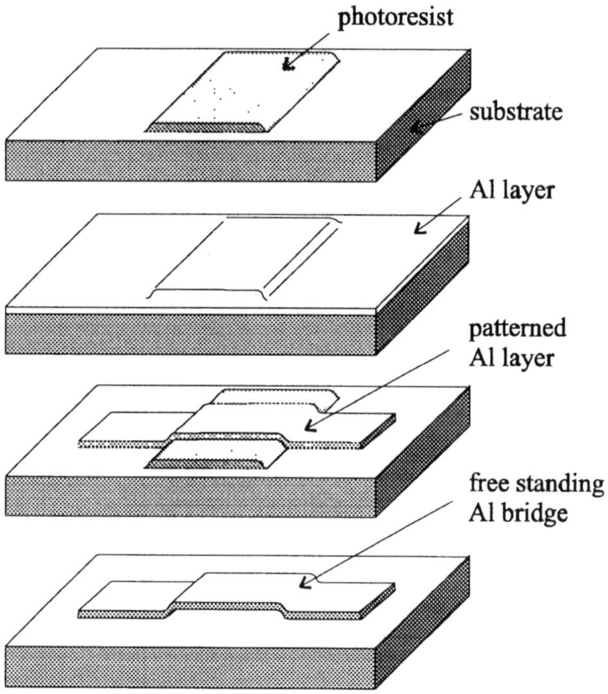

Fig. 1: Schematic fabrication sequence.

The conventional photoresists are optimized to achieve steep side walls for the optimum pattern transfer to the underlying layers. This property is not suitable in the proposed fabrication scheme. A poor step coverage, typical for layers deposited using evaporation, requires limited slope of the patterned sacrificial layer edges. Only then a conformal deposition of the structural layer can be accomplished. Multiple exposures of the same pattern with a small shift (< 1 μm) in the x and y directions and subsequent postbake yield the suitable photoresist edge shape.

3. Results and discussion

As a sacrificial layer a 1.4 μm and 2 μm thick photoresist has been tested. It was observed that for larger sacrificial layer regions (more than 100μm x 100 μm) and the thicker photoresist, outgasing of some spurious solvent from the first photoresist sealed underneath the Al layer, during development of the second photoresist, can cause delamination of Al structural layer. This can be easily eliminated by a postbake at 110°C for 30 min of the photoresist sacrificial layer before metal layer deposition. Fig. 2 shows the SEM photographs of test structures after patterning of the Al layer before the dry release step.

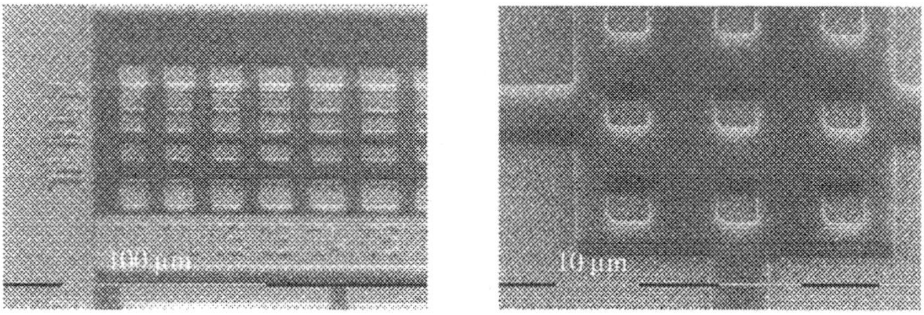

(a) *(b)*

Fig. 2: SEM photographs showing test structures after wet aluminium etching (before dry release in an oxygen plasma); (a) an array of double clamped beams, (b) a detail of a membrane.

Fig. 3 and 4 show different dry-released free-standing structures. Due to an internal stress in the deposited Al layer, a deflection of longer structures down to the substrate can be observed (see Fig. 4).

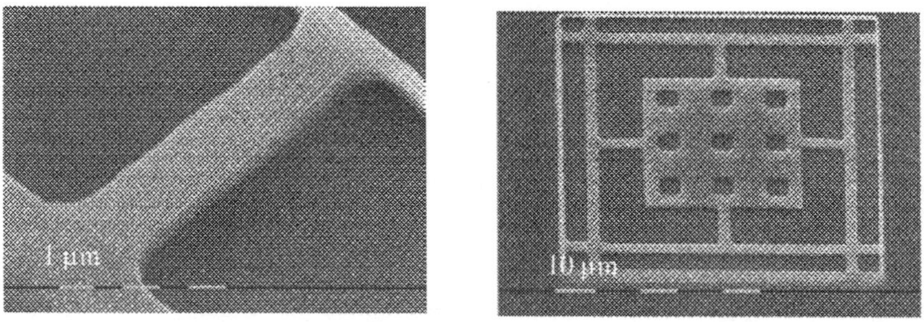

(a) *(b)*

Fig. 3: Free standing 0.5 μm thick Al test structures after dry release in oxygen plasma; (a) a 10 μm long double clamped beam, (b) a membrane.

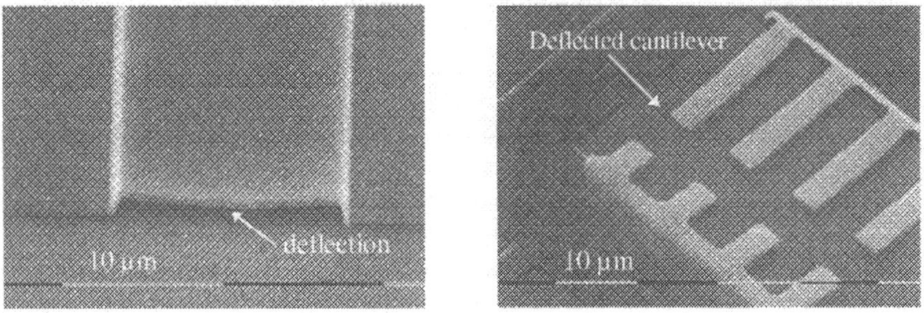

(a) *(b)*

Fig. 4: Internal stress in the Al layer causes a deflection of released structures; (a) a 15 µm wide bridge, (b) a 20 µm long cantilever.

The influence of access window dimensions and plasma power on the lateral etch rate during the dry-release in oxygen plasma has been investigated. It was found that the lateral under-etching rate is almost independent on the access window dimensions and is, at power of 1000 W, of about 1 µm/min (see Fig. 5). Also, for up to 20 µm of lateral underetching no saturation of the etch-rate was observed. This indicates that free-standing structures with lateral dimensions of at least 40 µm can easily be fabricated.

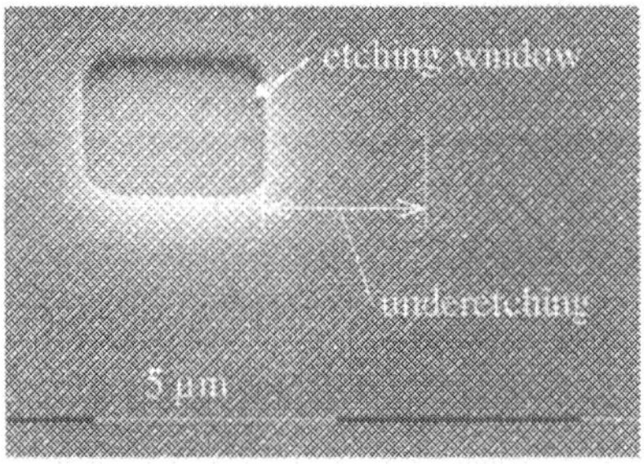

Fig. 5: A 5 min ashing in oxygen plasma yields lateral under-etching of about 5 µm.

4. Conclusions

A surface micromachining fabrication scheme based on organic sacrificial layers and dry-release in oxygen plasma is presented. Free-standing metal structures are formed using conventional photoresists as a sacrificial layer on which a metal structural layer is evaporated and patterned. Structures are then dry-released in an oxygen plasma. The presented dry-release technique is fully compatible with standard silicon IC processing and can be applied as an post-processing process module. The process eliminates any sticking problem and simplifies processing. The technique presented is directly applicable to any metal material which can be evaporated or sputter deposited. Applications include thermally isolated thermopiles, IR sources and different micromechanical structures.

Acknowledgments

The authors wish to thank the staff of Delft Institute of Microelectronics and Submicron Technology, especially J. Groeneweg, for technical assistance in fabrication of the devices. The work presented here has been supported in part by STW, project DEL 55.3733.

References

[1] P. Erikson, J.Y. Andersson, and G. Stemme, Thermal Characterization of Surface-Micromachined Silicon Nitride Membranes for Thermal infrared detectors, J. of Micromechanical Systems, Vol. 6, No. 1, March 1997, pp. 55-61.

[2] A.B. Frazier, C.H. Ahn, and M.G. Allen, Development of micromachined devices using polyimide-based processes, Sensors and Actuators, A45, 1994, p. 47.

[3] A.T.T.D. Tran, Y.H. Lo, Z.H. Zhu, D. Haronian and E. Mozdy, Surface Micromachined Fabry-Perot Tunable Filter, IEEE Photonics Technology Letters, Vol. 8, No. 3, March 1996, pp. 393-395.

[4] Z. Cui, and R.A. Lawes, A new sacrificial layer process for fabrication of micromechanical systems, Proc. 7th Micromechanics Europe Workshop, Barcelona, Spain, 1996, p. 38-41.

MINIATURIZED ISFET INTERFACE FOR THE USE ON A SERIAL PC PORT

G.R. Langereis, W. Olthuis and P. Bergveld
MESA Research Institute, University of Twente
P.O. Box 217, 7500 AE Enschede, The Netherlands
G.R.Langereis@EL.UTwente.NL

0. Abstract

A new interface for using an ISFET (Ion Sensitive Field Effect Transistor) was developed and realized. The advantages of the new design are the small size of the printed circuit board and the low power consumption. This resulted in an electronic circuit small enough to fit in a 25 pins sub-D connector without external power supply. As the circuit is connected to the serial port of the PC, easy interfacing is guaranteed. Furthermore, no hardware calibration is necessary, all calibration can be done by software.

Keywords: ISFET amplifier, interface, serial port

1. Introduction

Because the ISFET does not give a pH dependent voltage directly, it must be part of an electronic circuit for proper operation. This electronic circuit (commonly referred to as the ISFET amplifier) takes care of setting the bias point for proper pH sensing operation.
The proposed ISFET interface consists of a low power ISFET amplifier and an A/D converter. This A/D converter can be controlled and read out by a serial PC interface.

The total power consumption of the electronics is less than 8 mA. Since the standard serial RS232 port can supply up to 10 mA, the electronics could be supplied by the computer. The size of the printed circuit board of the proto type is small enough to fit in a 25 pins sub-D connector housing.

2. The electronics

The conversion from the sensor signal to a PC signal takes three steps. First, the ISFET, which behaves like a depletion type PMOS transistor from the electronical point of view, is forced into the unsaturated mode with a constant drain current and a constant drain-source voltage. This yields a pH dependent gate-source voltage [1] where the gate is being contacted via the reference electrode and the solution. The chosen drain current is 90 μA at a drain-source voltage of 0.4 Volt.

Next, this gate-source voltage is sampled by a 12-bit analog to digital converter with a serial output. The input range of this converter is set to 0 .. 5 Volt which results in a wide allowed range of ISFET offsets, while still having a satisfying feasible pH range. For example, a pH range from 0 to 14 will cover about 14×59 mV = 0.826 Volt. So, in a 5 Volt A/D converter window, still more than 4 Volt remains for differences in the ISFET offsets. This implies that with a 12 bits sample resolution, about 10 bits are used for the pH part and the other 2 bits for the offset part. So the chosen permitted offset range reduces the resulting pH resolution. This concession still leaves an acceptable resolution of 0.02 pH per bit (5 Volt divided by 2^{12} and 59 mV/pH).

The A/D converter has a TTL level output stage (0 and 5 Volt). However, the standard serial RS232 port uses -12 and +12 Volt stages. Therefore, a third step is necessary to convert the TTL levels to standard RS232 levels.

The total dissipated current of those three blocks could be limited to 8 mA. The standard RS232 port is capable of sourcing up to 10 mA through any pin, so the power supply for the electronics could be taken from the PC by just setting a free pin to the high level. This results in a power supply of 0 (ground) to 12 Volt. Because an ISFET amplifier requires a symmetrical power supply with respect to the reference electrode, this reference electrode is shifted to half the power supply. A consequence of the reference electrode not being at the computer ground is that the system might give grounding problems when used in combination with other (electrochemical) equipment.

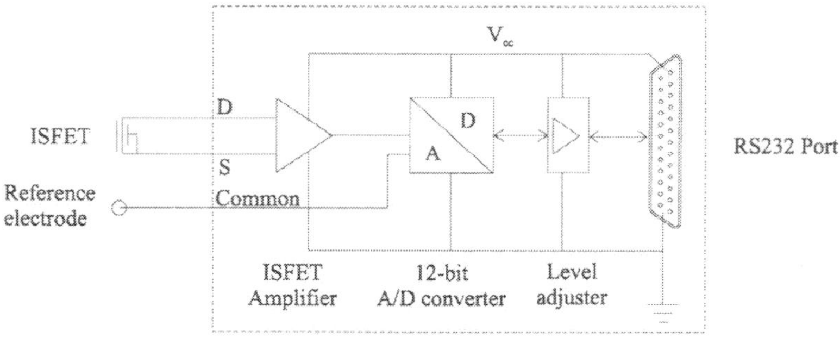

Figure 1: Schematic view of the electronic circuit

Figure 1 shows a block diagram of the whole converter. The components fitted on a single-sided printed circuit board of 3.5 × 3.5 cm (figure 2), when using SMD components this size might even be reduced.

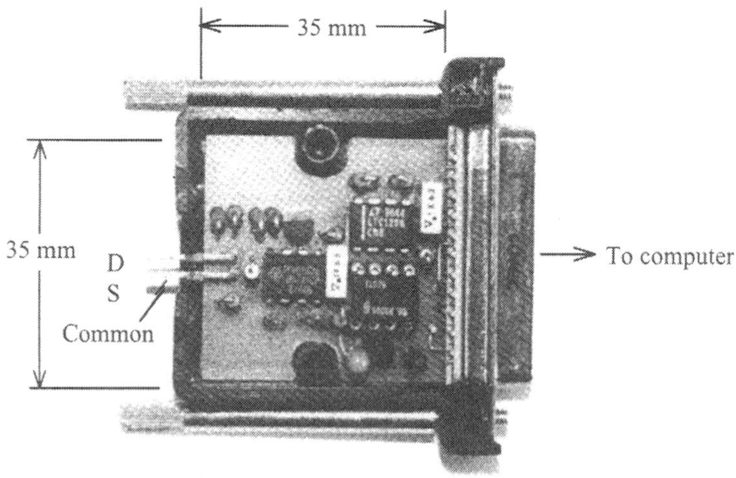

Figure 2: Top view of a completely assembled connector, S, D and Common are the connection pins for the source, drain and reference electrode respectively

Because of the low impedance characteristics of the ISFET, there are no special demands on the wiring between the probe and the converter. Using three leads of a cheap and flexible telephone wire (with almost no limits to the length) will be suitable. Notice that ISFETs should be used with the bulk short circuited to the source.

3. Software

To control and read out the interface, computer programs were written for the use under MS-DOS, MS-Windows and LabView. All programs offer an interactive calibration possibility for both one and two points calibration. These programs set the sample frequency to 10 Hz, not dependent on the used computer.

The advantage of the LabView application is that the program can be easily expanded to automated measurements set-up. For example, a drift measurement application program that stores the measured pH every minute during 24 hours can be programmed in a few minutes.

The simple MS-Windows program looks like a stand alone pH meter. All functions are accessible by menus. When the measured data need to be stored, it can be "polled" by another program like Excel using the standard MS-Windows Dynamic Data Exchange (DDE) features.

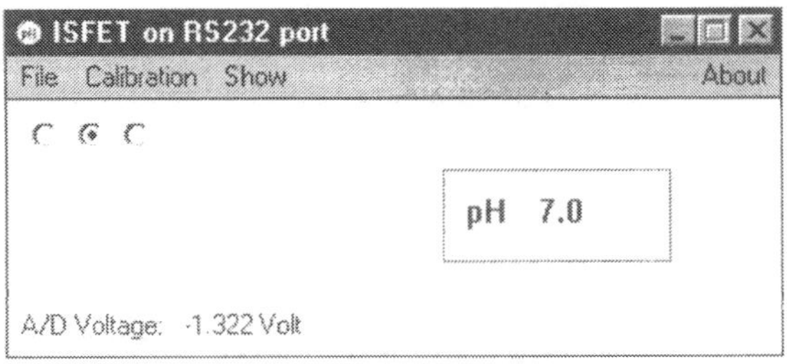

Figure 3: Screenshot of the MS-Windows program for using the converter

The screenshot, represented in figure 3, shows the MS-Windows program. Under the "Show" menu the monitoring of the A/D voltage and the calibration data can be toggled. The "Calibration" menu covers the single or two points calibration dialogs. The calibration result is automatically written to disk. When the program is started, the last saved calibration data is loaded. Using the "Settings" dialog, accessible under the "File" menu, the used COM port and the standard calibration buffers can be selected. Calibration data, being an offset and a slope, can be entered manually in this menu as well.

4. Specifications

A summary of the specifications of the ISFET to RS232 converter is given in table 1. It can be seen that neither the available resolution of the electronics (being 0.02 pH) nor the maximal allowed sample rate of the A/D converter (12.5 ksamples per second) is displayed by the software (0.1 pH and 10 samples per second respectively). This is done because in practical chemical applications these software criteria are sufficient. The overhead is used for reducing noise by software.

Table 1: Specifications of the interface

A/D Converter		
Resolution	12	bits
	1.22	mV/bit
Range	0 .. 5	Volt
Maximum sample rate	12.5	ks/s
Total system		
Size of prototype	3.5 × 3.5	cm
Supply current	8	mA
Resolution (at 59 mV/pH)	0.02	pH/bit
Printed circuit board	Single sided	
Software		
Operating system	MS-DOS, MS-Windows	
Displayed precision	0.1	pH
Calibration	1 or 2	Points
Sample rate	≈ 10	Hz

5. Experimental

The converter was tested using an ISFET with an Ag/AgCl reference electrode. This reference electrode was mounted on the backside of the ISFET stick in order to get a single probe pH sensor. A photograph of the system is displayed in figure 4.

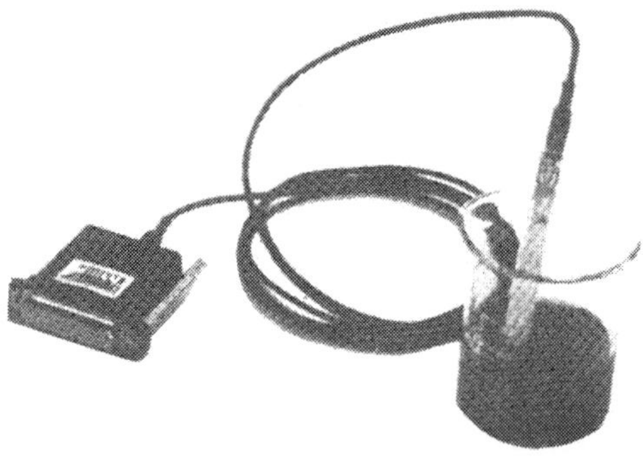

Figure 4: A complete pH measurement system

As a test, the ISFET with converter was compared to a conventional glass electrode. First, both the ISFET system and the conventional pH meter (a Radiometer PK80 with a GKH2401C glass electrode with internal reference) were calibrated using a two points measurement (pH = 4.01 and pH = 7.0). Directly after the calibration the probes were placed in a 100 mM acetic acid solution with 100 mM KNO_3 as a background electrolyte.

After applying several volumes of 1M NaOH and stirring, both measured pH's were written down. Figure 5 shows the recorded curve.

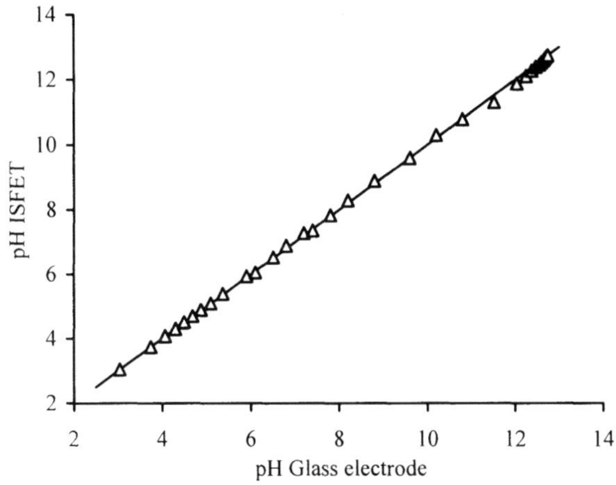

Figure 5: Comparison of a glass electrode to the ISFET system in acetic acid

It is obvious that there are no significant differences between the two pH meters. So the ISFET to RS232 converter can be used for standard pH measurement purposes.

6. Conclusions

The proposed ISFET interface offers a low-cost and small piece of hardware for sampling ISFET signals into a computer using the serial port. This interface can completely be controlled by the computer.

Besides being relatively cheap, the advantage of the device is in the field of the flexible test application building. Now the building of test applications using graphically programming techniques like LabView becomes more and more common [2], the demand rises for equipment that acquires data into the computer. The ISFET to RS232 converter appeared to be a flexible tool for acquiring pH where the calibration procedure can be implemented in software. An automated ISFET-drift measurement or titration set-up can easily be realized using graphical programming.

References

[1] P. Bergveld and A. Sibbald, in G. Svehla (ed.), Comprehensive analytical chemistry, Chapter 8, Vol. XXII, Elsevier, Amsterdam, 1988

[2] Gary W. Johnson, LabView graphical programming - Practical applications in instrumentation and control, McGraw-Hill Inc., 1994

A LOW POWER DETECTION METHOD FOR SMALL ULTRASONIC ECHO SIGNALS

E.A. Dijkstra*,#, W. Olthuis*, J. Holsheimer#, and P. Bergveld*

*MESA Research Institute and #Institute for Biomedical Technology
Electrical Engineering, University of Twente
P.O. Box 217, 7500 AE Enschede, The Netherlands
E-Mail: E.A.Dijkstra@el.utwente.nl

0. Abstract

Spinal cord stimulation is used for the management of chronic, intractable pain and vascular insufficiency. However, when stimulating at a constant voltage, the current density in the spinal cord changes due to the varying separation *d* between the spinal cord and the stimulating electrode, thus, preventing a stable clinical effect. In order to control the stimulation of spinal nerve fibers, a closed-loop system using ultrasonic delay time detection is being developed. This system is based on pulse-echo measurements using a piezoelectric transducer. The spinal cord stimulation system is an implanted system. Therefore, low power consumption (μW) is an important requirement. A low power electronic circuit is developed to detect the small echo signals (mV) from the spinal cord. The detection method is based on amplitude demodulation of the echo signal. The method is simulated in MATLAB, and the simulation results are compared to the results of a simulated matched filter method. The amplitude demodulation method shows results similar to the matched filter method.

Keywords: piezoelectric transducer, ultrasound, spinal cord stimulation, signal processing

1. Introduction

Spinal cord stimulation (SCS) is used for the management of chronic, intractable pain and vascular insufficiency. However, when stimulating at a constant voltage, the current density in the spinal cord changes due to the varying separation *d* between the spinal cord (floating in the cerebrospinal fluid, CSF) and the stimulating electrode, preventing a stable clinical effect. A schematic cross-section of the spinal cord is given in Fig. 1.

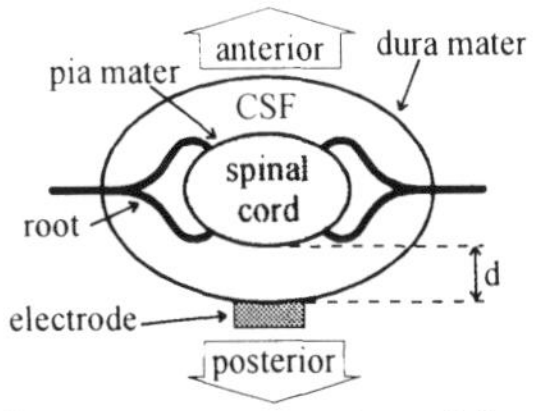

Fig. 1 Schematic cross-section of the spinal cord.

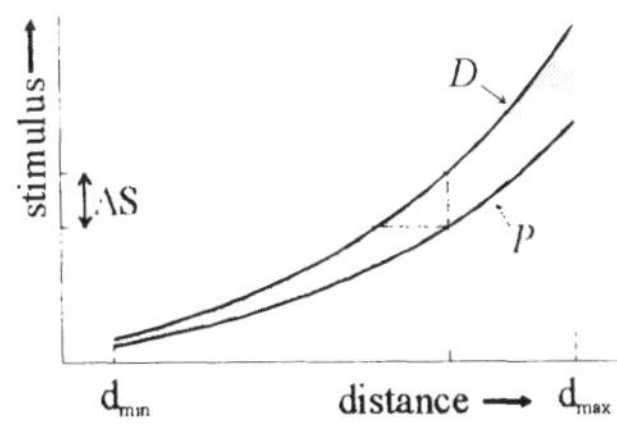

Fig. 2 The stimulus (applied to the electrode) as a function of the distance between the electrode and the spinal cord for optimal stimulation with the required stimulation window ΔS. The dashed area indicates the stimulation window ΔS, between the perception P and discomfort D thresholds.

In order to control the stimulation of spinal nerve fibers effectively, a closed-loop system using ultrasonic delay time detection is being developed. The current density in the spinal cord has to remain constant, which means that the stimulation voltage has to be between the perception P and discomfort D thresholds (stimulation window ΔS). See Fig. 2. The stimulation window is generally small (≈ 40% of P) [1]. In current SCS systems, the stimulus applied to the electrode is set at a constant value. A major improvement could be achieved when the delay time between the transmitted and the signal reflected by the spinal cord could be measured (linearly related to the distance d) and used to control the stimulus amplitude in a closed-loop system.

2. System Requirements

The distance measurement must be contactless, safe, and should be repeated regularly (e.g. once per minute). The distance between the electrode and the spinal cord varies roughly between 1 and 10 mm [2], and the resolution should be 0.5 mm, or less.

The SCS system is an implanted system and battery powered. The delay time measurement system should, therefore, have low power consumption (<40 μW), because a significantly increased power consumption reduces battery life. The transducer must be integrated on the electrode. This limits the size of the transducer to an area of 5×5 mm^2, and a thickness of 2 mm. Robustness is also an important requirement for an in-vivo used transducer.

3. Methods

A piezoelectric transducer using ultrasound was chosen to measure the delay time between the transmitted and the signal reflected by the spinal cord. The acoustic impedance of the spinal cord is close to that of the CSF, resulting in little reflection of the acoustic wave. The signal-to-noise ratio (SNR) of the echo signal is therefore small [3].

A low power electronic circuit is developed to detect the echo signals. The circuit must be able to detect the echoes with a reliability of 100%.

A. Measurement Setup

The measurement setup is shown in Fig. 3 and consists of a function generator (HP 33120A) connected to a piezoelectric transducer and protection circuit, consisting of a resistor of 1 kΩ and two Schottky diodes (D_1 and D_2) in anti-parallel. The protection circuit is connected to the input of the oscilloscope (HP 54520A), and limits the input voltage between –0.3 and 0.3 V. The protection circuit enables the scope to measure with a resolution of 2 mV/div.

The transducer is immersed in water and generates ultrasonic waves, which travel towards the "spinal cord" (modeled by silicone rubber) and are partially reflected. The spinal cord model is attached to a precision micromanipulator with a resolution of 1 μm. The manipulator is used to vary the distance d between the spinal cord model and the transducer.

The piezoelectric transducer is a PXE-5 element with an area of 4×4 mm^2 and a thickness of 0.3 mm. For all experiments a single sine wave with a period of 0.145 μs (7 MHz) and a peak-peak amplitude of 8 V was used as the excitation pulse. The total number of samples taken by the oscilloscope was 1024 and the sample rate was set to 50 MSa/s.

Ten sets of measurements for ten different distances d were made. Each set consisted of 101 pulse echo measurements. The distance d between the successive sets was varied by 0.5 mm.

B. Signal Analysis

The transducer signal was divided into two parts: the ringing signal (residual oscillation after excitation) and the echo signal, see Fig. 4. A single excitation pulse was applied to the transducer at t=0. The distance d was approximately 10 mm. The frequency components of the ringing and echo signals were in the same range. The amplitude of the ringing signal was much higher than the echo signal. The ringing lasted about 5 μs, which corresponds to a distance d of approximately 4 mm. Therefore, it is extremely difficult to detect echoes, when the distance d between the transducer and the spinal cord model is smaller than 4 mm, and the echo and ringing signal coincide.

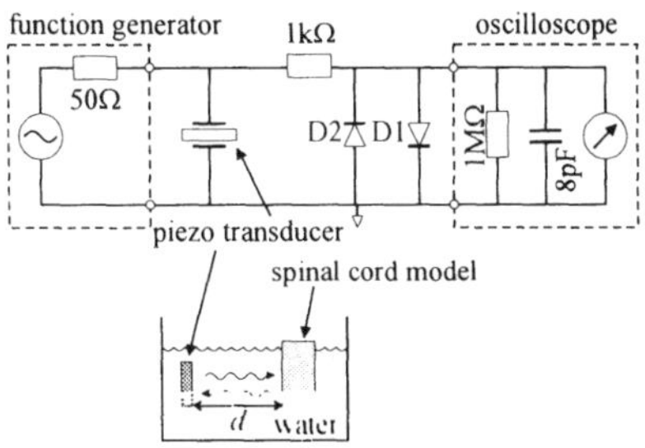

Fig. 3 Measurement setup for measuring echo signals with a piezoelectric transducer from a silicone rubber spinal cord model.

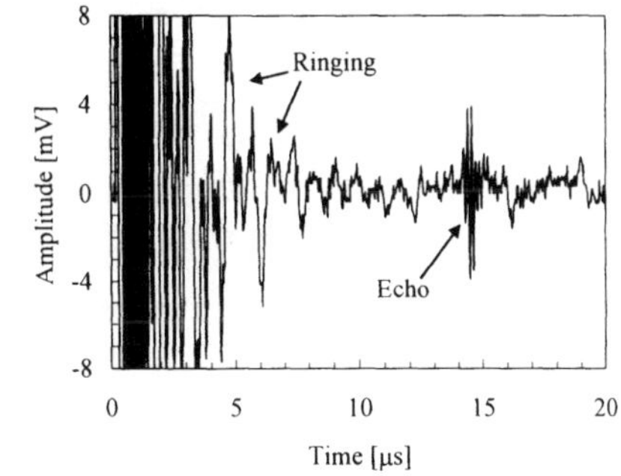

Fig. 4 The total transducer signal. The ringing part is shown on the left side and the echo signal on the right. The distance d between the transducer and the spinal cord model is approximately 10 mm.

The echo signals were analyzed in the time as well as in the frequency domain. The signal analysis was performed on two averaged signals (from 101 measurements each) for a distance d of approximately 5 and 10 mm. These averaged echo signals are shown in Fig. 5. The "5 mm" echo signal coincides with the last part of the ringing signal.

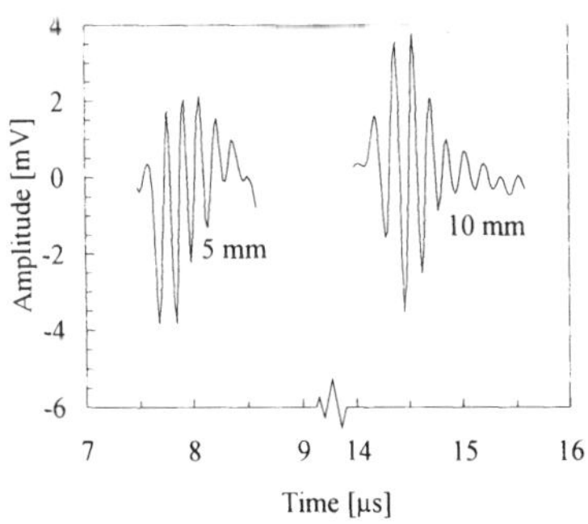

Fig. 5 Typical echo signals of the spinal cord model (reflector) at a distance of approximately 5 mm (left) and 10 mm (right).

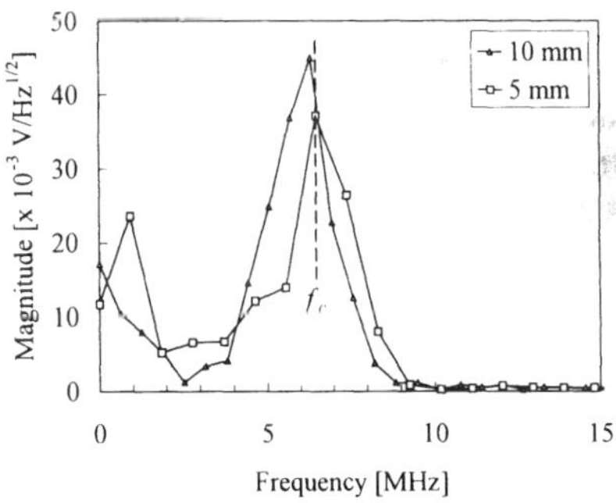

Fig. 6 Spectra of the echo signals of Fig. 5.

The spectra of the two echo signals were calculated and are shown in Fig. 6. The center frequency (f_c) of both signals is 6.5 MHz, and their bandwidths 2.5 MHz. The echo signal at a distance of 5 mm contains more low frequency components than the other echo signal due to ringing of the transducer.

C. Two Echo Detection Methods

A common way to detect echo signals with a poor SNR is the use of a so-called matched filter. This filter optimizes the SNR of the echo. However, the shape of the echo signal must be known. This shape depends on the angle of incidence/reflection on the reflector (spinal cord), and the distance between the transducer and the reflector (compare the echo signals shown in Fig. 5). The easiest method to create a matched filter is by using a digital signal processor (DSP). However, this requires AD-converters with a high sample rate (> 15 MHz) for accurate detection and, moreover, digital techniques are usually power consuming.

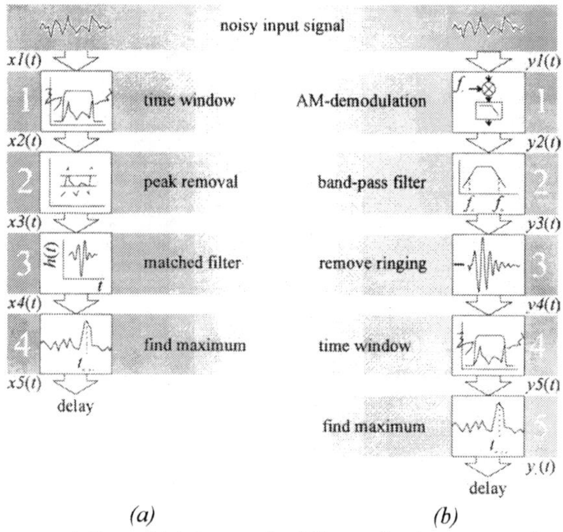

Fig. 7 Schematic representation of (a) the matched filter echo detection method and (b) the amplitude demodulation echo detection method.

The main frequency component of the echo signal is 6.5 MHz. Therefore, it is possible to use amplitude demodulation techniques to detect the echo signal. An advantage of this method is that it can be performed in an analog way.

Both detection methods were implemented in MATLAB, and the results were compared. A schematic diagram of both methods is shown in Fig. 7. The left side (a) shows the matched filter method and the right side (b) the amplitude demodulation method.

The Matched Filter Method

The entire echo signal $x_1(t)$ is passed through a time window $w(t)$:

$$w(t) = \begin{cases} 1 & 1.3\mu s < t < 18\mu s \\ 0 & elsewhere \end{cases} \tag{1}$$

The window corresponds to the time window in which an echo can be expected (1…10 mm corresponds to 1.3μs…13μs). The time window output can be written as:

$$x_2(t) = x_1(t) \cdot w(t) \tag{2}$$

Then some artifact peaks are removed:

$$x_3(t) = \begin{cases} x_2(t) & |x_2(t)| \le 4.5 \cdot 10^{-3} \\ 0 & elsewhere \end{cases} \tag{3}$$

This signal is passed through the matched filter. The mathematical equivalent of the matched filter is the cross-correlation. The impulse response of the matched filter is calculated from the "10 mm" echo signal of Fig. 5. The time corresponding to the maximum value of the matched filter output $x_4(t)$ is determined, and is the estimated time of arrival of the echo.

The Amplitude Demodulation Method

The echo signal is amplitude demodulated with a frequency of 6.5 MHz ($y_2(t)$). The next step is to filter the output of the demodulator by a band pass filter with a center frequency of 1 MHz and a bandwidth of 0.4 MHz ($y_3(t)$) (determined empirical). Due to ringing of the transducer there is a large signal component in the first 5 μs. Fig. 8 shows this reference ringing signal $r(t)$ when the input is a signal averaged over 101 measurements, and no echo is present. The signal $r(t)$ is subtracted from a single pulse measurement:

$$y_4(t) = y_3(t) - r(t) \tag{4}$$

Due to differences between the reference ringing signal $r(t)$ and the filtered signal $y_3(t)$ from a single measurement, i.e. phase lag and amplitude, the signal $y_4(t)$ contains still too high peaks below 4 μs to determine the echo reliably. Therefore, a time window $w(t)$ (determined empirical), shown in Fig. 9, is applied to $y_4(t)$:

$$y_5(t) = y_4(t) \cdot w(t) \tag{5}$$

The time corresponding to the maximum value of $y_5(t)$ is determined, and used as the estimated time of arrival of the echo.

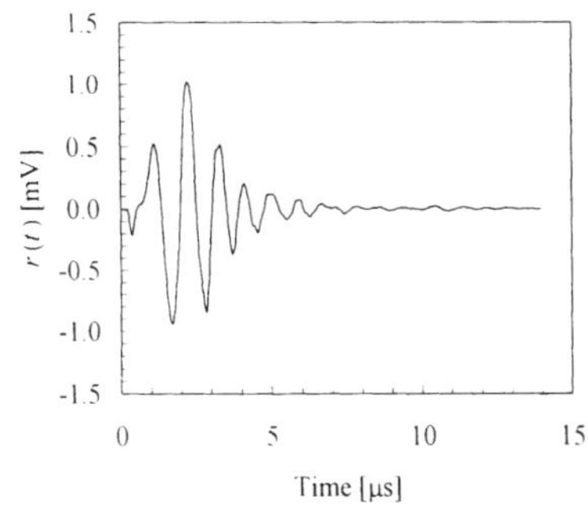

Fig. 8 Averaged signal output of the band pass filter due to ringing of the transducer (r(t)).

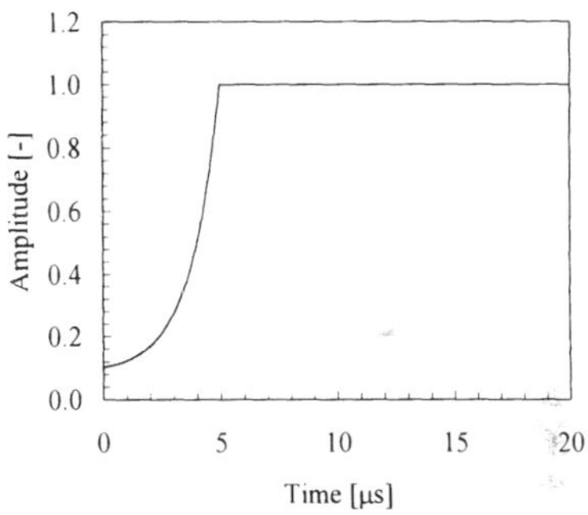

Fig. 9 Time window to reduce the influence of ringing.

4. Results

Both detection methods were used to estimate the delay times of the echo signals of the set of measurements, as described in section 3.A. For each distance 101 delay times were determined. The results of the matched filter and amplitude demodulation method are shown in Fig. 10 and Fig. 11, respectively. These figures show the estimated delay time as a function of the distance.

The delay time for each measurement was determined. Then a histogram of each set of measurements was made from the delay times. The classes have of width of 80 ns. The class containing most points is considered to be the estimated delay time. These points are represented by the diamonds in Fig. 10 and Fig. 11. The error bars indicate the maximum and minimum delay time. The numbers near the data points indicate the number of echoes detected in a time window of 240 ns, corresponding to a distance of 0.18 mm (3 classes of the histogram). Finally, a linear fit was made.

The matched filter (Fig. 10) has a slightly better performance for distances ranging from 5 to 7 mm. When the distance decreases below 5 mm the amplitude demodulation method (Fig. 11) shows a slightly better detection rate. However, both methods do not have a reliability of 100%.

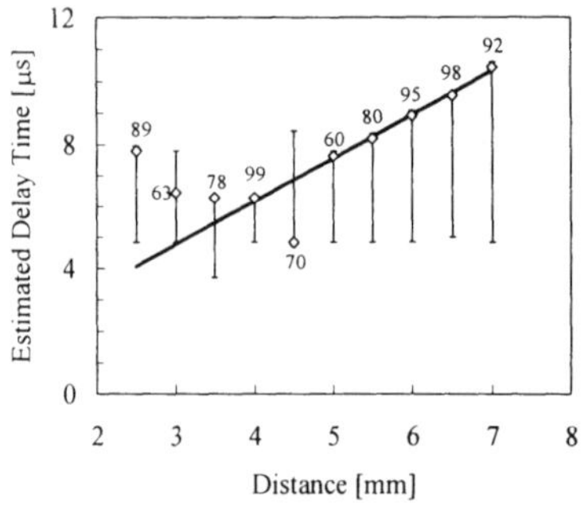

Fig. 10 Estimated delay time by the matched filter as a function of the distance d.

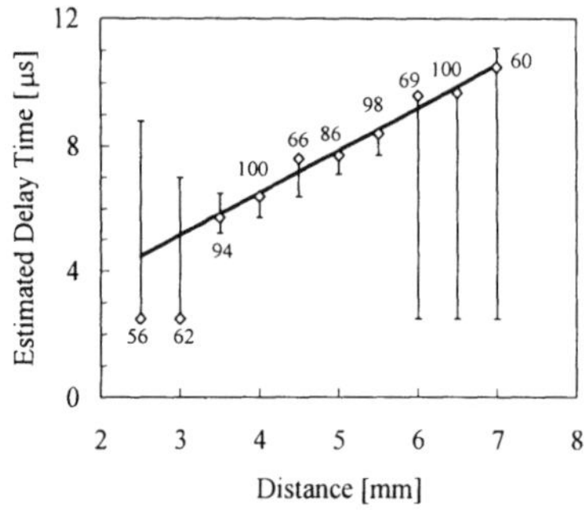

Fig. 11 Estimated delay time by the amplitude demodulation method as a function of the distance d.

Ringing of the transducer causes it to be extremely difficult to detect echoes for small distances d (<3 mm). The matched filter method overestimated the distance, where the amplitude demodulation method underestimated the distance.

5. Conclusions

The accuracy of the amplitude demodulation method is more accurate for small distances (<3..5 mm) than the matched filter method. The accuracy of both methods is similar for larger distances (≥5 mm).

A disadvantage of the matched filter method is that it requires digital signal processing techniques, while the amplitude demodulation method can be performed using analog circuitry. Digital techniques are usually power consuming, and become more inaccurate when the sample rate is decreased.

The amplitude demodulation method can be used for echo detection. However, the detection rate must be improved to make the method more reliable.

Acknowledgements

This investigation was supported in part by a grant from Medtronic Inc., Minneapolis, MN.

References

1 J. Holsheimer, J.J. Struijk, G. Barolat, and J. He, "Significance of the spinal cord position in spinal cord stimulation," *Acta Neurochir [Suppl]*, Vol. 64, pp. 119-124, 1995.

2 J. Holsheimer, J.A. Den Boer, J.J. Struijk, and A.R. Rozeboom, "MR Assessment of the Normal Position of the Spinal Cord in the Spinal Canal", *American Journal of Neuroradiology*, Vol. 15, pp. 951-959, 1994.

3 E.A. Dijkstra, J. Holsheimer, W. Olthuis, and P. Bergveld, "Ultrasonic Distance Detection for a Closed-Loop Spinal Cord Stimulation System," Proceedings 19th Conference of EMBS 1997.

DESIGN OF A SMART MICROACOUSTIC LIQUID SENSOR SYSTEM PART I: MICROACOUSTIC DEVICE

B. Jakoby, G.W. Lubking and M.J. Vellekoop
Electronic Instrumentation Laboratory/DIMES, Delft University of Technology,
Mekelweg 4, 2628 CD Delft, The Netherlands
email: jakoby@ei.et.tudelft.nl

0. Abstract

Microacoustic sensors have proved to be powerful devices due to a number of benefits like real-time electronic readout, small size, robustness, high sensitivity, and cost-effective fabrication. For sensing in liquids it has been shown that Love wave devices are are particularly well-suited. In this paper we illustrate the use of Love wave devices for liquid sensing applications and outline their proper design for utilization in a smart liquid sensor system.

Keywords: liquid sensing, Love waves, microacoustic sensors

1. Introduction

Microacoustic sensor devices are highly suitable for sensing physical quantities like mass density, viscosity, and acoustic properties of liquids. Using so-called chemical interfaces, they can also be used for chemical sensing where the concentration of a certain target compound in a liquid or gaseous environment has to be determined. A comprehensive overview on the state-of-the-art in microacoustic sensing can be found in [1].

Among the commonly used microacoustic devices, Rayleigh Surface Acoustic Wave (SAW) sensors have been proved to show high sensitivities, which is related to the concentration of the acoustic energy at the sensitive surface. However, they are not suited for liquid sensing applications since, due to the particle displacement component normal to the sensing surface, compressional waves are excited in the liquid leading to considerable damping of the wave. A suitable alternative is the utilization of shear polarized wave modes where Love waves have been theoretically shown to offer a very high sensitivity [8], which has been confirmed by experiments (see, e.g., [7, 3, 2]).

We use a Love wave delay line as feedback element in an oscillator. The oscillation frequency depends on the delay time which changes according to the quantity to be sensed. For (bio)chemical sensing [6] a selectively adsorbing film is deposited on the device; in the presence of the target molecule the mass density of the film increases, which leads to a corresponding change in the wave velocity and hence in the delay time of the device. For viscosity sensing we use the effect that a thin layer of a viscous liquid is entrained with the shear movement at the surface. This entrainment leads to a change in wave velocity and also to an increase in damping of the wave (this is related to viscous losses in the liquid) [5]. The principal geometry of a Love wave delay line is shown in Figure 1. The Love wave is excited and detected by interdigital transducers (IDTs) attached on a piezoelectric substrate. The Love wave is concentrated in a guiding layer which covers the substrate and the IDTs. To obtain a propagating

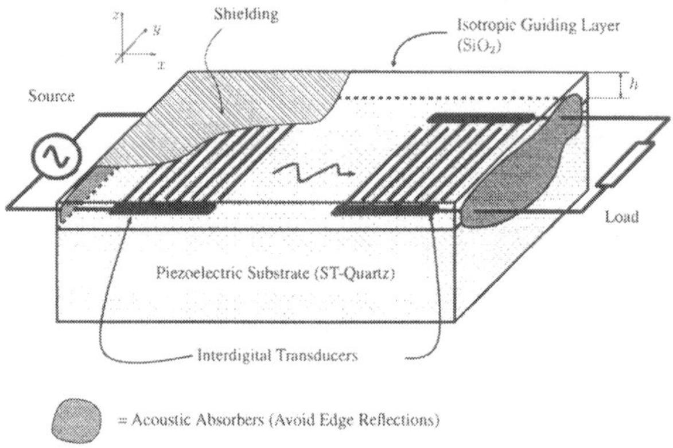

Figure 1: *Love wave delay line.*

Love wave, it is essential that the bulk shear velocity of the layer material is lower than that of the substrate. As layer material we used SiO_2 which has been deposited by Plasma Enhanced Chemical Vapor Deposition (PECVD). The substrate is ST-cut quartz where the propagation direction of the Love wave is orthogonal to the crystalline X axis.

The used oscillator circuit is shown in Fig. 2. The amplifier is equipped with Automatic

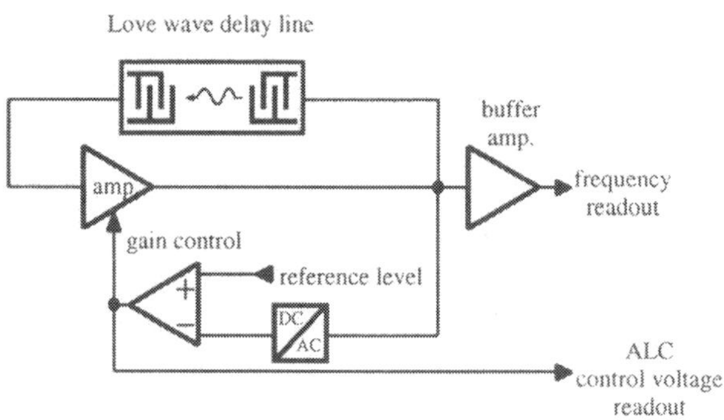

Figure 2: *Love wave delay line in the oscillator setup.*

Level Control (ALC). The control voltage of the ALC provides information about the damping of the delay line which can be used in viscosity sensing (see above). Using a dual setup with two oscillators where one contains a sensing delay line and one a reference delay line allows a first order compensation of temperature effects on the delay line.

In this contribution we discuss the essential issues in the design of this sensor system. We thereby concentrate on the device related aspects; the appropriate design of the amplifier is discussed in an accompanying paper in this Proceedings [9].

2. Sensitivity of the Delay Line

A crucial design parameter for the delay line is the thickness h of the guiding layer. In Fig. 3 we show the relative frequency sensitivity S_f of the oscillator with respect to mass loading at the sensing surface. S_f is defined as

$$S_f = \frac{1}{f}\frac{df}{d\rho_s}, \tag{1}$$

where f is the oscillation frequency and ρ_s denotes the surface mass density of the mass loaded upon the waveguide.

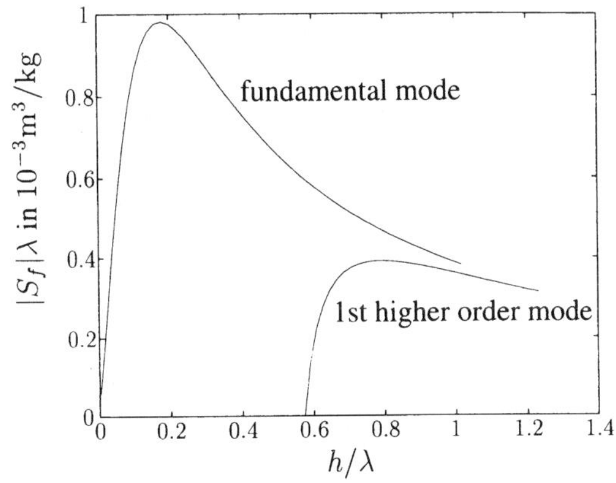

Figure 3: *Relative scaled frequency sensitivity $|S_f|\lambda$ for the fundamental and the first higher order mode on a fused quartz/ST quartz structure.*

Since the Love wave delay line is dispersive (meaning that the phase velocity v_p depends on the frequency ω), the frequency sensitivity S_f is *not* equal to the relative sensitivity S_v of the phase velocity v_p but they are related as[6]:

$$S_f = \frac{v_g}{v_{ph}} S_v = \frac{v_g}{v_{ph}} \frac{1}{v_p} \frac{dv_p}{d\rho_s}. \qquad (2)$$

Here v_g denotes the group velocity of the Love wave.

Fig. 3 shows that there is an optimum value h/λ to obtain a maximum for the scaled sensitivity $|S_f|\lambda$. Of course this optimum depends on the materials chosen; for the calculation we used the material parameters for ST quartz/fused quartz for the substrate/overlayer material[1]. The optimum value is given for the scaled sensitivity $S_f\lambda$ which means that using smaller wavelengths λ (corresponding to higher frequencies) will yield increasing values S_f. Practically the used frequency is bounded by spurious effects like crosstalk, which limit the actually achievable sensitivity. The scaling by λ is especially useful since the wavelength is linked to another important design parameter: the (electric) period of the IDTs which prescribes the wavelength at the center frequency of the delay line [10]. Hence for a given (chosen) IDT period, the wavelength is prescribed and hence Fig. 3 allows to determine the optimum overlayer thickness h and the resulting sensitivity S_f for that case.

3. Noise and Temperature Drift

The actually achievable sensitivity is limited by the stability of the oscillator circuit since the frequency change induced by the sensing effect must be distinguishable from the short term instability and long term frequency drifts. The latter is related to temperature variations of the sensor while the short term instability is determined by the noise characteristics of the entire circuit (see also [9]). For our devices short term instabilities in the order of one Hertz and below have been measured. These values are somewhat higher for liquid loaded devices, which indicates the influence of surface effects on the noise characteristics. Regarding temperature drift, temperature coefficients in the order of 25-30 ppm have been measured. The effect of temperature drift can be reduced by using the aforementioned dual (reference) setup which

[1]The material parameters for the actually used PECVD SiO_2 are not exactly known, the values for fused quartz should provide a reasonable approximation.

typically yields an improvement of about a factor ten. Additionally active temperature control can be employed. The control of the temperature drift is of special importance for chemical reactions involving slow kinetics.

4. Electric Characterization of the Device

For the design of the electronics the electric behavior of the delay line is essential. It can be characterized as a two port by using, e.g., the short circuit two port parameters Y_{ij}. The device is reciprocal ($Y_{12} = Y_{21}$) and, depending on whether the IDTs have chosen to be equal, often also symmetric ($Y_{11} = Y_{22}$). The equivalent circuit for the input admittance of a single IDT is given in Fig. 4. The static capacitance of the IDT is given by C_t, and ohmic losses

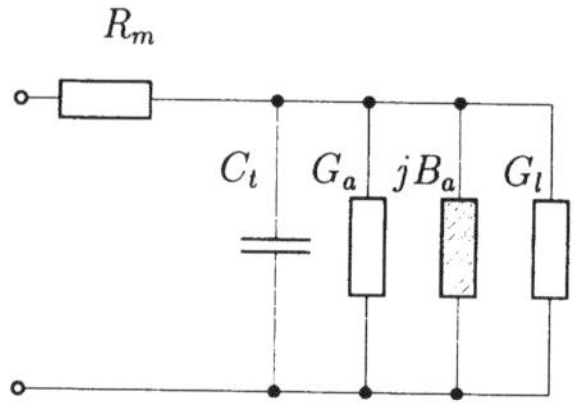

Figure 4: *Equivalent circuit for the IDT admittance.*

in the IDT metalization and dielectric (polarization) losses (see below) are represented by R_m and G_l, respectively. The (frequency dependent) radiation conductance G_a represents the electroacoustic power conversion taking place in the transducer; it is accompanied by a reactive component B_a which can be obtained from G_a via a Hilbert transform [10]. This equivalent circuit is valid for a single IDT; for two IDTs interaction between the two occurs due to launched and received acoustic waves. However, if an IDT is electrically short-circuited, it does practically not reflect the acoustic wave[2], which means that the above equivalent circuit for a single IDT can also be interpreted as describing the short circuit driving point admittance Y_{ii} of transducer number i ($i = 1, 2$). Moreover, since usually $Y_{12} \ll Y_{ii}$ ($i = 1, 2$), the input admittance into one transducer is approximately equal to the respective short circuit driving point admittance regardless of the electric load at the other IDT. The values of the equivalent circuit parameters of course crucially depend on the design of the IDT; in the following we provide typical order of magnitudes for the important parameters as they occur in our designs: Electrically the capacitance C_t is the dominating part in the input impedance; it is typically in the order of 10 − 30 pF. The radiation conductance is in the order of 1 mS and the loss resistor R_m typically amounts to a few ohms. The loss conductance G_l represents dielectric (polarization) losses in the materials surrounding the IDT. It can become important for example in the case of sensing in polar liquids or liquids containing ions. However, this influence can be eliminated by using electrical shielding (see below).

The transfer admittance Y_{12} is also determined by the radiation conductance [10]; its absolute value is given by $|Y_{12}| = \sqrt{G_{a,1}G_{a,2}}\exp(-\alpha L)$. Here $G_{a,1}$ and $G_{a,2}$ denote the radiation conductances of transducer 1 and 2, respectively. The factor $\exp(-\alpha L)$ accounts for acoustic losses (e.g., due to the viscosity of an adjacent liquid) of the wave traveling between the two IDTs, L denotes the center to center distance between the IDTs. Driving the input of the delay line with a voltage and loading it at the output with a high impedance, the modulus of the voltage transfer function is approximately given by

$$\left|\frac{V_2}{V_1}\right| \approx \sqrt{G_{a,1}G_{a,2}}\exp(-\alpha L)\frac{1}{\omega C_{t,2}}. \tag{3}$$

[2]This condition can be violated if there are considerable *mechanical* reflections from the IDT structure. These, however, can be suppressed by using the so-called split-finger design [10].

For liquid sensing applications an important issue occurs: The electric interaction of the liquid with the Love wave and the IDTs. Due to the piezoelectricity of the substrate material the Love wave carries an electric field with it. Hence the phase velocity of the wave would also be influenced by the electric properties (e.g., the permittivity) of the adjacent liquid. Furthermore, the permittivity of the liquid would also strongly influence the IDT capacitance C_t and dielectric losses, which would lead to a non-negligible loss conductance G_l as introduced above. All these effects would influence the phase shift of the signal traveling through the delay line which, in turn, changes the oscillation frequency. This represents an unwanted sensitivity to the electric properties of the liquid. A suitable shielding is obtained by depositing a metalization on top of the guiding layer. However, this shielding influences the electrical parameters of the delay line: The IDT capacitances increase and the radiation conductances decrease which can be explained by the fact that the electric field in the transducer region becomes more concentrated in the non-piezoelectric guiding layer. The increase of the capacitance is of importance because this means a reduction in the load impedance ($\approx 1/j\omega C_{t,1}$) of the amplifier. Since the maximum output current is limited and a certain voltage output amplitude has to be maintained to achieve a sufficient signal to noise ratio, this essentially reduces the frequency range that can be used. The reduction in the radiation conductance as well as the increase in the transducer capacitances furthermore cause an increase in the insertion loss of the delay line (see (3)).

The above discussed influence of the shielding on the electric parameters is also essentially determined by the overlayer thickness h: Increasing h makes the shielding more and more electrically invisible. To evaluate this influence numerically, we performed calculations giving the IDT capacitance and the radiation conductance (at the center frequency) versus the overlayer height h [4]. For the radiation conductance it has been found that its decrease due to shielding is less than 25% as long as $h/\lambda > 0.1$ [4]. The latter condition is anyway desirable to achieve a high sensitivity of the device. We note, however, that the radiation conductance shows an optimum at about $h/\lambda = 0.1$. Above this optimum, G_a slowly decreases (e.g., to about 50% of the optimum value at $h/\lambda = 0.25$) while below this optimum it drastically decreases.

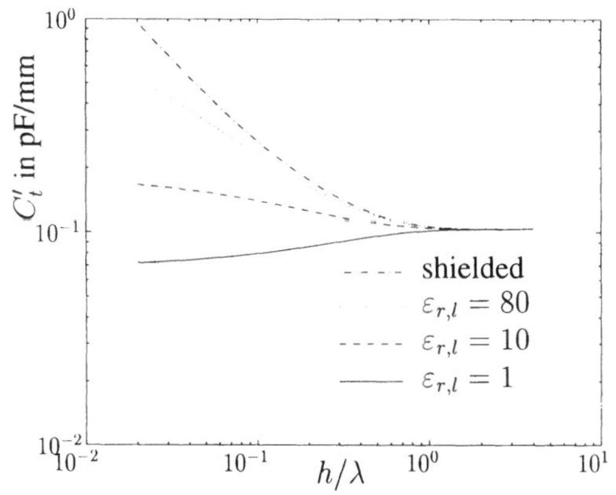

Figure 5: *Scaled capacitance* $C_t' = C_t/(Na)$ *(in pF/mm) versus relative layer thickness* h/λ *for a split finger IDT with* 50% *metalization ratio for different environment permittivities.* N *denotes the number of electric transducer periods and* a *is the acoustic aperture of the transducer.*

Fig. 5 shows the static capacitance versus h/λ for the unshielded case with adjacent air and substances showing relative permittivities 10 and 80 (approximately that of water). Finally, also the curve with shielding is shown. The air loaded case and the shielded case represent the bounding curves; capacitances for unshielded transducers with various environmental permittivities would yield capacitances between those two limits. It can be seen that shielding considerably increases the capacitance especially for low h/λ values but makes it independent from the permittivity of the liquid.

5. Conclusions

In this paper we summarized the device aspects in the design of a smart microacoustic liquid sensor system. The main design parameter of the delay line is the thickness h of the guiding layer. Although achieving a high sensitivity of course is a major issue in the design, one must not ignore the electric interaction of the sensing device with the supporting electronics (an amplifier in our case), which is essential to achieve a high signal to noise ratio in the sensor output signal. We presented the dependence of the device sensitivity as well as that of the electric parameters on the overlayer thickness and outlined the proper electric shielding of the device from the liquid. The resulting design of the amplifier and its measured characteristics are described in an accompanying paper [9].

Acknowledgments

This work was supported by the Brite-Euram project BE-95-1745:MIMICS.

References

[1] D. S. Ballantine, R. M. White, S. J. Martin, A. J. Ricco, E. T. Zellers, G. C. Frye, and H. Wohltjen. Acoustic Wave Sensors. Academic Press, San Diego, 1997.

[2] J. Du, G. L. Harding, J. A. Ogilvy, P. R. Dencher, and M. Lake. A study of Love-wave acoustic sensors. Sensors and Actuators A, 56:211–219, 1996.

[3] E. Gizeli, N. J. Goddard, C. R. Lowe, and A. C. Stevenson. A Love plate biosensor utilising a polymer layer. Sensors and Actuators B, 6:131–137, 1992.

[4] B. Jakoby and M. J. Vellekoop. Design of Love wave sensor devices for the operation in liquid environments. In Proc. IEEE Ultrason. Symp., Toronto, Canada, 1997.

[5] B. Jakoby and M. J. Vellekoop. Viscosity sensor based on Love waves. In Proc. Eurosensors XI, Warsaw, Poland, pages 919–922, 1997.

[6] B. Jakoby and M. J. Vellekoop. Properties of Love waves: Applications in sensors. Smart Materials and Structures, to appear in Dec. 97.

[7] G. Kovacs, G. W. Lubking, M. J. Vellekoop, and A. Venema. Love waves for (bio)chemical sensing in liquids. In Proc. IEEE Ultrason. Symp., pages 281–285, 1992.

[8] G. Kovacs and A. Venema. Theoretical comparison of sensitivities of acoustic shear wave modes for (bio)chemical sensing in liquids. Appl. Phys. Lett., 61(6):639–641, August 1992.

[9] G. W. Lubking, B. Jakoby, and M. J. Vellekoop. Design of a smart microacoustic liquid sensor system - part II: Electronics. In Proc. of the Dutch National Sensor Conference, 1998.

[10] D. P. Morgan. Surface-Wave Devices in Signal Processing. Elsevier, Amsterdam, 1985.

DESIGN OF A SMART MICROACOUSTIC LIQUID SENSOR SYSTEM - PART II: ELECTRONICS

G.W. Lubking, B. Jakoby, M.J. Vellekoop
Delft University of Technology, ITS, Electronic Instrumentation Lab./Dimes
Mekelweg 4, 2628 CD Delft, The Netherlands

0. Abstract

For the design of a micro acoustic delay line oscillator an adjusted design strategy is necessary in order to get optimal results. The here given description is aimed at the design of an oscillator which uses a so-called Love wave delay line [2]; the given considerations however are also valid for other types of delay line oscillators. The physical properties of the device are translated into electrical quantities. The consequences for the active circuit to be built and the practical solutions following from the specified demands are described. Experimental results obtained from the implemented circuits are given.

Keywords : Liquid sensing, delay line oscillator, design criteria

1. Introduction

The used sensing mechanism [3,4] is based one the alteration of the propagation properties of a Lovewave in a double layered structure of quartz and SiO_2 due to the presence of a fluid or gas above it; the propagation velocity decreases and the attenuation increases. The decrease in velocity is mostly correlated with mass loading while a high viscosity changes mostly the attenuation. In order to measure these changes we used the device as part of an oscillator where it determines the frequency By using level control in the oscillator there is also a voltage available which is correlated with the actual amplification factor and thus with the attenuation of the delay line.
The design of this so called delay line oscillator based sensor puts specific demands on the necessary electronic circuitry due to the properties of the device and the desired measuring sensitivity and resolution.

2. Electrical transfer of the delay line

We can use the equivalent circuit given in Fig.1 for describing the behaviour of the delay line. The current source I_t represents the acoustical transfer [1] and has the amplitude $V_i \times \sqrt{G_{a1}G_{a2}} \times e^{-\alpha L}$. G_{a1} and G_{a2} are the acoustic radiation conductances and C_{t1} and C_{t2} the static capacitances of the IDT's (inter digital transducer).

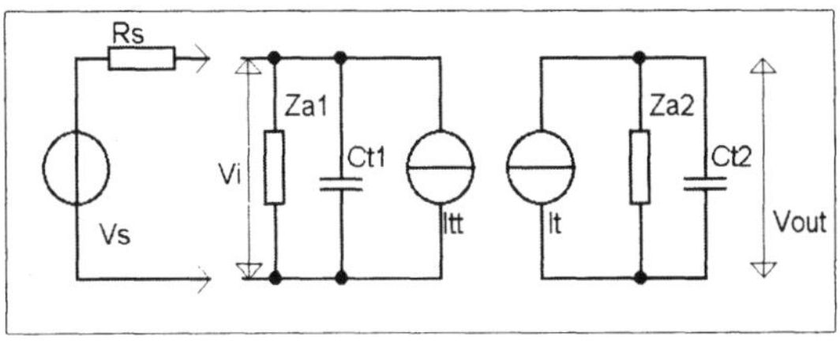

Fig.1 Equivalent circuit

Close to centre frequency $1/Z_{a1}$ en $1/Z_{a2}$ may replaced by G_{a1} and G_{a2}. The source I_{tt} represents the transfer in reverse direction. Assuming steering with a source with an impedance R_s, loading with a high impedance, a delay τ and a loss αL, the transfer function will be approximately:

$$\frac{V_{out}}{V_s} = \frac{1}{1+j\omega R_s C_{t1}} \times \sqrt{G_{a1}G_{a2}} \times e^{-\alpha L - j\omega\tau} \times \frac{R_2}{1+j\omega R_2 C_{t2}} \qquad (1)$$

because almost always $\omega R_2C_{t2} >> 1$ [5] and if we use a low impedance source $R_s << 1/\omega C_{t1}$ the transfer may be written as :

$$\frac{\sqrt{G_{a1}G_{a2}}}{j\omega C_{t2}} \times e^{-\alpha L - j\omega\tau} \qquad (2)$$

We use acoustic devices where piezoelectricity is responsible for the conversion of electrical - to mechanical energy; due to this principle we are faced with regeneration at the receiving side. The regenerated wave travels back to the source IDT where, again due to regeneration, a wave is launched which will be received also by the receiving IDT; this phenomena is called the triple transient echo. It should be noted that here we discuss only the echo which is caused by regeneration. Other parasitic effcts such as those due to mechanical reflections can cause comparable effects, will not be considered. They can be suppressed by using slanted edges and split finger IDT's and other mechanical measures [1].

For the transfer of the triple transient echo one can find:

$$\frac{R_s}{(1+j\omega R s C_{t1})^2} \times (G_{a1}G_{a2})^{3/2} \times \frac{-1}{\omega^2 C_{t2}^{\,2}} \times e^{-3\alpha L - 3j\omega\tau} \qquad (3)$$

For the ratio signal /echo we find:

$$\frac{\omega C_{t2}\sqrt{1+\omega^2 C_{t1}^{\,2} R_s^{\,2}}}{R_s G_{a1} G_{a2}} \times \varepsilon^{2\alpha L} \qquad (4)$$

3. The delay line oscillator principle

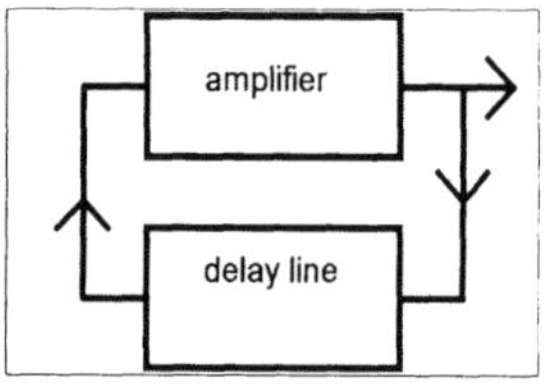

Fig.2 Delay oscillator set-up

The set up of a delay line oscillator can be drawn as shown in Fig.2. There are two terms which must be satisfied in a (delay line) oscillator:

_ the phase shift of the (closed) loop must be $2\pi n$; where n is an arbitrary integer

_ the loop gain must be ≥ 1.

The phase transfer of the delay line can be derived from (1) We assume a phase shift of the amplifier of φ_A ,an input impedance >> $1/\omega C_{t2}$ in the frequency band of interest and an output impedance of R_s. For all frequencies which satisfy

$$\omega\tau + \varphi_A + \pi/2 + \mathrm{arctg}\,\omega R_s C_{t1} = 2\pi n \quad \text{or} \quad f = \frac{n}{\tau}\left(1 - \frac{1}{4n} - \frac{\varphi_A}{2\pi n} - \frac{\varphi_O}{2\pi n}\right) \qquad (5)$$

the oscillation condition is fulfilled.(φ_A and φ_O in radians and $\varphi_O = \mathrm{arctg}\,\omega C_{t1} R_s$)

In order to meet the second demand the amplifier has a gain which exactly compensates for the attenuation of the line which in this case is :

$$\frac{1}{\sqrt{1+\omega^2 C_{t1}^{\,2} R_s^{\,2}}} \times \frac{\sqrt{G_{a1}G_{a2}}}{\omega C_{t2}} \times e^{-\alpha L} \qquad (6)$$

Up to this point we made an oscillator which can oscillate on *n* frequencies; they are called modes. In order to restrain our self to only one frequency we need magnitude selectivity. This type of delay line uses IDT's to get the acoustic wave into the solid; these transducers have a band pass character and the width of the pass band is determined by the number of finger pairs. So it seems that mode selection is inherent if we use this kind of delay line. For a number of reasons the pass band chosen so that a few modes will still be possible. These modes are attenuated slightly differently due to shape of the IDT transfer. By using a level control loop the action of this loop will not only fix the output level but also cause the choice of the mode with lowest attenuation.

The 'sensing' oscillator will be used for highly sensitive measurements of viscosity or mass adsorption where the frequency is correlated with the measurand. Frequency noise and thus short term instability, will determine the detection limit. For that reason we take a second look at (5) keeping in mind that this noise floor lies at about 10ppb of the oscillator frequency. The terms φ_A and φ_O are both caused by the amplifier; their typical values in the frequency band of interest are $\pi/4$ and $\pi/4$ respectively and their total contribution to the phase amounts (with n=100, also a typical value) about 0.25%; if we assume a temperature dependence of 0.3 % / ^{0}C (the temperature coefficient of a resistor on an integrated circuit) of the actual value, we end up with a temperature coefficient of 8000 ppb. The example given shows that although the delay line determines the oscillator frequency, the influence of the amplifier can certainly not be neglected. In the following part we describe how this influence can be kept to an almost negligible level.

4.0 Conditions to be met by the amplifier

Before stating guiding rules and specifications of the amplifier or also the active part of the oscillator we must look at some actual value's and facts of the delay line. The Love wave devices which we used have working frequencies in the range of 80 to 120MHz . The attenuation lies between 20 and 25dB in air and in liquid 6 to 10 dB higher. It should be noted that the increase due to liquid is not only caused by propagation losses but also by increased capacitance of the IDT's. From (6) we can see that both C_{t1} and C_{t2} effect the transfer. The actual value of the IDT capacitance is typical 15 pF in air and 30 pF in water; this is caused by the higher ε of liquids, especially of water. A solution to prevent this liquid depending capacitance is the use of a shield see [5]. Also in that case we are faced with an increase in capacitance but this capacitance is constant. In the IDT's 40 to 60 finger pairs are used; this gives a bandwidth of $0.88f_0/N$ with N the number of finger pairs and an acoustical length of about 100λ (thus n=100) is used. The just given values of the capacitance, have an impedance at 100MHz of 100 and 50 Ω. By using (4) we can make an estimation of the signal/echo ratio due to this capacitance's. If we would be able to make the source - and /or load impedance much lower than this 50 to100 Ω than this echo would not exist; formula (3) confirms this.
For the description of the electronic circuitry we shall first investigate the peripherals which are directly connected to the Love wave device, thus the input and output; after that the needed transfer characteristics, the control behaviour and at last the influence of temperature and other interferences. Throughout the discussion it will become clear that monolithic integration will have large advantages compared with a discrete design.

4.1 Input impedance

In this set up we have chosen for voltage sensing; its means that the connecting amplifier does not change the output voltage of the line. The consequence of this is that the input impedance of the amplifier must be much larger than R2//C2. Current sensing could also be used; it has the advantage that if we can realise an input impedance << R2//C2 a further increase of the triple transient suppression is possible. In practice however the just stated condition can not easily be met ; an impedance in the order of 15 Ω would be necessary. If we bear in mind that a connecting wire of 20 mm has already an impedance of about 10 Ω at 100Mhz it becomes obvious that this objective cannot be obtained and at the same time the phase shift of $\pi/2$, the second term in (3), would be replaced by a circuit dependent phase shift. More over, a good common mode rejection is difficult to realise in a current sensing circuit.

4.2 Output impedance, output voltage level

We have chosen for voltage steering. Voltage steering is done with a source with an output impedance (Rs) which is much lower than the load , in this case R1//C2. The classical reason for this is the same as before stated : an increase of the suppression of the triple transient echo but as we saw in the explanation above due the inherent induction of the connecting wires a total impedance of much lower than 25Ω is difficult to realise. If we would try to make an output impedance of the amplifier in the order of 10Ω, we need a high bias current. The output stage of the amplifier must steer a (capacitive) load with an amplitude which is as high as possible in order to get a good short term stability. In order to be able to do this we also need a high bias current. It is also clear that the phase shift arctg$\omega C_1 R_s$ will become smaller if we succeed in making R_s smaller. So by spending attention to lowering R_s also a decrease in circuit dependent phase shift is realised. We can make the conclusion that for a high output current we need a circuit which also can have a low output impedance; these combined advantages made us choose for the just described implementation.

4.3 Input noise

The short term instability of an oscillator is caused by noise and temperature effects.The noise has two sources : the Love wave device and the electronic circuitry. We will look at the noise of the circuitry. We assume that all the noise can be modelled by two independent sources u_n and i_n placed at the input of the amplifier. From the schematic shown in Fig3 we can derive that due the presence of C_{t2} the noise voltage at input can be written as $\sqrt{(u_n^2 + i_n^2/\omega^2 C_2^2)}$. For a differential pair with bipolar transistors the noise power density's are S_{nu}= 4kT($2r_{bb}+r_e+R_E$) and S_{ni}=4kT/4βr_e. Here is r_{bb} the total base resistance and r_e the (current dependent) diffusion resistance, β the current gain and R_E the total emitter resistance; f_T is supposed to be at least 10 times higher than the working frequency. It is possible to optimise the noise behaviour using an optimum value for r_e. In this case however this would lead to a very low value of r_e and a bias current of tens of mA would be needed and quite large transistors are necessary to get r_{bb} low enough. E.g. : with I_E = 2mA r_e=13Ω and with r_{bb} =25Ω and R_E=50Ω the noise power density S_{nu}=4kTx113 and S_{ni} =4kTx2.10^{-4}, which in a small frequency band around 100MHz will result in voltage density (impedance of C_2 ~ 100Ω) of 4kTx2. From this we see that the noise behaviour is dominated by the voltage source due to the 'high' value of C_2. The conclusion from this is : Take the bias current of the input stage as high as possible and use transistors which are as large as possible; it should be noted however that other demands will conflict with this criteria so that in the end we shall end up with a compromise.

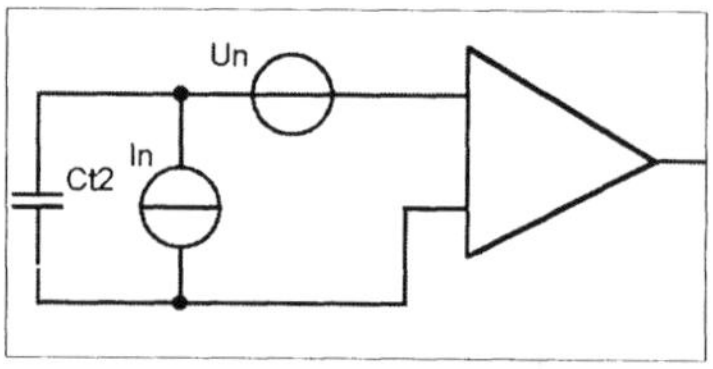

Fig.3 Noise sources

4.4 Control properties

We concluded earlier that the amplifier should have a (voltage)controllable gain with a control range of about 20 dB. There are various methods for varying gain but almost every method has disadvantages when one looks at noise behaviour, dynamic range, phase transfer, power consumption, behaviour close to transfer 0 and the control voltage/gain transfer. In this application the noise behaviour and the phase transfer are most important but also the dynamic range needs attention while a linear relation between the control voltage and the gain would very attractive. We want a circuit [6] with a linear noise transfer which means that referred to input noise and also the signal to noise ratio stays the same over the control range. The bonus for this is a short term stability which depends linearly on the loss of the line. Finally the change in gain should cause a negligible change in phase in order to avoid corresponding frequency changes.

4.5 Phase transfer

The frequency of the oscillator depends heavily on the phase transfer of the amplifier; a much longer line (e.g. n=1000) would ease the demands for this transfer but this line is almost impossible to make (both technologically and economically). As concluded earlier the phase shift at the working frequency was assumed to have a typical value in the order of $\pi/2$ and according to(5) this leads to a contribution of 0.25% in frequency. It clear that we should try to get this number as low as possible. We have two means to do that. Firstly we should use as little stages as possible in the amplifier and secondly the bandwidth of the amplifier should be as large as possible. So again it seems that we are faced with an open end demand. If one departs from the a total shift of $\pi/4$, the phase shift per stage is $\pi/4s$ where s is the number of stages. We used 5 stages so per stage $\pi/20$ is allowed. Assuming a first order transfer in one stage it means that the bandwidth per stage must be about 5 times the oscillator frequency (so in this case about 500MHz).

4.6 Temperature dependence tracking

From (2) it evident that apart from the here used Love wave delay line which has a temperature coefficient (TC) of around 25ppm, the TC of the amplifier ($\varphi_A+\varphi_O$) is a parameter which needs attention. The amplifier is made out of stages which use local feedback in a way that the transfer is mainly determined by a resistance or the quotient of two resistance's. The total phase transfer will be determined by the loop gain and phase transfer of each stage. The total phase transfer can be found by adding the phase shift of the subsequent stages. So if we assume that each stage contributes about the same shift we can say that the TC of one stage is also the TC of the total amplifier. In this design we used bipolar transistors; the RF behaviour in this transistors is amongst others determined by r_e which depends on the absolute temperature. This effect can be compensated. Other parameters depend on resistance's and capacitance's which can not easily be compensated. As a consequence of the strong feedback used in every stage the combined TC has a value in the order of 0.3%(see also section 3). In order to get rid of the temperature dependence we decided to use a dual configuration and couple the amplifiers so that they have the same temperature; in such a configuration the best temperature tracking is achieved by putting two identical circuits on the same chip. We decided for practical reasons to evolve this concept in two steps which means that the present experiments were done with two separate identical circuits in good thermal contact; the next step with two circuits on the same chip is in preparation.

4.7 Monolithic integration

From the discussion above it has become clear that we need a very wide band amplifier which uses as little power as possible. In order to get this bandwidth five stages are needed; amongst them is a gain controlled stage. The best way to cascade stages with little power loss is to keep the (parasitic) capacitance's as low possible. The best possibilities are offered by using monolithic integration.

In the paragraph about temperature dependency it became clear that using two circuits on the same chip would be preferred in order to get the necessary temperature tracking.
The scheme which is used for the controllable amplifier is not available on commercial chip. Finally short and rugged connections between amplifier and device can be made easier with a fully customised IC; additional capacitance on the input and output can be kept to a minimum.

4.8 The short term stability

Short term stability is probably one of the most important items when it comes to evaluating the performance of a measuring system. We used a not frequently applied method for getting a better inside in the involved parameters. Responsible for this instability are two noise sources: an 1/f source caused by the device and shot noise caused by the electronic circuitry. Within the scope of this article we shall not treat this subject extensively but an important conclusion is that the noise properties of the amplifier described above are so that its influence on the detection limit is almost negligible. Important in this field is the determination of the frequency where the contribution of the two sources is equal. Only if this frequency is accurately known one can specify the noise property's of the amplifier accurate. We measured at a 100MHz oscillator a standard deviation of 0.9 Hz; the gate time we used was 50 ms.

5. Measuring results

Shown are the transfer of the amplifier with respect to amplitude and phase; also the equivalent noise input voltage was measured and was 2.2 nV/√Hz. The amplifier was fed out a 50Ω source, loaded with 30pF between the differential outputs and biased for maximum gain, the power consumption is about 12.5mA at 12V so 150mW.

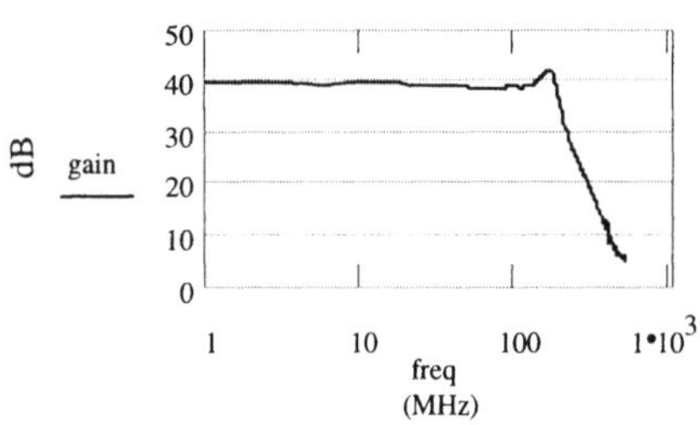

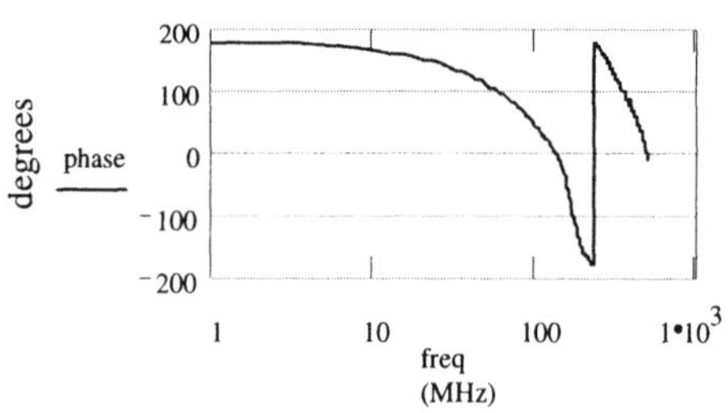

From the discussion under 4.8 became clear that the short term instability was better than 1Hz; this is the measuring result received by averaging over 100 measurements of the standard deviation of the frequency with a gate time of 50ms and a negligible dead time.

6. Conclusion

An amplifier and accompanying circuits can be built for a delay line oscillator which shows an excellent short term stability. It was made clear that the properties of the electronic circuitry have a strong influence on the final result and that monolithic integration of the circuit is necessary for getting optimal results.

7. References

[1] D. P. Morgan. Surface-Wave Devices in Signal Processing; Elsevier, Amsterdam, 1985.

[2] G. Kovacs, G. W. Lubking, M. J. Vellekoop, and A. Venema. Love waves for (bio)chemical sensing in liquids.In Proc.\ IEEE Ultrason.\ Symp., pages 281--285, 1992.

[3] B. Jakoby and M. J. Vellekoop. Design of Love wave sensor devices for the operation in liquid environments. In Proc.\ IEEE Ultrason.\ Symp., Toronto, Canada, 1997.

[4] B. Jakoby and M. J. Vellekoop. Viscosity sensor based on Love waves.
In Proc. Eurosensors XI, Warsaw, Poland, pages 919--922,1997.

[5] B. Jakoby, G. W. Lubking and M. J. Vellekoop.
Design of a smart micro acoustic liquid sensor system - part I: Micro acoustic device
In Proc. of the Dutch National Sensor Conference, 1998.

[6] K. Yamshita, T. Kinoshita, M. Maeda, T. Kaji, N. Maeda. Wide band and high-gain negative-feedback agc amplifier for high-speed lightwave digital transmission systems.
Electronic letters 9th May 1985 Vol.21 No.10

INFLUENCE OF SUBSTRATE VOLTAGE ON SENSITIVITY AND OFFSET OF (SPINNING-CURRENT) HALL PLATES

Sandra Bellekom
Laboratory for Electronic Instrumentation, Department of Electrical Engineering
Delft University of Technology, P.O.box 5031, 2600 GA Delft, The Netherlands
Phone: +31-15-2783342, Fax: +31-15-2785755, e-mail: sandra@ei.et.tudelft.nl

0. Abstract

Hall sensors are used for the measurement of magnetic fields and for the indirect measurement of mechanical quantities. Traditional Hall plates show a large, unpredictable offset, making them less suited for applications involving small magnetic fields. To reduce the offset of Hall plates, Munter introduced the so-called spinning-current principle [1]. Using this principle he successfully reduced the offset of Hall plates. The research into these spinning-current Hall plates concentrates on the influence of important parameters like temperature, substrate voltage, mechanical stress, etc. on the properties of these Hall plates. In this paper the influence of the substrate voltage on the sensitivity and offset of spinning-current Hall plates is discussed.

Keywords: (spinning-current) Hall plate, sensitivity, offset

1. Introduction

Magnetic sensors are used in many applications, for instance for the direct measurement of the earth magnetic field or for the reading of information from magnetic cards. Most magnetic sensors, however, are used in an indirect way. Combined with a permanent magnet they can be used as sensors for position, (rotation) speed or angle. As an example, a permanent magnet can be attached to a car acceleration pedal and a magnetic sensor can be used to measure the pedal angle. Such magneto-mechanical sensors have the advantage of being robust and insensitive to dust.

Hall plates are very simple magnetic sensors that can be produced in standard silicon IC technology. Fabrication in such a process yields cheap, small devices and allows the integration of electronics on the sensor chip. The drawback of these integrated silicon Hall plates has been their offset which is both unpredictable and a function of time. A lot of research has been carried out to explain the offset of Hall plates, but no complete theory exists.

Research into the different parameters that cause offset in Hall plates contains the study into influences of, for example, mechanical stress, bias current, temperature and substrate voltage. The effect of these parameters on the properties of traditional Hall plates is compared to the influence of these parameters on spinning-current Hall plates. This paper concentrates on one of these parameters: the substrate voltage. The influence on the sensitivity and the offset of (spinning-current) Hall plates is discussed.

2. Hall plates in silicon IC technology

The sensitivity of Hall plates depends on the mobility of the charge carriers in the device. In n-type material electrons and in p-type material holes are the charge carriers. Because the mobility of the electrons is larger than the mobility of holes, Hall-plates are realized in n-type material. The sensitivity also depends on the concentration of the charge carriers: a lower concentration gives a larger sensitivity [2].

In a CMOS process the n-well satisfies both conditions. The p-type silicon around the n-well is the p-substrate or p-epi layer. Using the junction between the p-epi (or substrate) and the n-well

as an isolation, separate Hall plates can be realized. Contacts to the Hall plate are made using an n^+ implantation, and to contact the substrate a p^+ implantation is used.
For bipolar processes the n-type epitaxial layer can be used as active material for the Hall plate. The p-substrate defines the lower boundary of the device and the deep p diffusions the side boundaries. Again n^+ areas are used to contact the device. Figure 1 shows Hall plates in both the CMOS and the bipolar process.

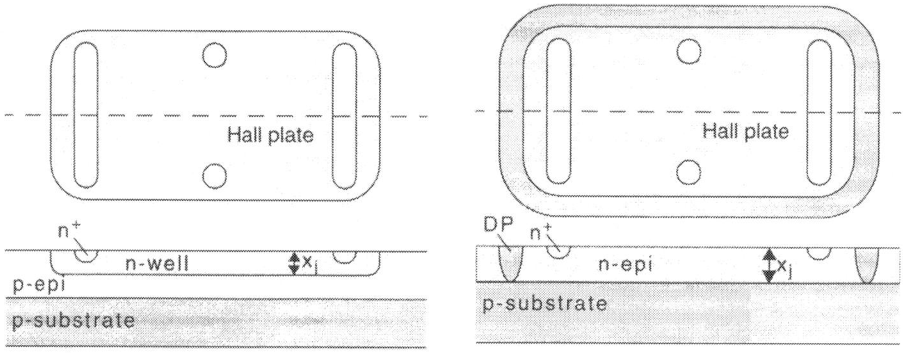

Figure 1: Integrated Hall plates in a CMOS process (left) and a bipolar process (right).

3. JFET effect

In IC processes the Hall plate's dimensions are no longer specified by the physical dimensions of the device. The isolation of the Hall plate is realized by a reverse biased pn-junction. The top boundary is defined by the interface between silicon and the oxide on top of the chip. All other boundaries are defined by the edges of the depletion layer between the n-type Hall plate and the p-type surrounding material. Because the width of these depletion layers depends on the bias voltage, so does the position of the boundary of the Hall plate. Because of the current through the Hall plate, the voltage across the depletion layer becomes a function of position, and so does the thickness of the Hall plate. This effect is called the JFET effect. Figure 2 shows the influence of the depletion layer thickness on the effective Hall plate thickness t.

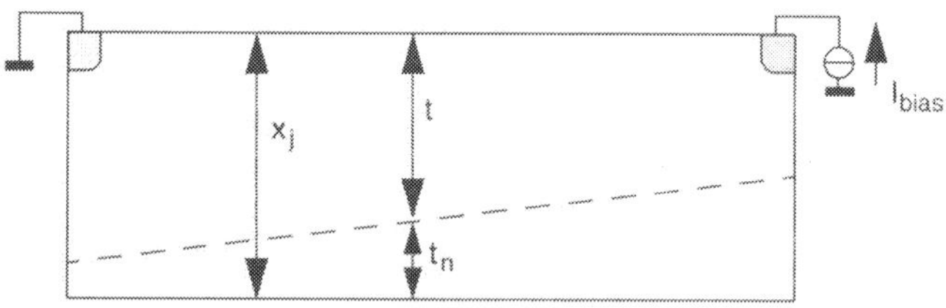

Figure 2: Reduction of the effective thickness t of the Hall plate because of the depletion layer thickness t_n.

The new effective thickness of the Hall plate, the thickness in the middle of the device, can be calculated:

$$t = x_j - \sqrt{\frac{2\varepsilon}{q} \cdot \frac{N_{sub}}{N_h(N_{sub} + N_h)} \cdot \left(V_{bi} + V_{sub} + \frac{1}{2}(I_{bias} \cdot R_{hp})\right)}$$

$$R_{hp} = \frac{2\rho}{\pi t} \cdot \ln\frac{D}{s}$$

$$V_{bi} = \frac{kT}{q} \cdot \ln\frac{N_h \cdot N_{sub}}{n_i^2}$$

In these equations ε is the dielectrical constant of silicon ($1.054\ 10^{-10}$ F/m), q is the elementary charge ($1.602\ 10^{-19}$ C), N_{sub} and N_h are the doping concentrations of respectively the p-type substrate and the n-type material in which the Hall plate is defined. I_{bias} is the applied bias current of the Hall plate, V_{bi} is the build-in voltage between those materials, V_{sub} is the applied substrate voltage and R_{hp} the Hall plate's resistance. The k is Bolzmann's constant ($1.38 \cdot 10^{-23}$ J/K), T the absolute temperature and n_i the intrinsic electron concentration of silicon ($1.45 \cdot 10^{10}$ cm^{-3}). The Hall plate's resistance depends on the geometry of the device, D is the diameter (of a circular Hall plate) and s the contact size, and of the material properties, ρ is the resistivity of the Hall plates material.

The equations clearly show a dependence of the effective thickness on the substrate voltage. Changing this voltage will influence t. Using the process parameters of the bipolar and CMOS processes it is possible to calculate the relative thickness change. Figure 3 shows the results of this calculation. From this figure the resistance and sensitivity change versus substrate voltage, as a result of changing effective thickness, can be determined since the sensitivity S of a Hall plate equals:.

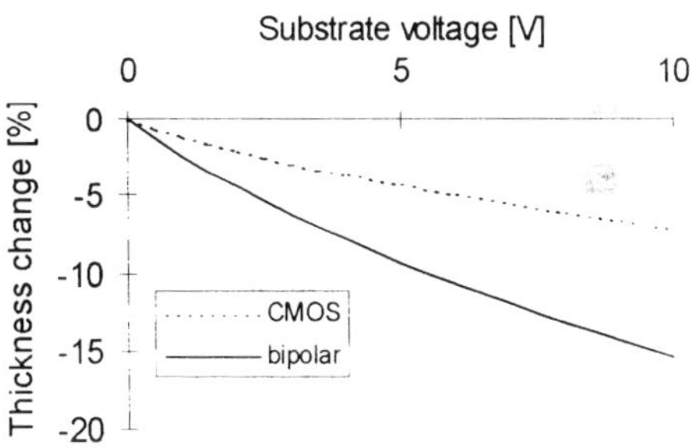

Figure 3: Calculated thickness change of bipolar and CMOS Hall plates versus substrate voltage.

$$S = I_{bias} G \frac{r_h}{N_h q t}$$

Here G is a geometrical correction factor and r_h the Hall factor. Both quantities can be measured and the measured and calculated values will be compared to check the JFET effect theory.

4. Spinning-current Hall plate

To reduce the offset of Hall plates spinning-current Hall plates were developed [1]. These are symmetrical Hall plates with 4, 8, 16 or more contacts. In these multi contact Hall plates the bias current I_{bias} is switched in, for example, eight directions through the Hall plate and the corresponding voltages V_H are measured, as shown in figure 4. The measured voltages represent one period of a (radial spatial) periodic signal. The spinning-current method uses this periodicity, it separates transduction effects according to their spatial radial periodicity. From the output signal a Fourier transform is calculated. The Hall effect itself does not depend on the bias current direction and is therefore present in the DC component. Offset voltages that contrib-

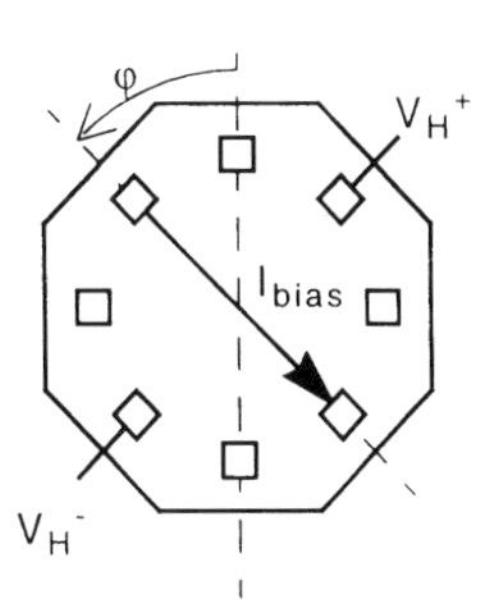

Figure 4: Lay-out of spinning-current Hall plate.

ute to the higher Fourier coefficients are eliminated.
The offset can be measured by placing the spinning-current Hall plate in a zero Gauss chamber. Figure 5a shows a possible output. The eight measured voltages are connected by lines to show the periodic behaviour. In a magnetic field, the output signal shown in figure 5b results. The shape of the output voltages versus current direction is the same, but the average level is changed. This averaged value is the residual voltage of the spinning-current Hall plate. The residual offset voltage of the spinning-current Hall plate can be 1000 times smaller than the offset voltage of a traditional Hall plate [3].

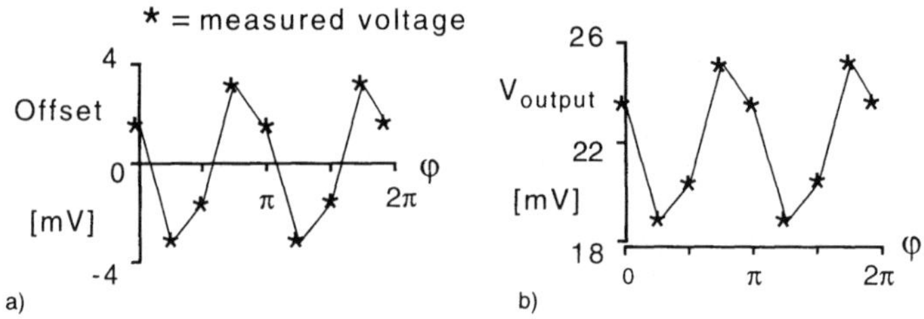

Figure 5: Measured voltages on a spinning-current Hall plate.
a) without magnetic field, b) with magnetic field.

Since the spinning-current method reduces all periodic offset causes, also offset resulting from resistance change will be reduced. The substrate voltage influences the resistance and therefore the offset of Hall plates, this influence should not be visible after applying the spinning-current method.

5. Experiments

Lay-out. The Hall plates used in the experiments are octagonal, with a diameter of 400 μm. The eight contacts are 4x4 μm^2 and the centre of the contacts is situated 5 μm from the rim of the Hall plate.

Measurement set-up. Using the experimental set-up shown in figure 6, the offset voltage and the resistance of the Hall plates are measured ($B = 0T$) as the substrate voltage is changed. The voltage in the current direction V_i is measured to determine the resistance of the Hall plate. The applied bias current is 0.5 mA.

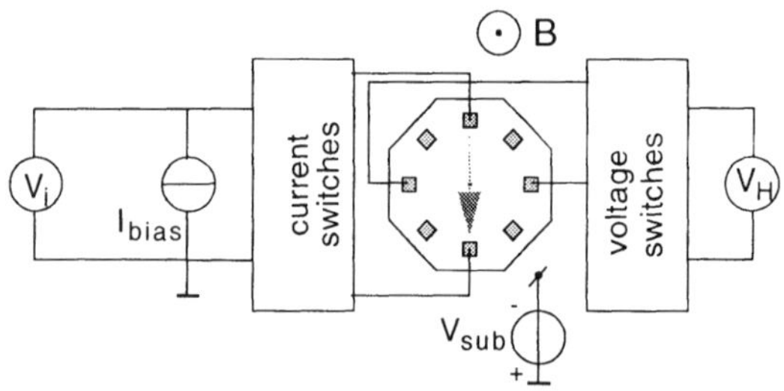

Figure 6: Offset and resistance measurement set-up.

To measure the sensitivity, the spinning-current Hall plate is placed between two Helmholtz coils which generate a magnetic field of about 20 mT in the centre.

Resistance and sensitivity. In figure 7 the measured and calculated resistance and sensitivity change versus substrate voltage is shown. The shapes of the measured and calculated graphs match quite well. The small deviation may have been caused by the used process parameters which are not accurately known. Also the equation used for the resistance only partly matches the actual geometry of the Hall plates. The measured CMOS sensitivity does very well coincide with the calculated one. Figure 7 shows that the JFET effect can predict the resistance and sensitivity change of bipolar and CMOS Hall plates quite well.

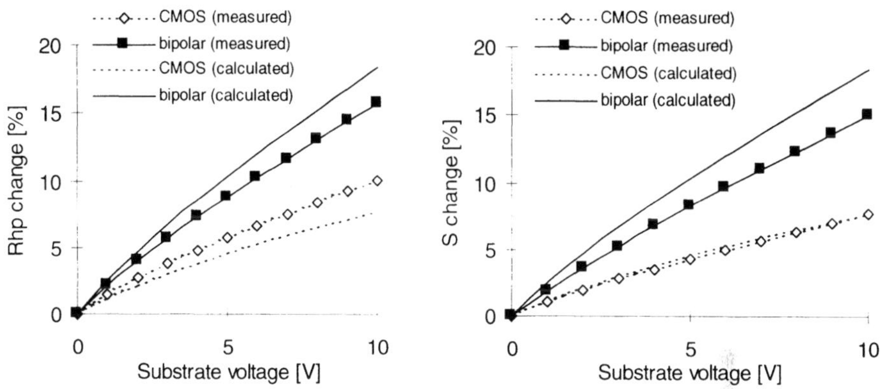

Figure 7: Measured and calculated resistance and sensitivity change versus substrate voltage for CMOS and bipolar integrated Hall plates.

Offset. Figure 8a shows the offset voltage without spinning-current and figure 8b shows the residual offset voltage when the current is made to spin (the vertical error bars show the standard deviation of the measurements). Though the measurements show that the offset of Hall plates depends very much on the substrate voltage, measurements also show that the residual offset of a spinning-current Hall plate only depends very little on this substrate voltage.

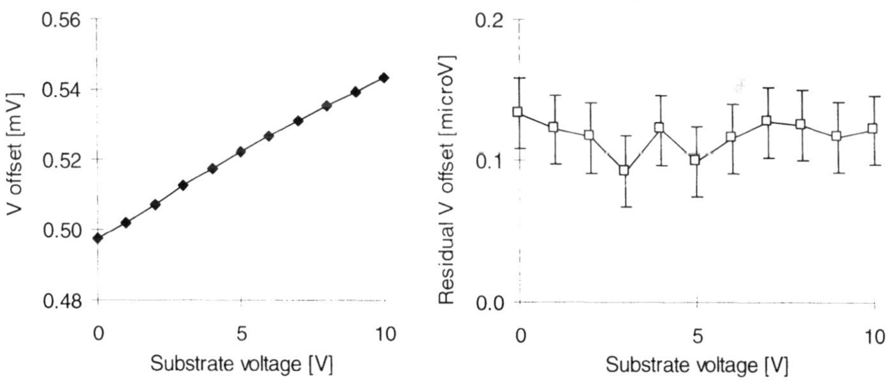

Figure 8: Measured offset voltage before (left) and after (right) applying the spinning-current method to a CMOS Hall plate.

6. Conclusion

Because of the JFET effect the substrate voltage influences the resistance and the sensitivity of bipolar and CMOS integrated Hall plates. The theoretically predicted resistance and sensitivity changes compare well with the measurements

Without the spinning-current method the offset of a Hall plate changes as the substrate voltage changes. However, as has been experimentally verified, the residual offset of a spinning-current Hall plate does not depend on the substrate voltage. The offset that is caused by substrate voltage variations is reduced significantly using the spinning-current method.

Acknowledgements

The author would like to thank Wilko Kindt, Paul Gennissen and Colin Ashruf for their support, the employees of DIMES (Delft Institute of Micro-Electronics and Submicron technology) for fabricating the devices, dr. P.M. Sarro for coordinating the fabrication process and Prof. Middelhoek for enabling the research and his useful suggestions. The research on spinning-current Hall plates is supported by STW, project DEL33.3216.

References

1. P.J.A.Munter, A Low-offset Spinning-current Hall Plate, Sensors and Actuators, A21-A23 (1990) 743-747.
2. Middelhoek S. and Audet S.A., Silicon Sensors, Academic Press, London, 1989.
3. A.A. Bellekom and P.J.A. Munter, Offset Reduction in Spinning-Current Hall Plates, Sensors and Materials, 5 (1994) 253-263.

HYBRID INTEGRATED ELECTRO-OPTIC SENSOR PLATFORM

A.J.T. de Krijger, K.W. Benoist, J.A.G. van Bezooijen, H.H.P.Th. Bekman,
TNO Physics and Electronics Laboratory,
Oude Waalsdorperweg 63, NL-2597 AK The Hague, The Netherlands

0. Abstract
Optical sensors are an important class of sensors. They can offer high sensitivity, high electro-magnetic immunity and compactness. An important sub-class is formed by the fiber-optic based sensors. They are used among others for smart structures, where the fiber is used in a distributed sensor configuration. In these systems the light generation and sensing is done in a remote unit. Another sub-class of optic sensors receiving increased attention are the electro-optic integrated sensors. Due to the integration of electronics and optical circuitry on one and the same silicon substrate miniature smart optical sensors can be realized. The price to pay for this marriage between electronics and optics is the loss of electro-magnetic immunity. In this paper we report on the development of a toolbox for electro-optic integrated sensors. Various components, part of the toolbox, like a laser-waveguide coupling, detector-waveguide coupling, optical waveguide structures are briefly discussed. The building blocks are combined in a basic sensor platform, thus showing the maturity of the toolbox approach. By choosing appropriate building blocks from the toolbox, in combination with a specific optical sensor circuit, specialized sensors can be realized.

Keywords: Integrated optics, semiconductor lasers, waveguides, sensors.

1. Introduction
The development of innovative micro systems for sensor applications is widely recognized as being important for a modern and competitive economy. Clear market signs point to the demand for miniaturized sensor systems for a large variety of markets like health care, safety and security, defense and process control. Obvious advantages for these systems include low weight, high reliability, low volume and power consumption. Silicon integrated optic micro sensors can be used for the detection of (bio-)chemical substances as well as for measurement of physical quantities like distance, pressure or temperature.

The aim of our research was to develop electro-optical sensors with integrated-optic technology. At the beginning of the project a few essential choices were made. As a start we based our integrated-optic technology on silicon technology, since silicon is a mature technology for integrated optic [1]. The extra advantage of silicon is that it has the potential for monolithic integrated electronics. Furthermore it was decided to pursue a hybrid approach in the first years, thus avoiding pain-staking problems with process compatibility's (between optical and electronic fabrication steps) and lack of optimized sub-assemblies. Each optical sensor principle is unique, but despite some differences various optical sensors share common modules. In our project we therefore started with the development of the identified common modules, our so-called toolbox elements. Similar approaches can be found in reference [2].

The work presented in this paper is part of the IOP Electro-optics cluster 3 project on silicon integrated optic microsensors, which is a joined activity of the TNO physics and electronics laboratory and the University of Twente. For other work performed in this cluster see references [3-5] and references therein.

2. Toolbox components

2.1. ZnO waveguide material

ZnO has been extensively studied in recent years because of its strong piezoelectric properties and subsequent application in both bulk and surface acoustic wave devices. Apart from its piezo-electric coefficients ZnO has also sizable thermo-optic and electro-optic coefficients [6]. One soon realized that ZnO is also a very interesting active optical waveguide material since ZnO is transparent from 0.4 to 2.0 micron, the wavelength range for which cheap laserdiode light sources are available. ZnO is, contrary to other active waveguide materials, like $LiNbO_3$, InGaAsP, AlGaAs, a very cheap material.

We have grown ZnO films on oxidized silicon substrates using a sputter deposition process. The films were grown in an RF-diode sputter machine at a temperature of 275°C by reactively sputtering a zinc target with oxygen at a pressure of 2×10^{-3} mbar. The growing rate is approximately 0.8 μm/hr.

The films were characterized by optical techniques, X-ray analysis and electron microscopy. We deduced that the grown ZnO films are polycrystalline and preferentially oriented such that the basal or (0002) planes of the hexagonal close-packed wurtzite crystal lattice parallel to the substrate (c-axis perpendicular to the substrate). The size of the crystallites in the c-direction as determined from an X-ray line-shape analysis is 30 to 60 nm. The as-deposited films have a large residual tensile strain along the c-axis which can be released by post-deposition annealing. The optical attenuation is typically 1 dB/cm @632.8 nm. The average ordinary refractive index at this wavelength is 1.9894 and the extra-ordinary index is 2.0080. For further details of the growth optimization and thermal annealing see reference [7].

2.2. Laser-waveguide coupling

The hybrid laser-waveguide coupling we developed is drawn schematically in Fig. 1. In a waveguide structure an anisotropic well is etched by a RIE process. The well depth is around 3-4 μm, depending on laser and waveguide geometry. The bottom of the well is metallized. With a home-made manipulation setup the laser is actively positioned in front of the waveguide. After the active alignment the laser is bonded to the substrate [8].

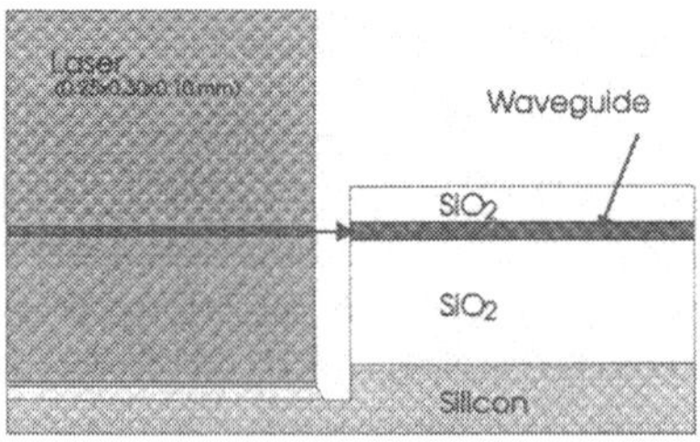

Figure 1. Schematic laser-waveguide coupling.

Coupling of laserdiodes to high contrast waveguides requires very accurate alignment between laser and waveguide. For a typical laserdiode and waveguide configuration an ideal coupling efficiency of 90% can be obtained. The alignment is most critical in the transversal direction. A transversal misalignment of 0.3 μm results in a reduction of the incouple efficiency to 45%. The results dictates that we use sub-micron alignment and that we use a special mounting technique since soldering techniques give insufficient transversal alignment precision.

The laser-waveguide coupling used in the present study makes use of sub-micron alignment with a piezo-electric manipulator and thermocompression bonding. In thermocompression bonding no liquid materials are involved; the bond is created solely by material transport in the solid phase. The temperature during bonding is far below the melting temperature of the materials used. As a result the thickness of the bonding layers should remain constant during the process. Therefore the critical transversal alignment is achieved solely by tuning the metallization thickness, and the alignment achieved before bonding remains unaltered in the bonding procedure [9].

The top layer of the laserdiode metallization consists of gold. Therefore it was most natural to employ a gold-gold thermocompression process. Gold is also attractive since the material is soft and it does not oxidize. The process is carried out under normal atmospheric conditions. We use a process temperature of 250°C, a pressure of 180 bar and a bonding time of more than 10 minutes. The temperature of 250°C is well away of the melting temperature of gold (1064°C). In normal thermocompression practice the bonding temperature is set at a temperature around 70% of the melting temperature (which would be around 663°C for gold). We have used only 250°C in order to protect the laser contact metallization integrity. As a result we had to use a relatively high bonding pressure, which is a disadvantage of the technique.

However in the mean time we have conducted lifetime experiments on thermocompression mounted laserdiodes. Two 850 nm laserdiodes are tested CW at a forward current of 30 mA at a temperature of 50°C, and a initial output power of 9 mW. After 3 years their output intensity has dropped to about 79%. Assuming a defect activation energy of 1 eV [10] and setting the failure criterion at 10% reduction in optical output power, the estimated lifetime at 30°C is 20-25 year. Thirteen 675 nm lasers were operated CW at a forward current of 50 mA (7-9 mW) at a temperature of 40°C. We determined a MTTF at 30°C of approximately 30 year (assuming a defect activation energy of 1 eV) [11].

The big advantage of this laser mounting scheme is that we have a very accurate control on the laser position after bonding. Furthermore it is also versatile since almost any laserdiode can be used.

2.3. Detector-waveguide coupling

Our objective was to realize an efficient and cheap waveguide-detector coupling. We therefore decided to glue a photodiode (dimensions 1 x 3 mm) on top of the waveguide. Two types of epoxies were used. An optical epoxy was used above the waveguide area and an electrical conducting glue was used next to the channel waveguide, since one of the contact of the photodiode surrounded its active area and faced the substrate. The epoxies were deposited with a syringe. The light we detected in the photodiode was either coupled out of the waveguide by the naturally present scatter centers or by a grating etched in the waveguide.

As scatter light is of low intensity, this signal can be used for optical feedback to stabilize the light intensity in the waveguide. The grating output coupling is more flexible. By tuning the grating parameters the output coupling efficiencies can be varied. Therefore the grating output coupling can be used for the optical feedback signal as well as for the sensor signal.

In figure 2 we show a detector-waveguide coupling with on the left a detector amplifier made with gate-array technology. On the right of this amplifier one recognizes the photodiode. A light beam enters the optical chip from the right and is coupled out of the slab waveguide by a grating underneath the photodiode. One can notice that all the light is coupled out of the waveguide, since on the left of the photodiode no light beam is visible.

Figure 2. Detector-waveguide coupling with a grating output coupler and detection electronics.

3. Sensor platform

A dummy sensor device was realized to demonstrate the viability of the proposed integration of the developed building blocks. Therefore the optical circuit (with dimensions of 6.5 x 12 mm) comprised of only a single straight waveguide, with a width of 4 μm defined in a RIE process. At one side of the optical chip a well was etched by reactive ion etching. The bottom of the well was metallized with Cr/Al/W/Au layers by e-beam evaporation. A 850 nm multi-quantum well laserdiode was coupled to the channel waveguide and mounted by thermocompression after lateral active alignment. The laserdiode was a 16 mW laser purchased from Philips Opto-electronics, with typical dimensions of 300 x 250 x 100 μm. The metallization at the bottom of the well served as one electrical contact. The other electrical contact was made by wire-bonding at the top the laserdiode chip. At the other end of the optical chip a silicon photodiode (dimensions 1 x 3 mm) was glued on top of the waveguide.

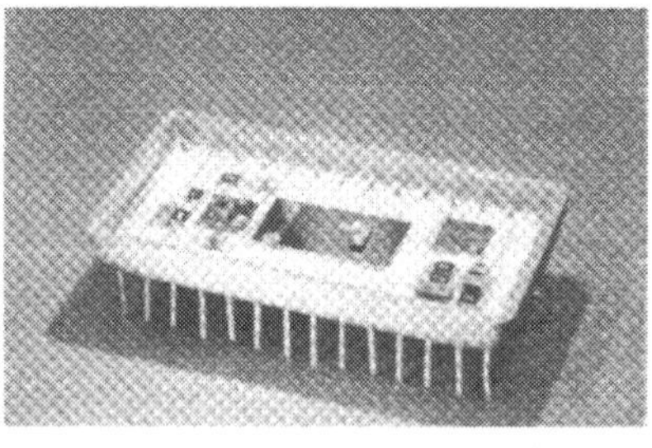

Figure 3. Integrated electro-optic sensor platform integrated in a DIL housing of 41 x 20 mm.

An electrical connection board was realized on a Al_2O_3 sub-carrier. On this carrier we glued the custom laser driver and detection circuit alongside the optical chip. In addition the custom electronics required a few external capacitors and resistors (smd components) for internal adjustment. The ceramic carrier with al its components was than glued in a standard DIL housing (28 pins, outside dimensions 41 x 20 mm), see Fig. 3. All electrical connections from the sub-carrier to the custom electronics and the DIL housing were made by standard wire bonding.

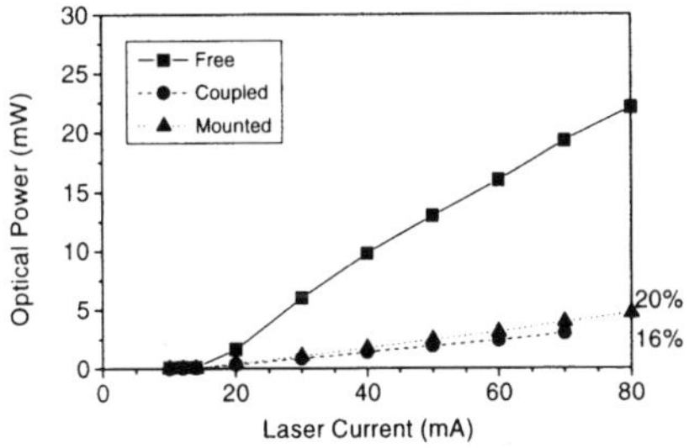

Figure 4. Laser power before and after coupling in 4 μm ridge waveguide.

3.1. Testing

The optical chip was characterized after laser-waveguide and detector-waveguide coupling. The first important parameters to be determined are the coupling efficiencies. Since the photodiode was only detecting scattered light we could simply measure the light at the end of the optical waveguide (length 9 mm). Correcting for reflection losses at the waveguide-air interface in front of the photodiode we deduced that the laser-waveguide incouple efficiency was about 20%, see Fig.4. Therefore at a maximum emitted optical output power of 16 mW, at the beginning of the waveguide we had an optical power budget in the waveguide of 3 mW. Simultaneously the detector-waveguide coupling efficiency was determined. Since we detected 30 μW of optical power, the optical detector-waveguide coupling efficiency was 1%.

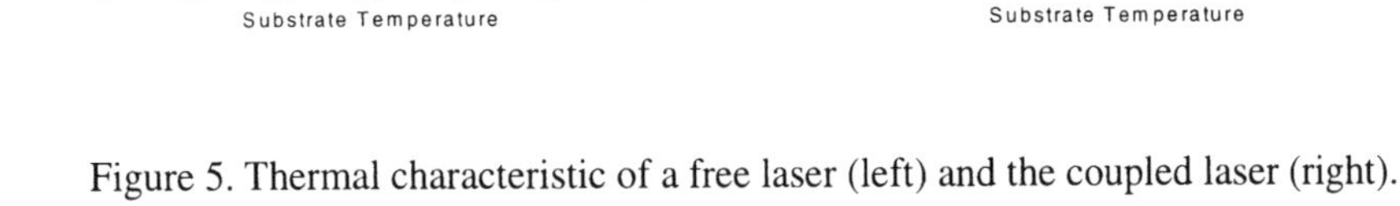

Figure 5. Thermal characteristic of a free laser (left) and the coupled laser (right).

Another important parameter for a stand-alone sensor is the influence of the environment on its operation. A major point of concern is the thermal stability of the very critical laser-waveguide coupling. The power coupled into the waveguide was measured as function of substrate temperature. The results are depicted in Fig. 5. In the same figure we compare the results with the thermal characteristics of a similarly mounted laserdiode emitting in free space. As can be noted the characteristics are very similar. The stability of the laser-waveguide coupling is therefore very good. We attribute the good thermal stability to our thermocompression mounting technique and the thin bonding layers used.

4. Conclusions

A ZnO sputter process for the realization of an active optical waveguide material, an efficient laser-waveguide coupling to high contrast waveguides with good thermal characteristics, and a simple detector-waveguide coupling has been developed for our integrated optic sensor toolbox.
After development of these building blocks the maturity of the technique was shown by the realization of the sensor platform. The versatility of the design allows the realization of sensor families.

Some of the developed toolbox elements will also be of value in other areas than the sensor market. An illustrative example is the laser-waveguide coupling, which can be used for e.g. integrated erbium-doped waveguide amplifiers (here we can apply our coupling technique to couple the pump laser to the doped waveguide) or blue-laser light generation (also for coupling the pump laser but now to the non-linear waveguide).

Acknowledgments

The authors would like to thank their colleagues at FEL-TNO: J.L. Joppe and R. Verboom for fruitful discussions and support. The optical chip in the sensor platform and the custom electronics used in the basic sensor platform were kindly provided by University Twente.
This work was in part financially supported by the Royal Dutch Navy and IC Technology Program (IOP Electro-Optics) of the Dutch Ministry of Economic Affairs.

References

[1] C.H. Henry, G.E. Blonder, and R.F. Kazarinov, J. Lightwave Technol. **7**, 1530 (1989).

[2] M.T. Gale, R.E. Kunz, and H.P. Zappe, 'Polymer and III-V transducer platforms for integrated optical sensors', Opt. Eng. **34**, 2396 (1995).

[3] P.V. Lambeck, R.G. Heideman, and T.J. Ikkink, 'Phasemodulated Mach-Zehnder interferometer for sensor applications', *10th Nordic Baltic Conference*, 1996, Medical and Biological Engineering and computing, **34**, 145 (1996).

[4] G.J. Veldhuis, and P.V. Lambeck, 'Highly-sensitive passive integrated optical spiral-shaped waveguide refractometer', Appl. Phys. Lett. **71**, 2895 (1997).

[5] G.J. Veldhuis, J.H. Berends, R.G. Heideman, and P.V. Lambeck, 'An integrated optical Bragg-reflector used as chemo-optical sensor', Pure and Applied Optics, to appear in jan. 98.

[6] R.G. Heideman, R. Lammers, and P.V. Lambeck, 'Electro-optical phase modulation in zincoxide (ZnO)', in *Proc. Nationale Sensor Conferentie*, 231 (1996).

[7] H.H.P.Th. Bekman, K.W. Benoist, and J.L. Joppe, Appl. Surf. Sc. **70/71**, 347 (1993).

[8] J.L. Joppe, A.J.T. de Krijger, and O.F.J. Noordman, Electron. Lett. **27**, 162 (1991).

[9] H.H.P.Th. Bekman, and A.J.T. de Krijger, 'A novel bonding technique for submicron precision mounting of algaas laserdiodes to silicon based waveguides', in *Proc. European Conference on Integrated Optics*, 11.4 (1993).

[10] R. Waters, 'Diode laser degradation mechanisms, a review', Prog. Quant. Electr. **15**, 153 (1991).

[11] A.J.T. de Krijger, and H.H.P.Th. Bekman, 'Hybrid coupling of laser light sources to silicon (oxy) nitride based waveguides', SPIE **3099**, 137 (1997).

A THEORETICAL AND EXPERIMENTAL STUDY OF THE PIEZOJUNCTION EFFECT IN SILICON AND ITS EXPLOITATION IN NEW SENSOR STRUCTURES.

Fabiano Fruett, Fredrik Creemer, Gerard C. M. Meijer and Paddy J. French
Faculty of Electrical Engineering, Delft University of Technology
Mekelweg 4, 2628 CD Delft, The Netherlands[1]
f.fruett@et.tudelft.nl

0. Abstract

The aim of this work is to start a theoretical study of the piezojunction effect, leading to two important applications. The first one is to develop strain sensors with maximum output. The second one is to reduce strain-effects using appropriate design, compensation structures and microelectromechanical systems. The piezojunction effect is a dominant source of the inaccuracy of many analogue circuits, such as bandgap references, DC amplifiers, monolithic temperature transducers and A/D converters. A cantilever structure chip with vertical bipolar transistors is used to analyze the behavior of this device under the different stress levels.

Key Words: Stress in silicon, silicon sensors, piezojunction effect.

1. Introduction

The piezojunction phenomenon quantifies the changes of bipolar transistor characteristics under mechanical stress. A strong influence of local mechanical stress on the electrical properties of pn junction has been observed [1, 2, 3, 4] The most important influence is the change of the base emitter voltage.

The mismatch of TCEs (Temperature Coefficients of Expansion) between package, glue and silicon, causes a strain on the chips at different temperatures, (see figure 1). This strain can affect the accuracy and reliability of the devices dramatically, thus makes the use of low-cost epoxy packaging for accurate devices impossible. As an example, it has been found that it is very difficult to improve the accuracy of bandgap references and monolithic temperature sensors without solving the problems due to the piezojunction effect.

On the other hand, this effect can be used to build pressure and force sensors, where the transistor can be a mechanical-electrical transformer [5]. The phenomenon can be explained by stress-induced changes in the energy band structure of the semiconductor. The studies and experimental results published in the literature have shown the anisotropy nature of the piezojunction effect. These studies considered specific current/strain combinations. However, these combinations do not cover the situation for the effects of, for instance, lateral strain in common bipolar transistors.

STW Project Number: DEL55.3908

In the first step of this work, the effect of stress on the electrical characteristics of the transistors will be measured for different alignments. In these experiments the I-V curves will be measured for several stress levels. Accurate measurement of the piezojunction effect is not easy, because the base-emitter voltages do not only depend on the strain but also on temperature and biasing current. Care must be taken to obtain the required discrimination of the effects to be measured, while eliminating cross effects. The stress will be applied as a modulated signal, and synchronization detection will give the discrimination of temperature drift effects.

The layout and structure of the chip will be designed as a cantilever structure, applying some deflection on the free end of the beam. The strain level of each point can be calculated and the I-V characteristic of each transistor along the bean will be measured.

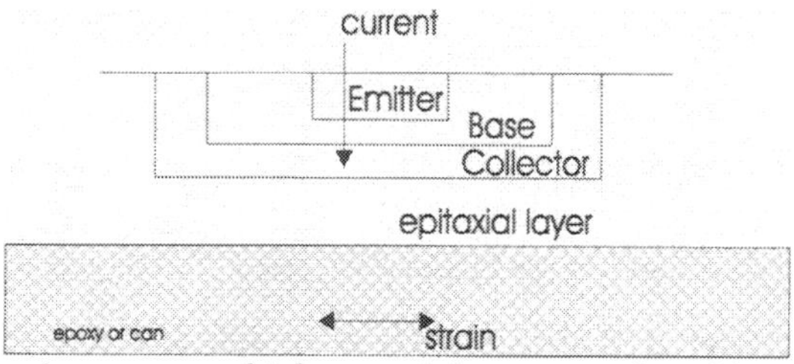

Figure 1. Electronic packages are multiple-material systems, where bi- and tri-material bonded edges and interfaces exist and often introduce stress concentrations. In vertical transistors the direction of the strain caused by the epoxy or can packaging is perpendicular to the direction of the electrical field and current.

2. Theory

2.1. Effect of strain on the silicon junction.

Bipolar transistors and diodes are very interesting devices to use as temperature sensors and band gap references due to the highly predictable and time-independent way in which the base-emitter voltage is related to temperature. In addition, their excellent fabrication compatibility, their small dimensions and their low power consumption make them very attractive in this kind of application. They have also some imperfections such as self-heating, mismatch, small drift after thermal cycling, small amount of non-linearity and piezojunction effect where the mechanical stress causes deviations in base emitter voltage. If the semiconductor is subjected to an anisotropy deformation, several effects can be observed. The main effect is the change in band structure, which has the following consequences:

- change in minority carrier concentration;
- energy level shift;
- change in carrier mobility;
- change in carrier lifetime;

The change in minority carrier concentration is the most important phenomenon caused by this band structure redistribution. Formula (1) illustrates the exponential nature of this phenomenon, where the change in minority carrier concentration is given by:

$$\frac{po_\sigma}{po} = \frac{no_\sigma}{no} = \exp\left(\frac{\Delta E_g}{kT}\right), \quad (1)$$

where:

$$\Delta E_g = \Delta E_v - \Delta E_c = \alpha . P\,, \quad (2)$$

where

ΔE_v	is the change in the energy level of the valence band;
ΔE_c	is the change in the energy level of the conduction band;
P	is the pressure over the junction;
α	is a constant and depends of the crystal orientation
po_σ , no_σ	are the minority-carrier densities under pressure, for n- or p-type material, respectively.
T	is the absolute temperature, and
k	is Boltzmann's constant

Figure 2 shows the simplified band diagram. The energy band structure of Si is more complicated and depends on the orientation in the crystal. The conduction levels are not changed by the same amount in all the directions. We will refrain from a detailed discussion of this subject here.

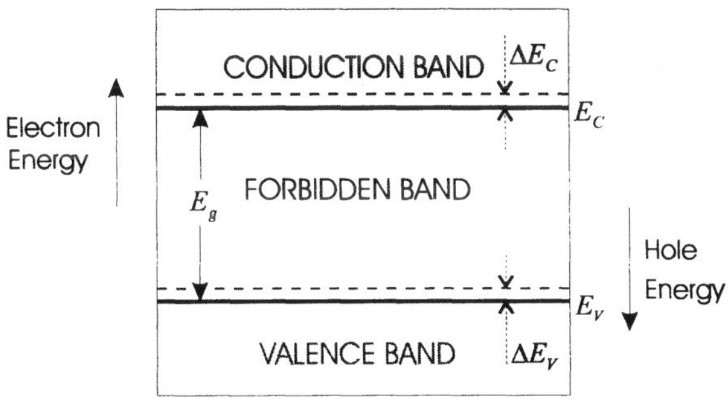

Figure 2. Simplified band diagram.

Considering a uniformly distributed stress σ, the stress-dependent relation for the collector current is:

$$I_{C\sigma} = I_S\left[1 + \exp\left(\frac{\alpha . P}{kT}\right)\right]\left[\exp\frac{V_{BE}}{Vt} - 1\right]. \quad (3)$$

Keeping the base-emitter voltage constant, the collector current will be increased under stress with an exponential dependence.

2.2. Beam bending technique

The chip test uses the beam bending technique where the silicon beam has one fixed end and another free end, it is shown in figure 3. The modulated displacement is applied at the free end using a simple lever.
The lever is controlled using a step motor. The strain at each point on the surface (co-ordinate x) of the beam, can be calculated using:

$$\varepsilon_p = \frac{3ycl(l-x)}{l^3}. \tag{4}$$

Where y is the deflection of the free end of the beam, c is half the beam thickness and l is the beam length.

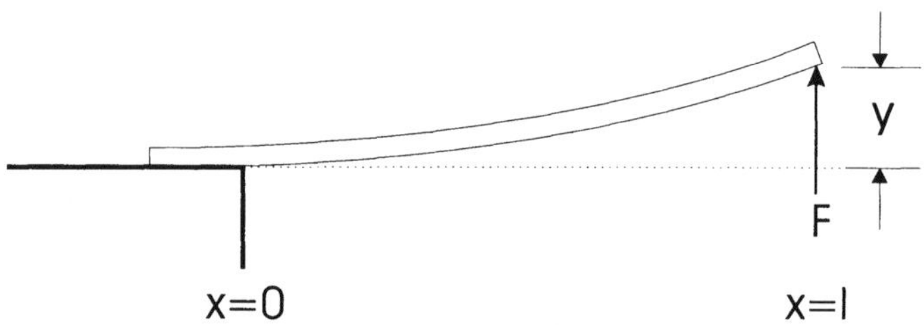

Figure 3. Beam bending technique.

3. Circuit Test Setup

A measurement setup is shown in figure 4(a) and figure 4(b)

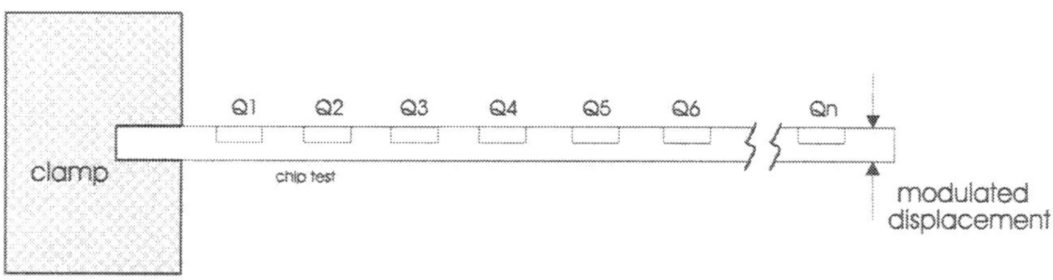

Figure 4(a). Mechanical setup for the test.

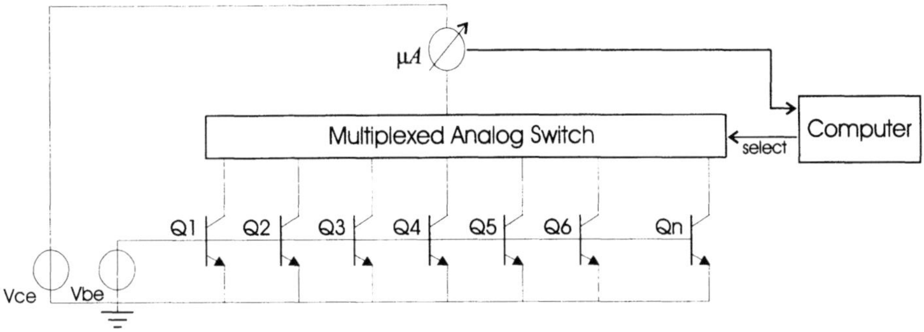

Figure 4(b). Circuit setup for the test.

The circuit setup for the tests is made using a multiplexed analogue switch which is controlled by an external signal (select). The base-emitter voltage is kept constant and the select signal (input of the multiplex) is changed. The collector terminal of each transistor is thus switched one by one with the microampere meter. The collector current through each transistor is measured and stored in the computer. After the multiplex has passed through all transistors the displacement is modified in a mechanical way with a controlled amplitude and a new sweep done. The displacement is modulated using a lever driven for by step motor.

These measurements have to be made in a short time period in order to avoid that small and undesirable temperature changes disturb the measurement.

4. Conclusions

The goal of this work is to form a start point for the development of the [2]theoretical study of the piezojunction effect in silicon and its exploitation in new sensor structures. In next steps methods will be developed to reduce the stress effects by releasing the stress and using compensated structures with electronics circuits and micromachine devices.

References

1. Y. Kanda, Effect of compressive stress on silicon bipolar devices, J. Appl. Phys., 44 (1973) 389-393.
2. Y. Kanda, Effect of stress on germanium and silicon p-n junctions, Jpn. J. Appl. Phys., 6(1967) 475-486.
3. J. J. Wortman, J.R. Hauser, R. M. Burger, Effect of Mechanical Stress on p-n . junction device characteristics, Journal of Applied Physics, 37 (1966) 3884-3892.
4. B. Puers, W. Sansen, Nem mechanical sensors in silicon by micromachining piezojunction transistors, Proceedings of Transducers '87 (1987) 324-327.
5. W. Sansen, P. Vandeloo and B.Puers, A force transducer based on stress effects in bipolar transistors, Sensors and Actuators, 3 (1982) 343-354.
6. P.J. french, Piezoresistence in polycrystalline silicon, PhD thesis, Southampton, UK, 1986

SACRIFICIAL OXIDE ETCHING COMPATIBLE WITH ALUMINIUM METALLIZATION

P.T.J. Gennissen, P.J. French
Lab. for Electronic Instrumentation, Dept. of Electrical Eng., DIMES
Delft University of Technology, The Netherlands
Po. Box 3500, 2600 GA Delft, Netherlands
Tel +31-15-2783342, Fax +31-15-2785755
P.Gennissen@et.tudelft.nl

0. Abstract

This paper reports on sacrificial oxide etching with very high selectivity to aluminium metallization using mixtures of 73% HF and IPA. Etch rate ratios up to 680 have been achieved even for (slow etching) thermal oxide. Thermal oxide etch rates up to 1.8μm/min. are reported. Thick polysilicon accelerometers with aluminium metallization and thermal sacrificial oxide have been made as well as full aluminium microstructures using plasma oxide as sacrificial layer.

Keywords: sacrificial etching, compatible processing, etch selectivity

1. Introduction

Traditionally one of the most difficult fabrication steps in surface micromachining has been the sacrificial etch step and subsequent drying of the free standing structures. Especially when aluminium is used as the interconnect layer it is difficult to selectively etch the sacrificial oxide without attacking the aluminium. Drying of the devices after sacrificial etching also needs special attention because the capillary forces of water can cause the free standing structures to stick to the surface.

Traditional methods to prevent metal attack during sacrificial oxide etching are the protection of the metallization using resist [1], the use of alternative etch mixtures like BHF and glycerine [2] or 'pad-etchants' [3]. However all these methods have disadvantages: Resist protection is only possible when etch times are limited to 10 minutes due to adhesion problems and etching of the resist itself [4]. Special etch mixtures can offer reasonable selectivity, but have a low oxide etch rate HF vapour etching is an additional method which offers a very high selectivity to metal and avoids the sticking problem since no fluid is used to etch the oxide. A disadvantage of this method is the non standard set-up of the process.

A new approach is to etch in very high HF concentrations. In these solutions the concentration of H_3O^+ ions, which are responsible for the attack of aluminium is relatively low, while the concentration of HF, responsible for oxide etching is high. This technique approaches the method of HF vapour etching without the need of special equipment. After sacrificial etching the devices need to be kept wet before they are dried in a special way to prevent sticking. Traditionally the sacrificial etchant is removed by a dilution rinse. In order to keep the devices wet when they are taken out of the wet etch, the fluid in the etch bath needs to have a low-surface tension. IPA and acetone are commonly used for this purpose.

2. Experiments

In our experiments we used the highest commercially available HF concentration (73% manufacturer: Fluka). In order to lower the etch rate and to lower the surface tension of the etchant we diluted the HF with various parts of IPA. The etch rate of thermal oxide and aluminium was measured and the etch selectivity was calculated. The same experiments were performed for mixtures of 40% HF and IPA. The results are summarised in table1. Figure 1 shows the etch rates as a function of the IPA:HF ratio.

Table 1: Etch rates of thermal oxide and aluminium in IPA/HF mixtures

IPA : HF	73% HF			40%HF		
	Ox. (nm/ min)	Al (nm/ min)	selectivity	Ox. (nm/ min)	Al (nm/ min)	selectivity
0 : 1	1500	2.2	680	844	68	12
1 : 1	833	12	69	400	122	3
2 : 1	167	18	9	233	145	1.6

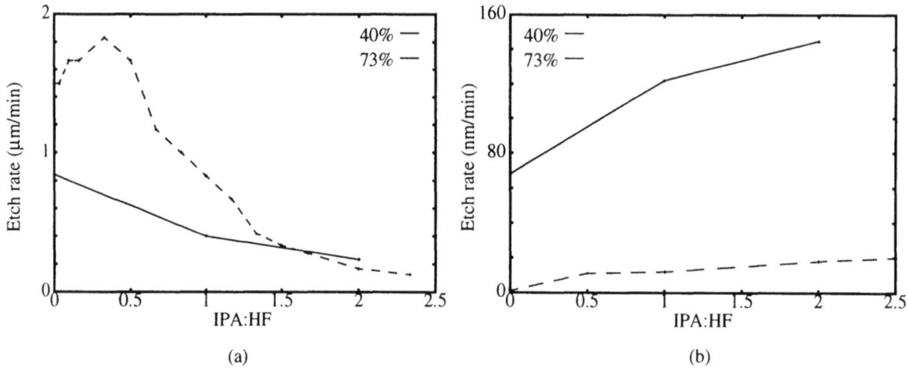

Figure 1: Etch rate for IPA:HF mixtures for thermal oxide (a) and aluminium (b)

The highest selectivity is achieved for pure 73% HF. However, addition of IPA to the HF is preferred since this makes it possible to take the etched sample out of the etchant without the need of dilution rinsing. For 40% HF concentrations (in water) the aluminium etch rate is higher due to the increased H_3O^+ concentration. The oxide etch rate is mostly dependent on the HF concentration in the etchant: 40%HF has almost the same oxide etch rate as 73% HF diluted with 1 part IPA. The initial increase in oxide etch rate for 73% HF with low IPA additions as shown in figure 1 (a) can be explained by the increased wetting capability of this mixture, compared to straight 73%HF.

3. Examples

To demonstrate the usefullness of the1:1 HF:IPA etching mixture accelerometer structures were made using epipoly [5] as structural material and TEOS oxide as sacrificial material. The unprotected aluminium bondpads on the accelerometer were not attacked at all during the 6 min. sacrificial etch which was needed to release the structure. After the 6 min etch the samples were taken out of the etch mixture and put IPA for 15 min, followed by a second 15 min IPA rinse. Finally the samples were put into cyclohexane for 15 min and freeze dried on a -5°C cold plate under a nitrogen flow at atmospheric pressure. In 5 min all the cyclohexane was sublimated and the samples were ready for packaging. A SEM photograph of a vertical capacitive epipoly accelerometer structure with bondwires is shown in figure 2.

Figure 2: SEM photograph of a free etched vertical accelerometer. The aluminium bondpad is not attacked during the sacrificial etch step, which makes wire bonding possible.

4. Conclusions

When high (73%) HF concentrations are used for sacrificial oxide etching an etch selectivity up to 680 to aluminium interconnect can be achieved. Dilution rinsing before taking the sample out of the etchant is not needed when a mixture of 73%HF and IPA is used. This saves chemicals and simplifies the etching process. The samples should not be rinsed in water, since water addition to the HF will result in rapid attack of the aluminium interconnect.

Acknowledgements

The authors wish to thank the staff and processing crew of DIMES Technology Centre for assistance in fabricating the devices. This work is supported, in part, by the Dutch technology foundation (STW), project DEL 44.3578.

References

[1] C. Linder, N.J. de Rooij, 'Investigation on free-standing polysilicon beams in view of their application as transducers', Sensors and Actuators A, 21-23 (1990), pp.1053-1059

[2] H.A.C. Tilmans, K. Beart, A. Verbist, R. Puers, 'CMOS foundry based micromachining', MME'95, Denmark, pp. 183-187

[3] J. Bühler, J. Funk, P. Steiner, P.M. Sarro, H. Baltes, 'Double pass metallization for CMOS aluminium actuators', Transducers '95, June 25-29, Sweden, Vol. 2, pp.360-363

[4] J.F.L. Goossen, B.P. van Drieënhuizen, P.J. French, R.F. Wolffenbuttel, 'etching problems of sacrificial layersin the presence of aluminium interconnect, proc. Eurosensros X, September 8-11, Leuven, Belgium, pp. 271-274

[5] P.T.J. Gennissen, M. Bartek, P.J. French, P.M. Sarro, 'Bipolar -compatible epitaxial poly for smart sensors: stress minimization and applications', Sensors and Actuators A 62 (1997) pp.636-645.

FREE STANDING BEAMS MADE BY SINGLE STEP ELECTROCHEMICAL ETCHING IN HYDROFLUORIC ACID

Hiroshi Ohji, Piet J. Trimp, and Patrick J. French
Electronic Instrumentation Lab., Department of Electrical Engineering, DIMES,
Delft University of Technology, P.O. Box 5031, 2600 GA Delft.
Tel.: 015-278-6432, Fax: 015-278-5755, e-mail: ohji@ei.et.tudelft.nl

0. Abstract

This paper presents a new technique of micromachining using single step electrochemical etching in hydrofluoric acid(HF). Using this etching technique, free standing beams with height, width and length of 40μm, 2μm and 250μm, respectively, are made of single crystal silicon. Basic concept of this etching technique is a combination of anisotropic and isotropic etching mode. First, trenches are made at a certain current density. After the desired depth is obtained, the current density is increased to enlarge the trench width without effecting the width of existing trenches. Finally, the trenches are connected under the beams. Thus free standing beams are obtained. After connection of the trenches, electropolishing occurs under the beams, so smooth silicon surface and desired gap length are also obtained.

Keywords: electrochemical etching, free standing beam, porous silicon, wet etching

1. Introduction

The two main techniques commonly used in micromachining are bulk micromachining and surface micromachining. Both of them have their own advantages and disadvantages. In bulk micromachining, structural parts are made of single crystal silicon which has good mechanical properties, however, the structural shape suffers from crystal orientation. For example, the anisotropic etching for (100) wafer is stopped at (111) planes which are slanted 54.7 degrees. Alternatively, in surface micromachining, the structural parts are made of poly silicon which has mechanical properties inferior to those of single crystal silicon. Gap between the structural parts and the silicon surface is not large which leads to sticking problems. Electrochemical etching solves many of these problems.

The electrochemical etching in hydrofluoric acid(HF) is known as a technique of porous silicon formation which has been studied since the 1950's[1][2]. Recently porous silicon has been used for micromachining or micro devices because of its interesting properties. For example, micro porous silicon can be removed by 1% KOH at room temperature so it is utilized as a sacrificial layer of surface micromachining[3][4]. However, it is difficult to define the porous silicon area with mask because the micro porous silicon is formed isotropically. Alternatively, macro porous silicon which pores are formed anisotropically has been studied by Lehmann[5][6]. Lehmann found that initial pit made by KOH can be used as a starting point for macro pore and some devices using macro porous silicon were fabricated[7][8][9]. However, fabrication of mechanical structures using this etching technique has not been well studied. Recently trench structures using V-shaped groove as a initiation have been demonstrated[10]. In this paper, the fabrication of free standing beams by a combination of anisotropic and isotropic mode are presented.

2. Experimental set-up

The sample is etched in 5% HF under 3V anodic bias[10]. A sketch of the electrochemical etching set-up is shown in Fig. 1. The set-up consists of outer HF bath and inner sample holder. The inner sample holder has a hole to illuminate the back side of the sample with a white light of which the intensity can be varied. Copper strip is put around the hole to make a electrical connection with sample. The sample is fixed on the copper strip by silver epoxy. A platinum grid is used for a counter electrode. Positive voltage is applied between the sample and the counter electrode. Current is adjusted by the light intensity and monitored by a multimeter.

3. Fabrication process

Process flow for electrochemical etching is shown in Fig. 2. Start material is n-type (100) silicon with 10^{14} cm^{-3} phosphorus doping concentration. The effects of the crystal orientation were studied by Chuang in which (100) silicon wafer is good for the vertical etching[11]. Silicon nitride was deposited on the top of the surface and patterned to define initial V-shaped groove by KOH. Vertical walls were formed by electrochemical etching using the set-up shown in Fig. 1. After desired depth was obtained, current density was increased by increasing the light intensity. The width of the trenches increased and trenches were connected under the beams. Thus free standing beams made of single crystal silicon were fabricated with only one mask.

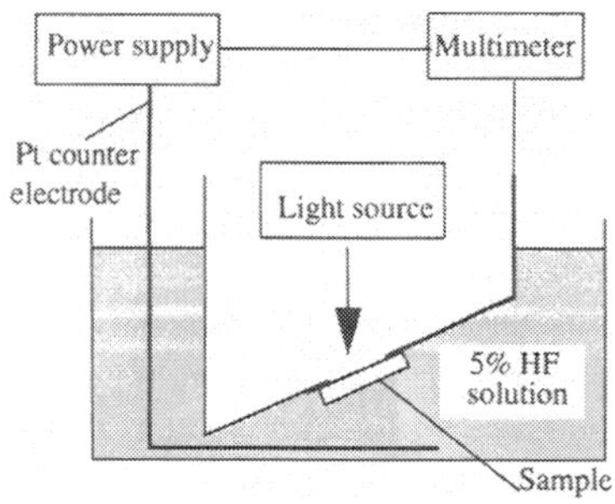

Fig. 1 Electrochemical etching set-up

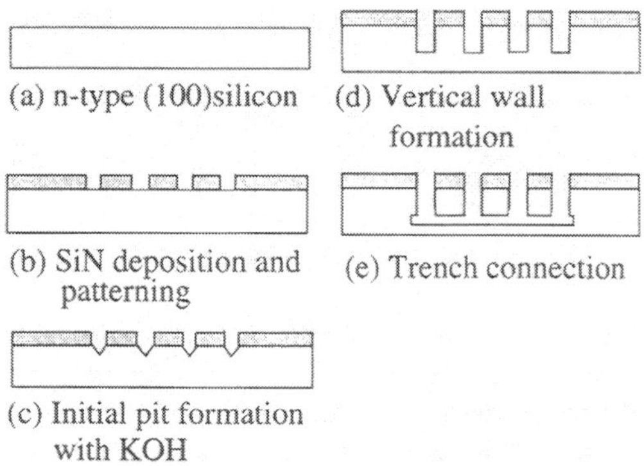

Fig. 2 Process flow for electrochemical etching

4. Single step electrochemical etching

Anisotropic mode

The etching mechanism of this process is as same as that of macro porous silicon formation[12][13]. Figure 3 shows the schematic view of the anisotropic etching mode. In n-type silicon, electronic hole is a minority carrier so the holes which are consumed by the etching are supplied with illumination from back side of sample. The electronic holes generated at the back side of the sample diffuse to the front side of the sample and reach the pore tip, therefore straight pores or trenches can be achieved.

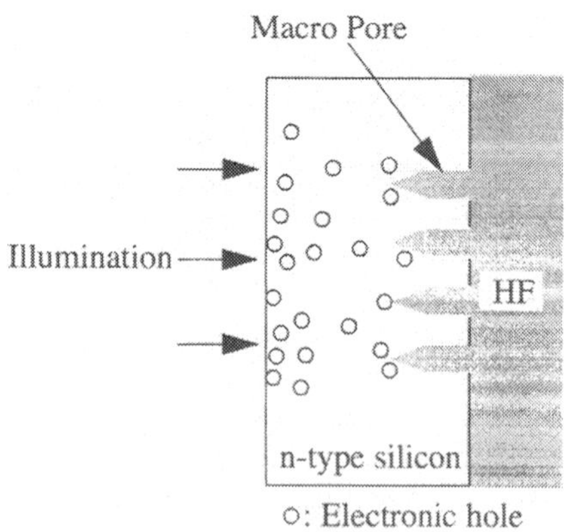

Fig. 3 Schematic view of macro pore formation

Isotropic mode

In this electrochemical etching, pore diameter is controlled by adjusting the light intensity during the etching which is an advantage of this etching technique. If the light intensity is increased, the more electronic holes are generated. Additional holes are used not to increase the vertical etch rate but to enlarge the pore diameter. Etch rate and pore diameter as a function of current density are shown in Fig. 4 and 5, respectively. Initial pits which were used in these experiments were 4μmx4μm square with a pitch of 8μm. The etching takes place under the initial pits and the straight pores are formed. The etch rate stays around 1.6μm/min. against the increase of the current density as shown in Fig. 4. On the other hand, pore diameter is increased with increasing of the current density in Fig. 5. These results indicate that the pore diameter is controlled by adjusting the light intensity. If the current density is decreased gradually during the etching, the pores with taper walls can be achieved and shown in Fig. 6. The taper angle as a function of reduction rate is shown in Fig. 7. Good relationship can be seen between the taper angle and the reduction rate of the current density.

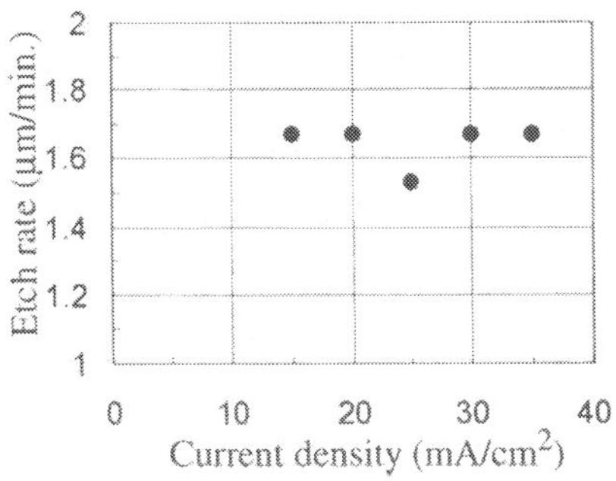

Fig. 4 Etch rate as a function of current density

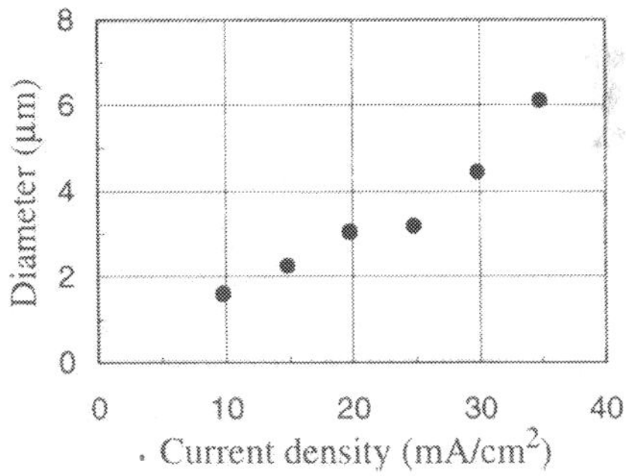

Fig. 5 Pore diameter as a function of current density

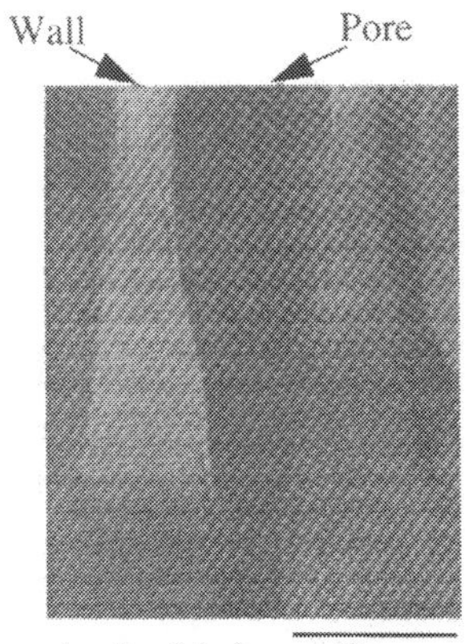

Fig. 6 Pore with taper wall

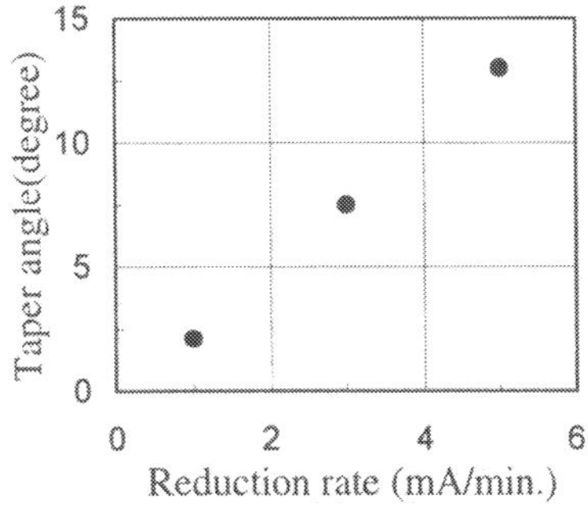

Fig. 7 Taper angle vs. reduction rate of current density

Using this advantage, free standing beams are fabricated by this etching technique. First, the trenches are formed at a certain current density. After the desired depth is obtained, the current density is increased by increasing the light intensity. The width of the trenches is increased under the beams without effecting the width of existing trenches. The connection of the trenches under the beams can be achieved and free standing beams are obtained with only one mask. If the current density is larger than the critical current density, electropolishing occurs at the silicon surface under the beams[14]. Thus the desired gap length and smooth surface are also obtained.

5. Results and discussion

Using this single step electrochemical etching, free standing beams made of single crystal silicon are obtained with only one mask. Figure 8(a) shows two types of free standing beams and close up view of beam tip and clamping point is shown in (b) and (c), respectively. In first step, the current density was 26mA/cm^2 for 30 minutes which was used to form the trenches. In the following step, the current density was raised to 40mA/cm^2 for 7 minutes to connect the trenches. Initial V-groove of the upper beams in Fig. 8(a) was 4μm wide and 8μm pitch and that of the lower beams was 2μm wide and 4μm pitch shown in Fig. 8(d). The free standing beams with height, width and length of 40μm, 4μm(upper), 2μm(lower) and 250μm, respectively, were made of single crystal silicon. Clear gap between the beams and substrate can be seen in Fig. 8(b). In Fig 8(c), clamping point of the beam is not straight and lateral etching can be seen in the anchor. Surface of the large anchor was covered with SiN mask, so the electronic holes which are generated under the anchor diffuse to the clamping point of the beams. The current density around the clamping point is locally higher than any other area of the beams during the trench formation. The lateral etching around the clamping point is caused by this higher current density[15]. This problem should be solved and a two-mask method has been proposed elsewhere[10]. Figure 9 shows the base of lower beams in Fig. 8(a) which looks like a knife edge. This is caused by the transition time in current density. When the light intensity is increased the trench width also increases. Once the current density is stabilized, the etching returns to an anisotropic etch with the new trench width. If the trenches are connected before the constant condition, the excess electronic holes which are not consumed by the etching remain on the silicon surface. The electropolishing occurs under the beams due to the excess electronic holes. Figure 10 shows the silicon surface under the beams. The smooth surface can be seen in Fig. 10 because of the electropolishing.

6. Conclusions

In this paper, the single step electrochemical etching in hydrofluoric acid by way of new technique for micromachining has been presented. Using this technique, the fabrication of free standing beams made of single crystal silicon has been demonstrated. After making the trench structure, the current density was increased to connect the trenches under the structures. As a result of this connection, free standing beams with height, width and length of 40μm, 2μm, and 250μm, respectively, were fabricated with only one mask. Investigation are continuing to improve the profile of the under side of the structures.

Acknowledgment

The authors would like to thank the staff and processing crew of DIMES for assistance in preparing the samples. In addition we wish to thank T.Ozeki and Dr. K.Tsutsumi of Mitsubishi Electric Corporation for their encouragement.

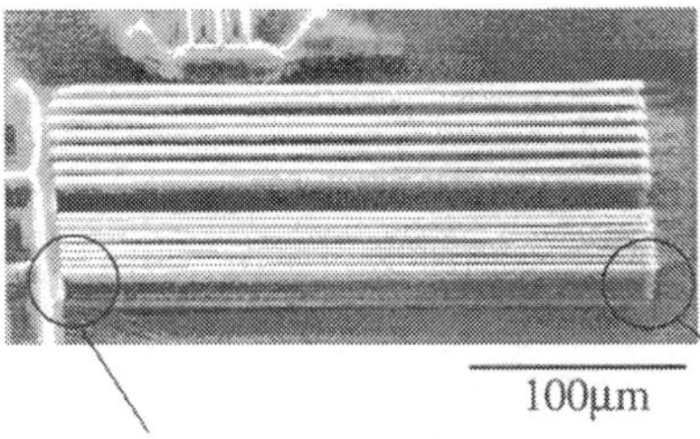

Fig. 8(a) Free standing beams

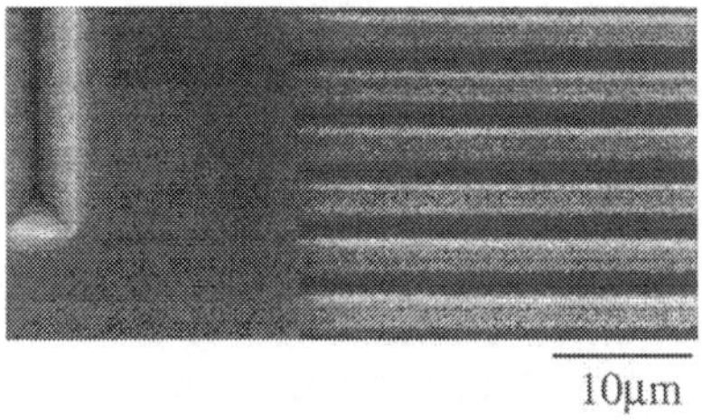

Fig. 8(d) Initial V-shaped grove

Fig. 8(b) Close up view of clamping point

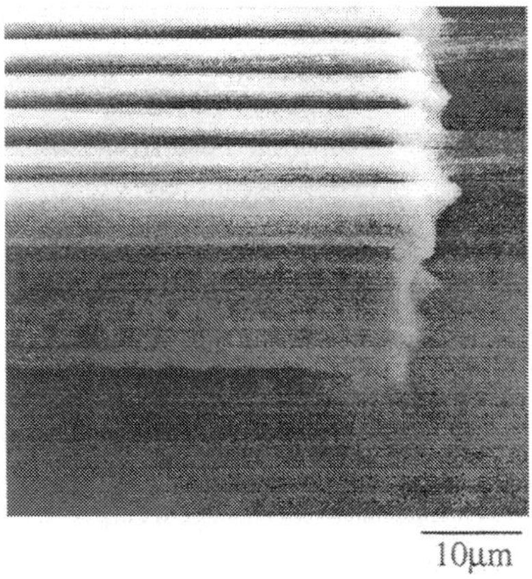

Fig. 8(c) Close up view of beam tip

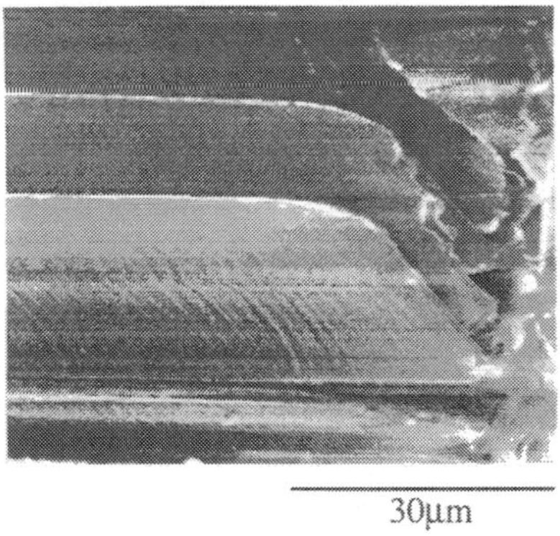

Fig. 9 Base of the free standing beams

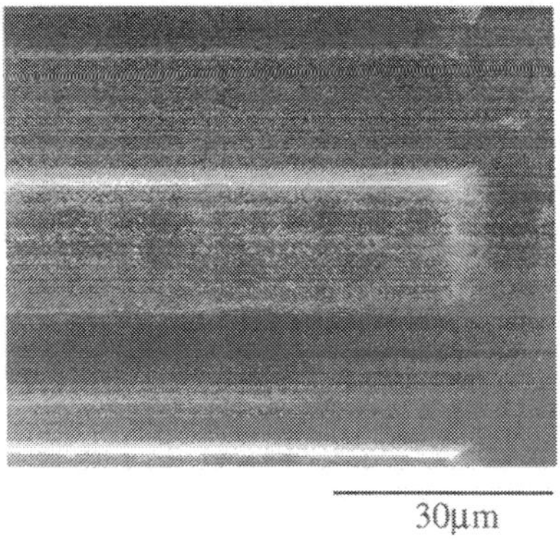

Fig. 10 Silicon surface under the beams

References

[1] A Uhlir, "Electrolytic shaping of germanium and silicon", *Bell Tech. J.* **35**, pp. 333-347, 1956.
[2] D. R. Turner, "Electropolishing silicon in HF acid solutions", *J. Electrochem. Soc.* **105**, pp. 402-408, 1958.
[3] M. Esashi, H. Komatsu, T. Matsuo, M. Takahashi, T. Takishima, K. Imabayashi, and H. Ozawa, "Fabrication of catheter tips and sidewall miniature pressure sensors", *IEEE Trans. Electron Devices* **ED-29**, pp. 57-63, 1982.
[4] T. E. Bell, P. T. J. Gennissen, D. DeMunter, and M. Kuhl, "Porous silicon as a sacrificial material", *J. Micromech. Microeng.* **6**, pp. 361-369, 1996.
[5] V. Lehmann, and U. Gosele, "Porous silicon formation: A quantum wire effect", *Appl. Phys. Lett.* **58**, pp. 856-858, 1991.
[6] V. Lehmann, "The physics of macropore formation in low doped n-type silicon", *J. Electrochem. Soc.* **140**, pp. 2836-2843, 1993.
[7] V. Lehmann, and H. Foll, "Formation mechanism and properties of electrochemically etched trenches in n-type silicon", *J. Electrochem. Soc.* **137**, pp.653-659, 1990.
[8] S. Ottow, V. Lehmann, and H. Foll, "Processing of three-dimensional microstructures using macroporous n-type silicon", *J. Electrochem. Soc.* **143**, pp. 385-390, 1996.
[9] V. Lehmann, "Porous silicon - a new material for MEMS", *IEEE MEMS Workshop '96, San Diego, USA*, pp. 1-6, 1996.
[10] H. Ohji, S. Lahteenmaki, and P. J. French, "Macro porous silicon formation for micromachining", *Proceedings SPIE, Micromachining and Microfabrication Process Technology III, Austin, Texas, USA*, September 1997, pp.189-197.
[11] S. F. Chuang, S. D. Collins, and R. L. Smith, "Preferential propagation of pores during the formation of porous silicon: a transmission electron microscopy study", *Appl. Phys. Lett.* **55**, pp. 675-677, 1989.
[12] M. J. J.Theunissen, "Etch channel formation during anodic dissolution of n-type silicon in aqueous hydrofluoric acid", *J. Electrochem. Soc.* **119**, pp. 351-360, 1972.
[13] R. L. Smith, and S. D. Collins, "Porous silicon formation mechanisms", *J. Appl. Phys.* **71**(8), pp.R1-R22, 1992.
[14] X. G. Zhang, S. D. Collins, and R. L. Smith, "Porous silicon formation and electropolishing of silicon by anodic polarization in HF solution", *J. Electrochem. Soc.* **136**, pp. 1561-1565, 1989.
[15] P. Steiner, and W. Lang, "Micromachining application of porous silicon", *Thin Solid Films* **225**, pp.52-58,1995.

DEVELOPMENT OF A DISPOSABLE BIOSENSOR CHIPCARD SYSTEM

Gijs Dirks, Geert Besselink, Wouter Olthuis and Piet Bergveld
MESA Research Institute, University of Twente
P.O. Box 217, 7500 AE Enschede, The Netherlands

0. Abstract

The proposed system provides a fast, simple and cheap method for on-site determination of protein concentration in aqueous solutions. This paper explains how the ion-step method is performed on Ta_2O_5 ISFETs that are incorporated in a disposable chipcard. An external pump mechanism (stamps) eliminates the need for micro-machined components on the chipcard.

Keywords: ISFET, chipcard, ion-step, biosensor

1. General introduction

The ion-step method is a measuring procedure for the determination of ISFET-bound charge, which has been developed several years ago [1]. Binding of proteins to the bare IFET gate, or to a membrane covered ISFET, introduces shifts in numbers of fixed charges, due to the charged nature of the proteins. This change in charge is proportional to the protein concentration, and can be detected with the ion-step method. An important advantage of the ion-step measurement method is the fast response, which nearly eliminates the influence of drift on the measurement [2].

Achievements have been made toward the development of a laboratory set-up for an ISFET based heparin sensor and protein sensor [3,4]. Recent research has resulted in improved procedures for modifying the ISFET gate with proteins, which can serve as ligand [5]. This enables us to prepare ImmunoFET devices.

These achievements however, require laboratory personnel to do measurements. Logically the next research step is to simplify the measurement procedures. These simplifications lead to the disposable chipcard idea. The use of the proposed system should be fast quantitative measurements at the point-of care, or bedside monitoring. It should be possible for non-laboratory personnel to do the measurements easily.

The intend to make the system disposable has led to a new concept for construction of the system which, in contrary to μTAS systems, does not contain expensive micro-machined pumps and valves. Instead, the chipcard only contains a surface-modified ISFET, a reference electrode, plastic channels and two pods (containers) containing the two required salt solutions (Figure 1). The card can be inserted into a computer-controlled device, which controls two mechanical stamps. A computer controls the stamps as well as measures and processes the ion-step response. The stamps can change the ion concentration near the ISFET gate (the actual ion-step) by pushing the different buffer solutions out of the pods through the channels.

The next paragraphs will give an introduction to the possible construction of the proposed disposable biosensor chipcard system.

The *results & discussion* paragraph will show an ion-step response, measured on a chipcard with a bare ISFET (non modified ISFET). The reproducibility of the responses of bare ISFETs is a good indication for the applicability of the fluidic system to do measurements with modified ISFETs.

2. System design

The chipcard system can be divided into three subsystems: (1) The chipcard itself (2) a measurement device in which the chipcard can be inserted and (3) a computer to control the measurement as well as process the results. An important aspect of the system is the ability to perform more than one ion-step on a chipcard, so averaging the results is an available option to enhance accuracy.

The next three paragraphs will describe the three previously mentioned subsystems.

2.1 The chipcard.

The chipcard is built on a polycarbonate board (60x70 mm). The ISFET can be mounted in the board with epoxy resin (Hysol). The channels are made of 'plain paper copier sheets' (3M) on which the desired structure is printed. On both sides of the sheets, double-sided tape is applied and the sheet is cut into the required pieces. The channels can be constructed by aligning the pieces to the ISFET gate area. After aligning, the channels are covered with another sheet to close the system. This will result in channel cross-sections of about 1000x150 μm. A vent in the waste container is necessary to sustain atmospheric pressure in the waste.

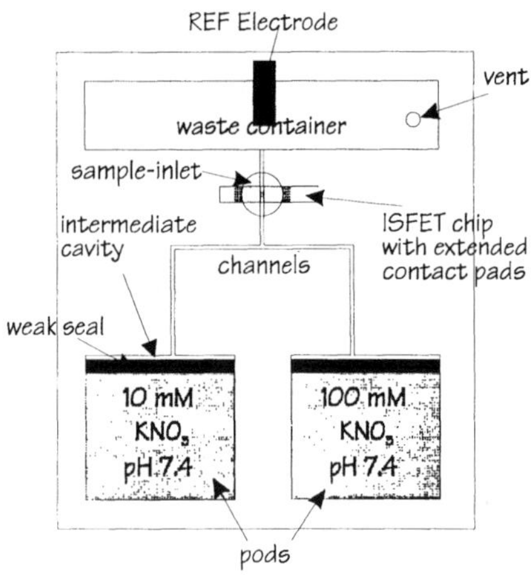

Figure 1 *The cipcard. The sample inlet has not yet been realized.*

The pods with KNO_3 solutions are made according to the 'Polaroid direct ready' photographic technology. They are sealed flexible bags (containing 400 μl) of which one side (the weak seal) will burst when a certain pressure is applied to the pod.

The weak seals are connected to intermediate cavities (the interfaces between the pods and the channels) with silicone rubber (Dow Corning, Silastic 734 RTV) (Figure 2)

The amount of air included in the cavity determines whether air bubbles will disturb the measurement. Experiments have shown that reducing the amount of air included at the start of the measurement to approximately 2 μl was enough to eliminate this problem.

The Ta_2O_5 ISFETs are made according to standard procedures. The source, drain and bulk contacts have been placed on the chip in such a way that the width of the gate can be aligned along with the flow direction. The contacts are made big enough for probes (pogo pins) in the measuring apparatus to contact the chip directly, so there's no need for bonding on the chipcard (Figure 3).

An important part of the chipcard is the reference electrode and its position. In theory, any electrode shows an ion-step response. To avoid the interference of this response with the ISFET response, the reference electrode can be placed in a channel with a constant electrolyte concentration. Alternatively, it can be placed in the waste container where the flow rate is relatively small.

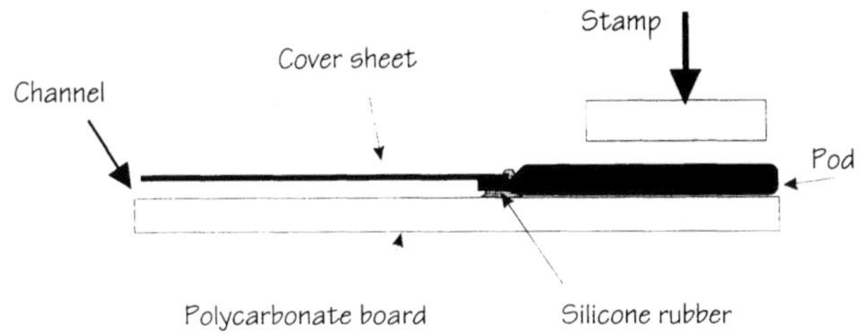

Figure 2 *Connection of a pod to a channel. Note that only in top view (figure 1) the intermediate cavity as mentioned in the text is visible. The grayed rectangle and the arrow represent the mechanical stamp.*

2.2 Measurement device

The measurement device mainly consists of two stamps driven by DC-motors (MAXON RE013-020-13) and an interface, which will be discussed briefly in the next paragraph. To avoid the need for pressure- and conductivity-sensors, which could be used to control flow and detect the opening of the pods, motor current sensing is used. Both angular speed (which represents flow) and momentum (which represents pressure) are reflected in the motor current.

When the stamps start moving at the beginning of the measurement procedure, their function is to open the pods. The bursting of the pods can be detected from a drop in the motor current (caused by the sudden drop in the motor load). A band-pass filter converts this current drop to a voltage peak which drives a Schmitt-trigger that triggers a relay to stop (short-circuit) the motor.

When both pods have opened, and the stamps have stopped, the channels are completely filled with buffer solution. The load the stamp system experiences from then on is nearly constant. Consequently, a feed-forward stamp-control system is good enough to ensure the same flow rate for every ion-step made hereafter.

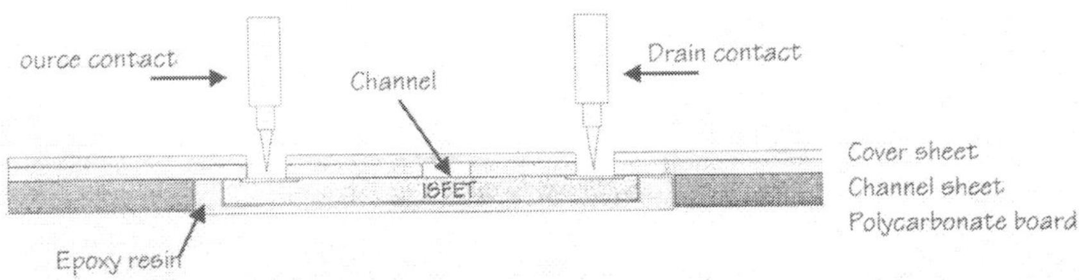

Figure 3 *Cross-section of the chipcard with the ISFET. The probes are within the measurement apparatus and make contact with the ISFET as soon as the card is inserted.*

2.3 Computerized controller

The system is controlled with Labview running on a Windows 95 platform. An electronic interface is controlled through a DAQ-board (National Instruments, lab PC-1200).

The interface comprises the amplifiers for the motors, the pod-burst detectors and stamp position checks. Also a passive version of a non-linear high-pass filter [6] to cut off the DC component of the ISFET signal, which can vary between –1 and 0 Volt is integrated in the interface.

3. Measurement procedure

The experiments performed on chipcards start with the burst detection of the pods (Figure 4), after which a stabilizing time for the ISFET and the high-pass input filter (cut-off frequency = 0.1 Hz) is pursued.

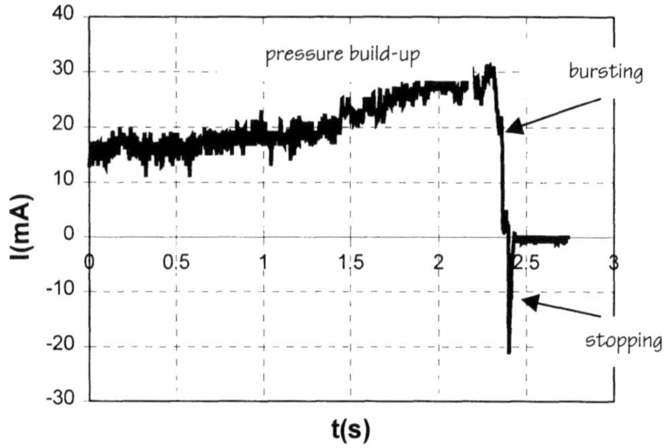

Figure 4 *Current measured during burst detection. The negative pulse after bursting is caused by stopping the motor.*

After bust detection the changing of the buffer concentration starts by activating the stamps. The stamps are driven by 40 ms voltage pulses of 20 Volt. This pulse results in the transport of 50 μl KNO_3 solution. There is an unknown delay between the stamping rate and the flow rate because of the flexible intermediate cavity (and channels).

The stamps have been optimized for a short mechanical time-constant (15 ms) as well as for stiffness, to ensure a fast and tight response of the fluidic system to the applied voltage to the motor. The forces the stamps can exert are much bigger than the resistance of the channels, so the stamps can be considered flow sources rather than pressure (effort) sources.

A measurement sequence is presented in Table 1.

Table 1 *measurement sequence.*

burst detection
stabilizing of the ISFET and amplifier for 5 sec
pre-pumping 100 ms. with both stamps
change the ion concentration by switching the active stamp every second. (40 ms pumping, 960 ms pause for each stamp)
process results
move stamps back up

4. Results and discussion.

Experiments have been done with two positions of Ag/AgCl reference electrodes. The position was either (1) within the channel with the 100 mM KNO_3 solution or (2) in the waste container close to the end of the channel.

The fist case showed ion-step response amplitudes of approximately 60 mV (rise time 150 ms). These responses could not be explained with known ion-step models [3,4]. They may be explained from the liquid junction potential between the two KNO_3 solutions [7], that is being transported along the ISFET when ion stepping.

The second case showed ion-step responses with amplitudes and rise-times similar to the theory and laboratory set-up results (Figure 5). However, the response is disturbed by a second response. This second response can be explained from the position of the reference electrode, where it experiences an ion-step.

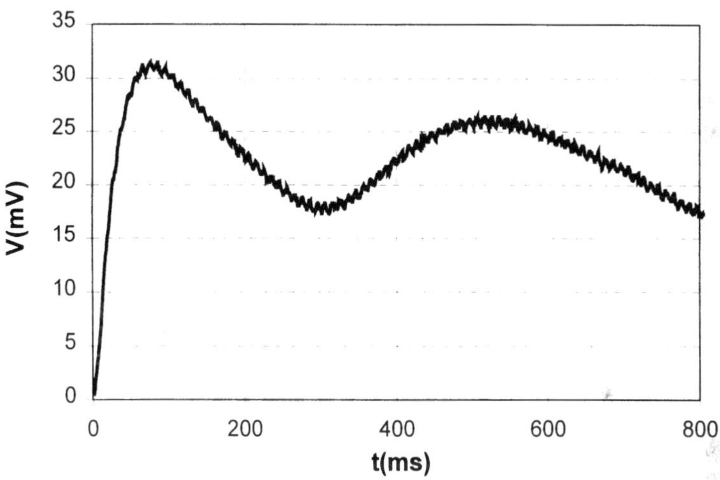

Figure 5 *Ion step (10→100 mM KNO_3) response of a bare Ta_2O_5 ISFET on a chipcard. The second response (maximum) is caused by the reference electrode.*

Acknowledgements

This project is financially supported by the Dutch Technology Foundation (STW).

References

1. R.B.M. Schasfoort, A new approach to ImmunoFET operation, Thesis University of Twente, Enschede, the Netherlands, ISBN 90-9003145-6 (1989)
2. P. Bergveld, Stimulus-Response Measurements with ISFETs, Biocybernetics and Biomedical Engineering, 16 (1996) 5-21
3. J.C. van Kerkhof, The development of an ISFET-based heparin sensor, Thesis University of Twente, Enschede, the Netherlands, ISBN 90-9007514-3 (1994)
4. J.C.T. Eijkel, Potentiometric detection and characterization of adsorbed protein using stimulus-response measurement techniques, Thesis University of Twente, Enschede, the Netherlands, ISBN 90-9008615-3 (1995)

5. G.A.J. Besselink, Covalent coupling of polyamines to aminobead-covered ISFETs, published elsewhere in this book
6. A. Volanchi, Dynamic surface tension measured with single nucleation site electrodes, Thesis University of Twente, Enschede, the Netherlands, ISBN 90-365-0852-5 (1996)
7. J.Koryta, 'Ions, Electrodes and membranes', Wiley Publishers, England, ISBN 0-471-93080-6 (1991)

QUARTZ-TO-SILICON FUSION BONDING FOR MICRO ACOUSTIC WAVE APPLICATIONS

A. Berthold, P.M. Sarro, M.J. Vellekoop

Delft University of Technology
DIMES, Electronic Instrumentation Laboratory
PO Box 5031, 2600 GA Delft, The Netherlands
E-mail: axel@ei.et.tudelft.nl

0. Abstract

A new process for the fabrication of piezoelectric quartz thin films on silicon is investigated. With this process, new silicon-implemented acoustic wave delay lines for sensor applications can be realized. The process consists of low temperature fusion bonding of a piezoelectric quartz wafer to a silicon wafer followed by a back-lap of the quartz to a thickness of between 10 and 50 μm. In this contribution we report on the first step: the fusion bond process.

An annealing temperature of 140°C was found to be sufficient for further micromachining of the wafer stack.

Keywords: micro acoustic wave detector, quartz-to-silicon bonding, low temperature bonding

1. Introduction

In situ monitoring of the concentration of compounds in a chemical solution receives increasing attention in medical science and in the production of foods and chemicals. It opens up the way for automatic process control.

The use of detectors based on physical principles avoids typical problems encountered with biochemical sensors using chemical sensitive layers such as ageing, clogging and the fabrication of a reproducible and stable biochemical interface.

Micro acoustic-wave sensors are often used for the determination of liquid properties such as density and viscosity [1,2,3,4]. An acoustic-wave delay-line consists of two interdigital thin film metal transducers (IDTs) fabricated on a piezoelectric crystal. In order to realize acoustic-wave devices on (non-piezoelectric) silicon, the use of piezoelectric thin films such as zinc oxide, aluminum nitride or PZT has been reported [5,6,7,8]. However, these films often exhibit stress, aging, pinholes, or poor reproducibility which affects the performance of the device. Piezoelectric quartz crystal wafers are known for their high stability and reproducibility of piezoelectric and mechanical properties. Therefore, the realization of piezoelectric quartz thin films on silicon would be very attractive for silicon-implemented micro acoustic-wave detectors. Lamb wave sensors are an example of devices which could benefit from this technique, as will be discussed in the following section.

2. Thin-film piezoelectric-quartz on silicon for Lamb-wave sensors

Lamb-wave delay-lines consist of a membrane in which interdigital transducers are realized. The membrane is typically made of silicon or silicon nitride by using silicon bulk etching techniques. For the generation and detection of the acoustic Lamb waves, a piezoelectric film is required. The fabrication sequence of the Lamb-wave device proposed in this paper is shown in Fig.1.

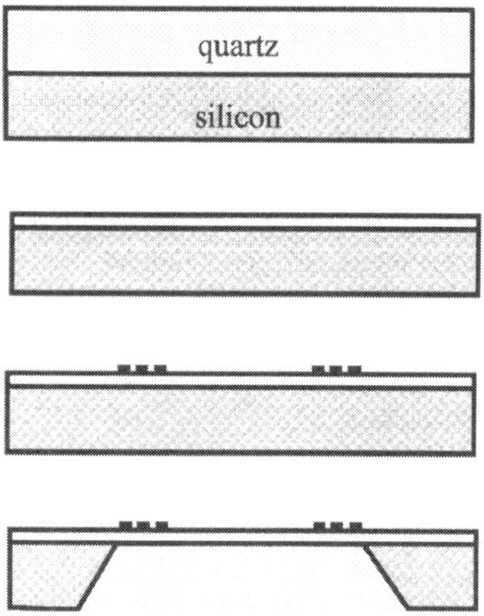

Fig. 1 Fabrication sequence of the lamb wave device: Quartz and silicon are fusion bonded after which the quartz is lapped back to a thickness of 10 to 50 μm. The metal IDTs are fabricated by a metal deposition followed by an etch step. The last step consists of silicon bulk etch process from the back.

Starting materials of the Lamb-wave device are a piezoelectric ST-cut quartz wafer and a silicon wafer which are bonded together at low temperatures. In the following section this is described in more detail. The wafer stack is annealed in order to obtain a high bond strength, after which the quartz wafer is lapped back to a thickness of between 10 and 50 μm. The next step is metallization of the quartz wafer, followed by an etch step to form the IDT structures. In order to construct the quartz membranes, the silicon bulk material is etched from the back. This can be done by using KOH-etching. The masking material on the silicon is silicon dioxide, the etching will stop automatically when the quartz material is reached.

3. Process and results

We have bonded quartz wafers to bare silicon wafers. For the experiments, we used ST-cut quartz wafers, 100 mm diameter with a thickness of 500 micron and standard silicon wafers, p-type, 2 - 5 Ωcm, 100 mm diameter with a thickness of 525 micron and a roughness of 0.2 nm. The process consists of three main steps which are described in the following section. The bond process has been applied to silicon-to-silicon [9] and silicon to oxidized silicon [10]. In the process describes here, the oxide etch time has been adjusted to the etch rate of quartz.

Surface treatment:

The wafers were etched in an ammonium fluoride etch mixture for 1 minute in order to remove the native oxide at the silicon wafers and slightly etch the quartz wafers. The wafers are rinsed in deionized water for 5 minutes followed by immersion in 100% nitric acid for 10 minutes and a second water rinsing for 1 hour. The wafers are now hydrophilic and ready for prebonding.

Prebonding:

Immediately after spin drying the wafers are prebonded by bringing them into contact using a small force (weight) at room temperature. After prebonding the bond strength is sufficient for further handling of the wafer stack.

Bonding:

In order to obtain strong bonding, the prebonded wafer stack is annealed in a vacuum furnace at 140°C for 3 hours at a pressure of 100 mbar. The thermal ramp, both up and down, does not exceed 30 degrees per hour.

Due to the different thermal extension coefficient of both materials, the annealing of the wafer stack is a critical step in the production of thin film quartz on silicon devices. The prebonding takes place at room temperature. During annealing, bending of the wafer stack is observed. After cooling down to room temperature, the stack is flat again. The stress in the quartz and in the silicon is expected to be very low, therefore.
The bond strength was checked with the razor blade test [11]. The bonding showed to be sufficient strong for reliable further processing (e.g. micromachining of silicon).

4. Conclusions

A piezoelectric-quartz to silicon fusion-bonding process for the fabrication of silicon-implemented micro acoustic wave delay-lines is presented in this paper. The low annealing temperature enables merging materials with different thermal extension coefficient.

This opens up the way for the fabrication micro acoustic devices using a thin film of piezoelectric single crystalline quartz with its stable properties instead of sputtered layers such as zinc oxide, aluminum nitride or PZT.

Further work will be concentrated on back lapping of the quartz and micromachining of the wafer stack.

Acknowledgments

The authors would like to thank staff of the DIMES Technology Centre, ICP-group for technical support.

This research is supported by the Technology Foundation STW, applied science division of NWO and the technology program of the Ministry of Economic Affairs. Project: BIOMAS, no. DST66.4351.

References

1 W.P. Mason, Measurement of the viscosity and shear elasticity of liquids by means of a torsionally vibrating crystal, Proc. A.S.M.E., 1947, 359-370

2 J.W. Grate, S.J. Martin, R.M. White, Acoustic wave Microsensors Pt. I, Anal. Chem. 65 (21), 1993, 940-948

3 J.W. Grate, S.J. Martin, R.M. White, Acoustic wave Microsensors Pt. II, Anal. Chem. 65 (22), 1993, 987-996

4 M.J. Vellekoop, G.W. Lubking, P.M. Sarro, A. Venema, Evaluation of liquid properties using a silicon Lamb wave sensor system, Sensors and Actuators A43, 1994, 175-180

5 D.R. Evans, M.F. Lewis, E. Patterson, Sputtered ZnO surface wave transducers, Electron. Lett. 7 (18), 1971, 557-558

6 H.M. Liaw, W. Cronin, F.S. Hickernell, The SAW characteristics of sputtered aluminum nitride on silicon, Proc. IEEE Ultrasonic Symp., Baltimore, USA, 1993. 267-271

7 Ph. Luginbuhl, S.D Collins, G.-A. Racine, M.-A. Gretillat, N.F. de Rooij, K.G. Brooks, N. Setter, Flexural-plate-wave actuator based on PZT thin films, Proc. IEEE MEMS, Nagoya, Japan, 1997, 327-332

8 M.J. Vellekoop, Microfabrication of acoustic-wave devices, Proc. SPIE Symp. on Micromachined Devices and Components III, Austin, USA, 1997, 90-97

9 A. Berthold, M.J. Vellekoop, IC-compatible wafer-to-wafer bonding, Sensors and Actuators A 60, 1997, 208-211

10 A. Berthold, M.J. Vellekoop, IC-compatible wafer-to-wafer fusion bonding with SiO_2 insulating layer, Eurosensors XI, Warsaw, Poland, 1997, 1373-1376

11 W.P. Maszara, G. Goetz, A. Caviglia, J.B. McKitterick, Bonding of silicon wafers for silicon-on-insulator, J. Appl. Phys. 64 (10), 1988, 4943-4950

CYCLODEXTRINS AS RECEPTORS ON SURFACE ACOUSTIC WAVE DEVICES

Hubertus F.M. Nelissen, Menno R. de Jong, Fokke Venema, Martinus C. Feiters, and Roeland J.M. Nolte
Department of Organic Chemistry, Nijmegen SON Research Center,
University of Nijmegen, Toernooiveld 1, 6525 ED Nijmegen, The Netherlands
e-mail : bartn@sci.kun.nl

0. Abstract

Cyclodextrins (CDs), which bind organic substrates in their cavities, can be used to give mass sensitive sensor devices a selective coating. We have tested this principle for the detection of toluene vapour by a Quartz Crystal Mircrobalance (QMB) with crystals coated with an amorphous layer of silylated CDs. We also describe the covalent linking of a CD monolayer to a quartz crystal for applications in a Surface Acoustic Wave (SAW) sensor.

Keywords: Cyclodextrin, Quartz Crystal Microbalance, Surface Acoustic Wave

1. Introduction

Cyclodextrins are cyclic oligomers of α-D-glucose. They are torus shaped and contain a cavity which can be used as a receptor to bind various organic substrates. The most common CDs are α-, β-, and γ-cyclodextrin, which differ in the number of glucose units (6, 7 and 8, respectively) and the size of their cavity (Figure 1). The difference in cavity size is reflected in the binding properties of the different cyclodextrins. α-CD with the smallest cavity only shows affinity for small molecules, whereas β- and γ-cyclodextrin can also bind larger molecules.

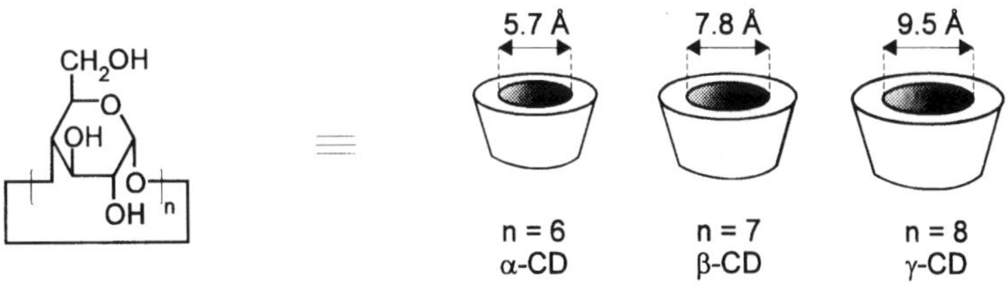

Figure 1 : Structure of α-, β- and γ-Cyclodextrin.

We have recently reported on the functionalisation of cyclodextrins with different groups [1-3]. In the synthesis we protect all primary OH-functions with *tert*-butyldimethylsilyl chloride. The resulting silylated cyclodextrins are soluble in organic solvents, which simplifies their purification, and allows us to introduce a functional group on one of the secondary OH-functions. The secondary hydroxy groups are located on the broad side of the cyclodextrin, which is the most accessible side for binding of substrates. Especially of interest to us are cyclodextrin derivatives with functional groups that can have additional interactions with a substrate. This would enable us to increase the sensitivity for certain substrates.

We want to use these modified cyclodextrins as receptors on a mass sensitive sensor device for the detection of volatile organic compounds in gas-phase and for pesticides in water. In principle four mass sensitive devices are available [4] of which the Quartz Crystal Microbalance (QMB) [5] and the Surface Acoustic Wave (SAW) [6] are the most frequently used. The setup of both devices is shown in Figure 2. With a modified SAW device it is possible not only to detect volatile organic substrates but also organic pollutants in aqueous solution. This setup uses the so called Love Waves [7].

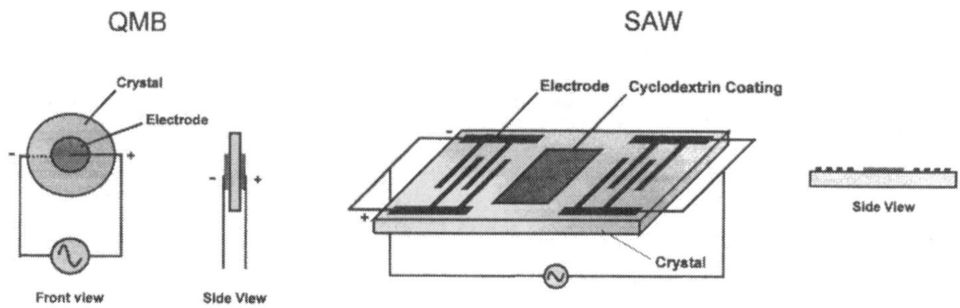

Figure 2 : Setup of the Quartz Crystal Microbalance (QMB) and Surface Acoustic Wave (SAW).

2. Results and discussion

In order to test the possibility of using CDs as a recognising coating we used an amorphous coating of α-, β- and γ-CD on a QMB to detect volatile organic compounds. The cyclodextrins were silylated as described above and applied as a solution in dichloromethane after which the solvent was allowed to evaporate.

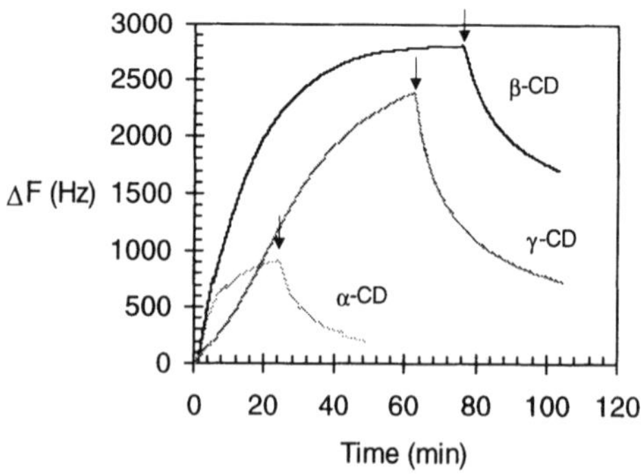

Figure 3 : QMB response of various coatings to toluene; Arrows indicate the moment that fresh nitrogen gas is passed over the sensor to clean it.

Upon binding of substrate in the CD-cavities the mass of the crystal increases and the resonance frequency of the device changes. The experiments show that the coatings respond differently to toluene vapours as a result of the difference in cavity size as can be seen in Figure 3. β-CD gives the highest response due to a perfect match in cavity size and the molecular dimensions of toluene. The cavity of α-CD is too small to accomodate toluene and as a result its response is low. The response of γ-CD is higher but not as high as that of β-CD, because the γ-CD-cavity is too large. The fact that the adsorption and desorption rate are slow for toluene compared to *e.g.* ethanol is not well understood but may have to do with the lack of order in the coating.

In an amorphous coating not all CD-cavities are accessible for binding of vapour molecules. Organisation of the CD-units in a monolayer would increase the accessibility, because all cavities are exposed to the gas stream. The mass increase of binding of substrates on the monolayer, however, is too small to be detected by a QMB device. Therefore the more sensitive Surface Acoustic Wave (SAW) transducer will be used.

To secure the stability of the sensors, the CDs are covalently linked to the surface of the SAW device. This involves the modification of the quartz surface of the SAW device with reactive amino groups, which can subsequently be reacted with a functionalised cyclodextrin (Figure 4).

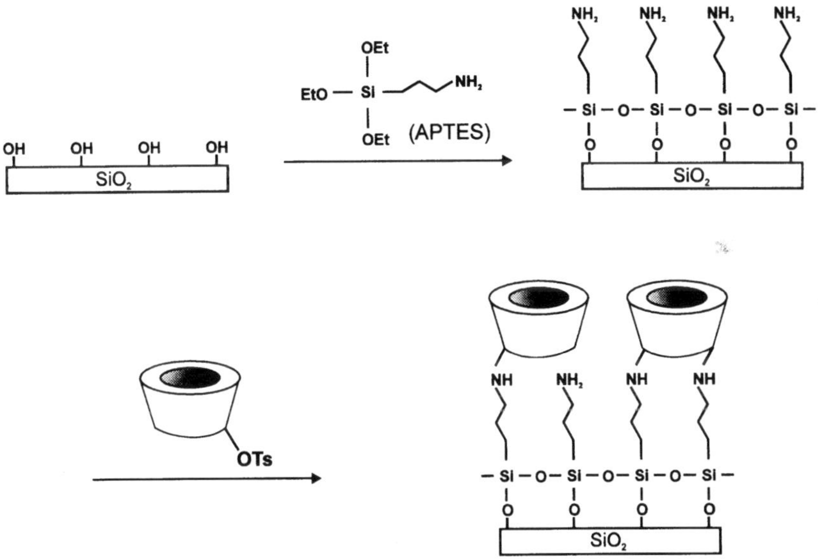

Figure 4 : Reaction scheme for the modification of glass substrates with cyclodextrins.

To test the proposed reaction scheme, glass slides were cleaned with piranha solution (H_2O_2/H_2SO_4, 1:2, $^v/_v$, 1 hour). After drying for 2 hours in a stove, the slides were dipped in a 1% solution of aminopropyltriethoxy silane (APTES) in toluene for 1 hour. The slides were washed with toluene to remove physisorbed APTES and dried in a stove again for one night to complete the formation of the siloxane network. The contact angle of the glass did not change after this treatment, as can be expected since their is only a small change in polarity going from a silanol to amino covered surface. By means of reflectance IR we studied the presence of alkyl groups on the surface by looking at the CH stretch vibrations. Unfortunately we did not see any signals in this region. This might be due to the fact that even when a complete monolayer is formed, the amount of CH_2 groups is below the detection limit of the apparatus. The characterisation of the APTES coated glass as well as the product of its reaction with tosyl-CD are in progress.

3. Conclusions

We have demonstrated that the size of the cyclodextrin cavity can be a factor determining the selectivity of the coatings of mass sensitive sensors. We also describe a strategy towards more ordered mono-layered coatings which are expected to lead to more sensitive and selective systems and higher adsorption/desorption rates.

Acknowledgements

The authors thank the Netherlands Technology Foundation (STW) for financial support.

References

1. F. Venema, C.M. Baselier, E. van Dienst, B.H.M. Ruel, M. C. Feiters, J. F. J. Engbersen, D. N. Reinhoudt and R. J. M. Nolte, Synthesis and Binding Properties of Novel Cyclodextrin Dimers, Tetrahedron Letters, 35 (1994) 1773-1776.
2 F. Venema, C.M. Baselier, M. C. Feiters and R. J. M. Nolte, Synthesis and Conformational Behaviour of Novel Cyclodextrin Hetero-dimers, Tetrahedron Letters, 35 (1994) 8661-8664.
3 E. van Dienst, B.H.M. Snellink, I. von Pickartz, M.H.B. Grote Gansey, F. Venema, M.C. Feiters, R.J.M. Nolte, J.F.J. Engbersen and D.N. Reinhoudt, Selective functionalisation and flexible coupling of cyclodextrins at the secondary hydroxyl face, Journal of Organic Chemistry, 60 (1995) 6537-6545.
4 J.W. Grate, S.J. Martin and R.M. White, Acoustic Wave Microsensors Part I, Analytical Chemistry, 65 (1993) 940A-948A.
5 F.L. Dickert and A. Haumschild, Sensor Materials for Solvent Vapor Detection—Donor-Acceptor and Host-Guest Interactions, Advanced Materials, 5 (1993) 887-895.
6 M.S. Nieuwenhuizen and A. Venema, Surface Acoustic Wave Chemical Sensors, Sensors and Materials, 5 (1989) 261-300.
7 G. Kovacs, M.J. Vellekoop, R. Haueis, G.W. Lubking and A. Venema, A Love Wave Sensor for (Bio)chemical Sensing in Liquids, Sensors and Actuators A, 43 (1994) 38-45.

A LOW COST SMART INTERFACE FOR ACTIVE SENSORS

Xiujun Li and Gerard C.M. Meijer

Faculty of Electrical Engineering, Delft University of Technology
Mekelweg 4, 2628 CD Delft, The Netherlands
email: lixj@duteela.et.tudelft.nl

0. Abstract A low cost interface for active (voltage- and current-generating) sensors, based on the use of a microcontroller and an universal transducer interface, has been presented. The proposed interface acts as an asynchronous voltage- or current-to-digital converter employing a current-to-voltage converter, a voltage-controlled oscillator (VCO), switch-control circuit and a microcontroller. A prototype of the interface which can measure voltage and current signals has been built and tested. The measurement results are presented.

Keywords: smart interface, sensors, voltage and current measurements.

1. Introduction

Recently, some interesting novel sensor interfaces have been presented in which the measurand is converted into a period-modulated output signal [1 – 3]. The advantage of this way of modulation is that powerful low-cost processing circuits, such as microcontrollers can directly process these signals without needing an extrinsic A/D converter. The combined features of such an interface and a microcontroller enable smart signal processing. Using the memory and processing facilities of the microcontroller many nonidealities of the analog front end in the interface can be eliminated or significantly reduced.

A Universal Transducer Interface (UTI) which recently has been presented in [1] has especially been designed to support passive sensing elements, such as capacitive, resistive and resistive-bridge sensing elements.

In this paper, we present a low-cost interface based on the use of the UTI, but modified in order to process also signals of current and voltage-generating sensors, such as thermo-couples.

2. Measurement Concept

2.1 The capacitance measurement using UTI

The Universal Transducer Interface (UTI) is a sensor-signal-to-time converter, based on a period-modulated oscillator, which provides interfacing for the capacitive sensors, Platinum resistors, Thermistors, resistive bridges and potentiometers [1].

To perform the calibration of offset and gain of the interface, a reference signal and a constant part (including offset voltages) are measured in exactly the same way as the sensor signal during two additional phases. During the first phase, the offset of the interface is measured. During the second phase, the reference signal, which can be the resistance or capacitance, is measured, and during the last phase the sensor signal itself is measured. These

This project is supported by the Dutch Technology Foundation (STW) and Smartec B.V., The Netherlands.

phases are automatically controlled and outputted by the UTI itself. The output signal of the UTI is very simple and has a discrete amplitude (V_{CC}), so it is compatible with microcontrollers.

As an example, Figure 1 shows a setup for the measurement of small capacitance by using the mode C23 of the UTI, and the output signal of the UTI with a complete cycle consisting of three phases.

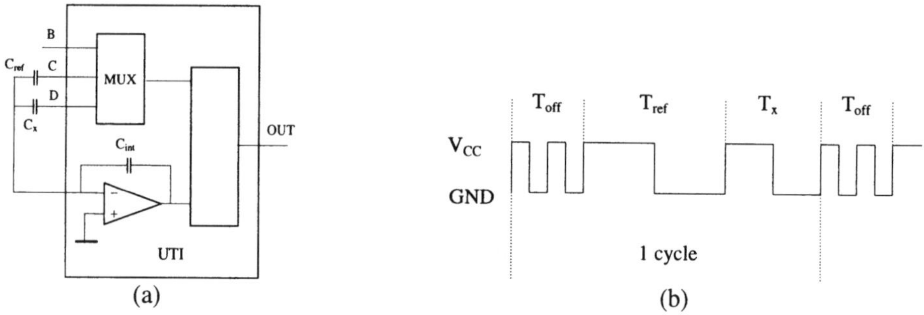

Figure 1. (a) A setup for the measurement of capacitance using the UTI, (b) The output signal of the UTI for a 3-phase mode.

The time interval of each phase is proportional to the measured signal, and the time intervals of the three phases are given by:

$$\begin{aligned} T_{off} &= NKC_0 \\ T_x &= NK(C_x + C_0) \\ T_{ref} &= NK(C_{ref} + C_0) \end{aligned} \quad , \tag{1}$$

where C_x is the capacitive sensor signal to be measured, C_{ref} the reference capacitance, C_0 a constant part (including offset capacitances etc.) and K the transfer coefficient. The factor N represents the number of internal oscillator periods in one phase. The microcontroller samples the output signal of the UTI by counting the number of internal clock cycles fitting in each phase. This results in the digital numbers N_{off}, N_{ref} and N_x. The capacitance ratio C_x/C_{ref} is calculated by the microcontroller:

$$M = \frac{N_x - N_{off}}{N_{ref} - N_{off}} = \frac{C_x}{C_{ref}} . \tag{2}$$

This ratio does not depend on the constant part and the transfer coefficient. In fact, the interface is calibrated for the offset and gain. Even in the case of drift or other slow variations of the offset and gain, these effects are eliminated also.

The three phases are time-multiplexed, as depicted in Figure 1. The offset phase with two short intervals is also used as a synchronization signal for the microcontroller.

2.2 Voltage measurement

Based on the use of the UTI, Figure 2 shows a schematic diagram of an interface for the measurement of the voltage and current signals. This interface acts as an asynchronous voltage or current-to-digital converter employing a current-to-voltage converter, a voltage-controlled oscillator (VCO), switch-control circuit and a microcontroller. The VCO with a period-modulated output signal is formed by the UTI, capacitor C_s and switches S_x, S_{ref}. The microcontroller is used to measure the output signal from the UTI, to process data, to control the switch-control circuit and to output the measured sensor signal. For the measurement of voltage and current signals, the UTI is set as a charge controlled oscillator, which converts the charge of the capacitor C_s to a output period.

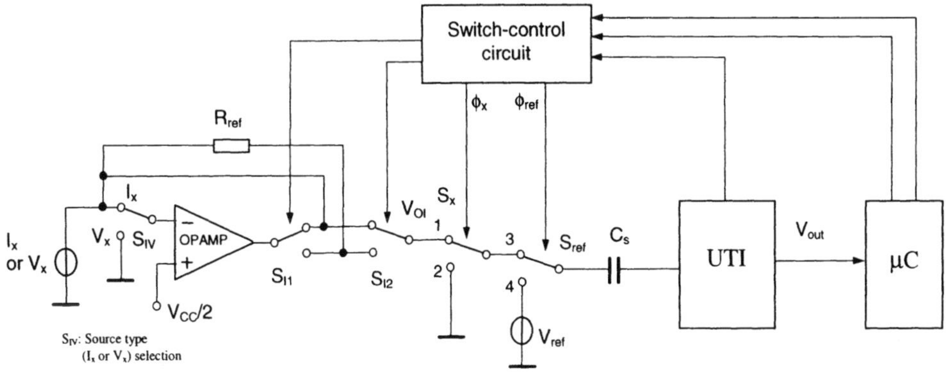

Figure 2 A schematic diagram of an interface for the measurement of the voltage and current signals by using the UTI.

When a voltage signal V_x is measured, the measurand V_x is connected to the V_{OI} (see Figure 2). As presented in section 2.1 and ref. [2], for the measurement of voltage signal V_x, the three-signal technique [4] is also applied to ensure the accuracy and reliability of the circuit by eliminating the gain and offset parameters of the interface for the voltage signals.

To perform the three-signal technique at the voltage level, a reference signal (V_{ref}) and a constant part (including offset voltages) are measured in exactly the same way as the signal to be measured ($V_{OI} = V_x$) during the total of three phases.

In Table 1, three different periods of the output signal form the UTI, T_{off}, T_x and T_{ref}, and the corresponding switch situations are listed.

Table 1 The output of the UTI and switch situations.

S_X	S_{ref}	T_{OUT}
2	3	$T_{off} = 2\frac{V_{cc}}{I_{ch}}C_{off} + 4t_d$
ϕ_x	3	$T_x = 2\frac{C_s}{I_{ch}}V_x + T_{off}$
2	ϕ_{ref}	$T_{ref} = 2\frac{C_s}{I_{ch}}V_{ref} + T_{off}$

where V_{CC} is the power supply voltage, I_{ch} is the amplitude of the charge/discharge current, t_d is the delay time in the oscillator loop of the UTI. These periods of the output signals are measured by the microcontroller.

With table 1, the measured voltage V_m is found from

$$V_m = \frac{T_x - T_{off}}{T_{ref} - T_{off}} V_{ref} = V_x . \tag{3}$$

Formula (3) shows that the inaccuracies of the capacitance C_s and the charge/discharge current I_{ch} do not affect the measured result because of the applied three-signal technique.

Moreover, In ref.[1, 2] it is shown that the effect of other nonidealities sources which cause additive or multiplicative errors in the UTI is also eliminated by the three-signal technique.

2.3 Current measurement

In Figure 2, an OPAMP, an accurate resistor R_{ref}, and switches S_{II} and S_{IV} form a current-to-voltage converter for the measurement of the current signal I_x with two different conversion factors. The output voltages V_{OI1} and V_{OI2} of these two converters are given by

$$\begin{aligned} V_{OI1} &= \frac{1}{2}V_{CC} + V_{os} - R_{ref} \cdot I_x \\ V_{OI2} &= \frac{1}{2}V_{CC} + V_{os} \end{aligned} \tag{4}$$

where V_{os} is the offset of the OPAMP. It is same as the measurement of the voltage signal, two additional signals, T_{off}, and T_{ref}, are also measured in same way. Then the measured current I_m by the interface described in Figure 2 is found from

$$I_m = \frac{T_{OI2} - T_{OI1}}{T_{ref} - T_{off}} \cdot \frac{V_{ref}}{R_{ref}} = I_x \tag{5}$$

where T_{OI1} and T_{OI2} represent the output periods of the UTI corresponding to the voltage signals V_{OI1} and V_{OI2}, respectively.

From equation (5), it is shown that the effects of V_{CC} and V_{os} as well as the effects of the additive or multiplicative errors of the rest of the interface, are eliminated. In fact the three-signal technique has been extended to a four-signal technique to eliminate the additional parameter ($V_{CC}/2 + V_{os}$) of the I–V converter.

3. Measurement Results

A prototype has been implemented based on the circuit shown in Figure 2. The sampling and selection switches were implemented with a simple quad bilateral switch (CD4066). A simple dual OPAMP (TLC272AC) with an offset of 5 mV (maximum value) was used for the current-to-voltage converter. The switch-control circuit was implemented using standard CMOS gates. The UTI is set to work as an oscillator with a oscillating frequency of about 20 kHz to 50 kHz, depending on the measurand and the measurement phase. A microcontroller of the type INTEL D87C51FC with a 5 MHz counting frequency, is employed to measure the output period of the UTI, to process the data, to control the logic circuitry and to communicate with the outside digital world. The interface is powered by a single 5 V supply voltage.

Experimental results show that the interface is able to measure voltages in the range of 0 V to 5 V, with an accuracy and resolution of $\pm 64\times10^{-6}$ and 20×10^{-6}, respectively and to measure currents in the range of 0 mA to 20 mA, with an accuracy and resolution of $\pm 145\times10^{-6}$ and 60×10^{-6}, respectively. The measurement time is about 100 ms.

4. Conclusions

In this paper a low-cost smart interface for the measurement of DC voltage and current signals has been presented. The use of the microcontroller, UTI, and the concept of the three-signal and a four-signal measurements enables the elimination of many undesired nonidealities of the complete interface. A prototype of the interface has been built and tested. A relatively high degree of the resolution and accuracy has been achieved for the measurement of voltage and current signals at a measurement time of about 100 ms. The proposed interface is very suited to be implemented in low-cost integrated CMOS technology.

References

[1] F.M.L. van der Goes, "Low-cost smart sensor interface," Ph.D. Thesis, Delft University of Technology, Apr. 1996.
[2] X. Li, G.C.M. Meijer and Erik J. Schnitger, "A novel smart interface for voltage-generating sensors," In Proc. IMTC/97, Ottawa, Canada, May 19-21, 1997, pp. 1472-1475.
[3] X. Li, G.C.M. Meijer, "A low-cost and accurate DC current measurement system," In Proc. ICECS'97, Cario, Egypt, Dec. 15-18, 1997.
[4] G.C.M. Meijer, J. van Drecht, P.C. de Jong and H. Neuteboom, "New concepts for smart signal processors and their application to PSD displacement transducers," Sensors and Actuators, vol. A35, pp. 23-30, 1992.

SMART OPTICAL SENSOR SYSTEMS IN CMOS FOR MEASURING LIGHT INTENSITY AND COLOUR

G. de Graaf and R.F. Wolffenbuttel
Delft University of Technology, Dept. of Electrical Engineering
Laboratory for Electronic Instrumentation / DIMES
Mekelweg 4, 2628 CD Delft, The Netherlands
Email: G.de Graaf@et.tudelft.nl

0. Abstract

An optical sensor system operating in the visible part of the spectrum, with digital outputs has been developed in a standard CMOS process. The photodiodes operate in the charge integrating mode. Two methods for on-chip A-to-D conversion, namely voltage switching and capacitor switching have been realised. The output signal is a bitstream or a digital word proportional to the intensity of the incident light. Information about the intensity and the spectral distribution of the impinging light is obtained by switching between two PN junctions at different depths in the silicon and read-out in two measurement cycles.

Keywords: CMOS compatible smart sensor, smart colour sensor, integrating mode read-out, charge redistribution AD converter.

1. Introduction

Sensors with on-chip signal processing open new fields of application in measurement systems [1]. When combining the sensor with read-out electronics, the main advantage is lower unit cost of the total system, provided that a standard process is used [2]. Advantages of on-chip integration are: small size, reduced parasitics effects and EMC susceptibility, lower costs for wiring and generally an improved reliability. When using CMOS technology there is also the advantage that complex circuits can be added with a small increase in cost and that these circuits are low-power [1][2]. Many smart sensor systems in CMOS or BICMOS have been developed recently and good quality photodiodes can also be fabricated in MOS technology. Therefore optical sensor systems with a frequency output in CMOS are now already commercially available [3].

The very large dynamic range of silicon photodiodes generally requires a high performance A-to-D converter. A charge read-out (integrating-mode) principle can conveniently be applied in the sensor system presented here. The principle of integrating-mode read-out of PN-junction detectors is well known [4]. Several papers have described the application of this principle for high sensitivity silicon image detectors [5][6]. A Light-to-Frequency converter in bipolar technology based on current switching and current integration has been reported before [7], but a realisation in CMOS is more efficient in terms of power consumption and the chip area needed. The next paragraph shows that photodiodes with a significantly different spectral response and good electrical properties can be fabricated in a standard CMOS process.

2. Sensor structure

The basic sensor structure has been presented before in bipolar technology [8][9] and is shown in fig. 1. Few modifications are required for fabrication in a standard 1.6 μm poly-silicon gate CMOS process. For application as a combined intensity-colour detector, two stacked PN junctions are used: The deep junction is formed by the P-epilayer and the N-well. The shallow junction is formed by the N-well and a P+ implanted layer that is normally used for the drain/source contacts (SP). Due to the wavelength-dependent absorption of light in silicon and the large difference in depth, these diodes will have different spectral responses. Fig. 2 shows the measured spectral response of the upper and the lower junction.

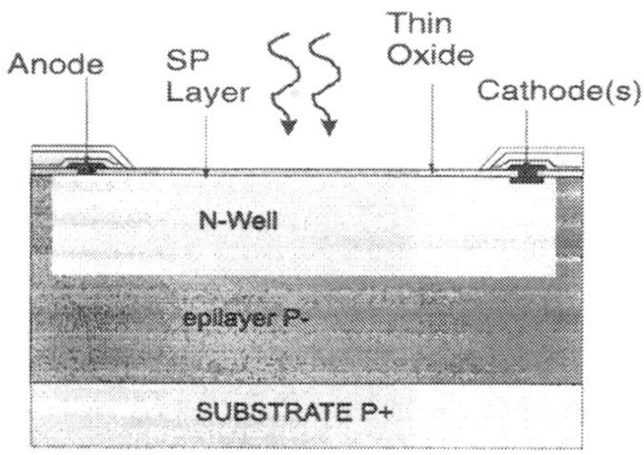

Fig.1 Sensor structure.

The response of the upper junction peaks at 460nm and the lower at 730nm. Fig. 3 shows the electric equivalent of the structure of Fig. 1.

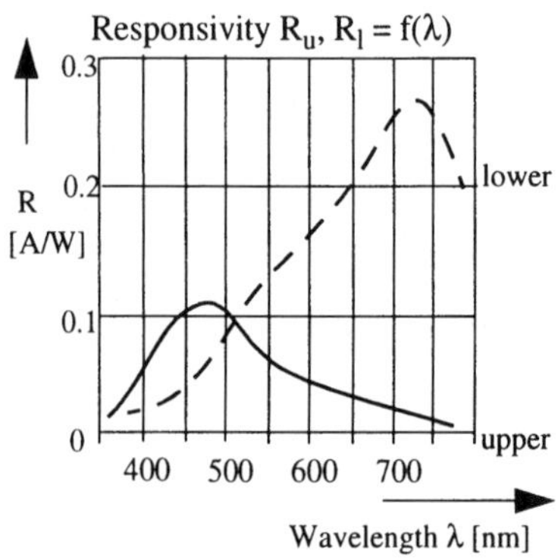

Fig.2 Spectral response of both junctions.

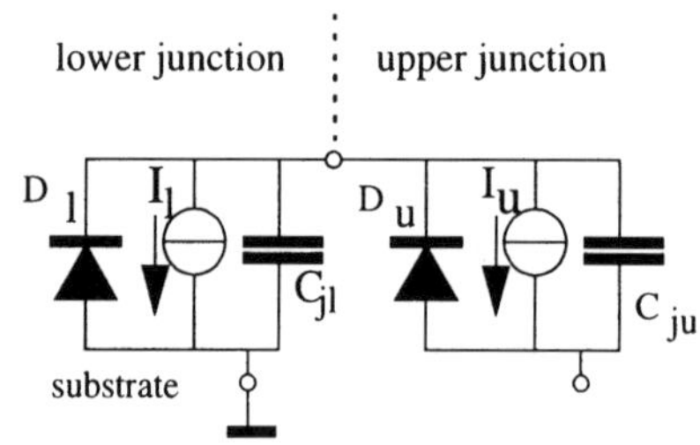

Fig.3 Electrical equivalent of the photodiode structure.

The photocurrents of the lower- and upper PN junction are represented by the current sources I_l and I_u respectively. Capacitors C_{jl} and C_{ju} are voltage dependent, since these represent the junction capacitance of the respective photodiode. The relation between the reverse voltage and the junction capacitance of a diode is given by:

$$C_j = kA_j(V_j)^M = C_0(V_j)^M \quad [1]$$

Where A_j denotes the area of the junction, k is a constant, M is a constant depending on the doping profiles of the junctions or using C_0 representing the junction capacitance at the built-in voltage. Some design constraints are imposed on the read-out of the PNP structure shown in fig. 1. Firstly, since the anode of the lower diode is always connected to ground via the substrate, both photodiodes share the n-well as a common cathode. Secondly both diodes must operate at reverse bias and the substrate contact cannot be used for photocurrent read-out. Consequently it is not possible to measure each current separately, but the sum of both photocurrents can be measured at the n-well contact and the current of the upper junction is measured at the SP contact. Since MOS transistors perform very well when used for voltage sensing- and voltage- or capacitor switching, a charge integrating read-out mode can be conveniently applied in CMOS as will be described in the next paragraph.

3. Basic read-out principle

Fig. 4 shows the basic read-out circuit. If switch S1 closes, the photodiode is reverse biased at voltage V_d. When the switch opens again the photocurrent will discharge the junction capacitance C_{ju}. The relation between the photocurrent I_{ph} and the voltage V_j is given by:

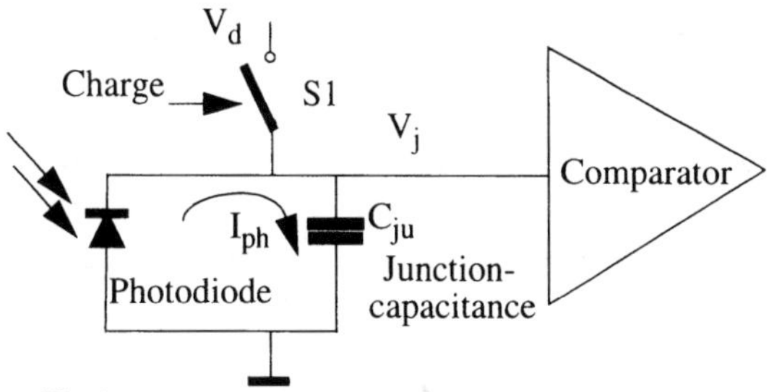

Fig.4 Integrating mode read-out principle.

$$-I_{ph} = \frac{\delta Q}{\delta t} = C_j\frac{\delta U_j}{\delta t} + U_J\frac{\delta C_j}{\delta t} = kA_j(V_j)^M(M+1)\frac{\delta V_j}{\delta t} \qquad [2]$$

The discharge time from a reference level V_d to a low level threshold voltage V_{th} can be calculated as:

$$t_d = \frac{kA_j}{I_{ph}}\left[(V_d)^M - (V_{th})^M\right] = \frac{khc}{\lambda\eta_e P}\left[(V_d)^M - (V_{th})^M\right] \qquad [3]$$

With:
P = optical power [W]
η_e = external quantum efficiency
λ = Wavelength
h = Planck's constant
c = speed of light

Therefore, the discharge time t_d is inversely proportional to the light intensity P. Note that the discharge time does not depend on the diode area A_j, and the voltage across the junction will be non-linear. Based on the charge integrating read-out principle, two basic A-to-D converters have been realised. These are described in the next two paragraphs.

4. Voltage switching A-to-D converter

Fig.5 shows the block diagram of the voltage switching A-to-D converter principle, as used with the sensor described before. Two conversion cycles are needed to measure both currents, one cycle integrates the sum of both photocurrents, while the other integrates the photocurrent provided by the lower junction only. We can distinguish four different states. During the first conversion S2 is closed. During one period of the master clock S1 will close and both junctions (C_{ju} and C_{jl}) will be charged to V_d (+5V). After the opening of S1 the voltage across both photodetectors (switched in parallel) is monitored by the comparator, while $C_{ju}+C_{jl}$ is discharged by I_u+I_l until the lower threshold voltage of the comperator is reached. At the next clock cycle S2 opens and the same procedure starts again but now only the lower junction C_{jl} is monitored while discharged by I_l only, after this the two conversion cycles will repeat. The

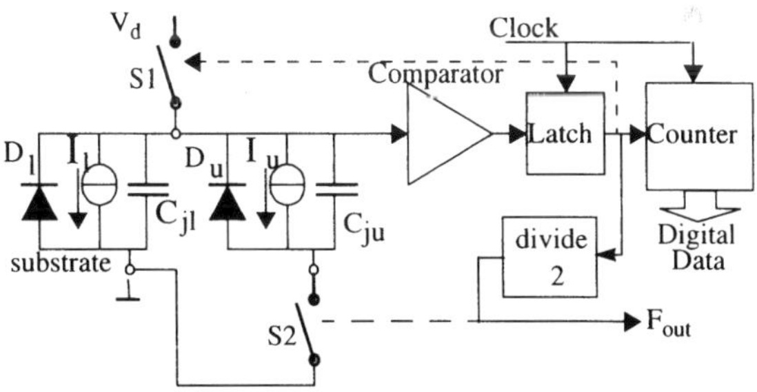

Fig.5 Voltage switching A-to-D converter.

discharge times of state 2 and state 4 follow from eqn.3 and are measured by counting the number of master clock periods into two 16-bits counters. The maximum conversion time T_{max} is thus determined by the number of bits (M) of the counters and the clock frequency T_{cl}:

$$T_{max} = 2^M \cdot T_{cl} \qquad [4]$$

The maximum conversion time can also be derived by substituting the leakage currents of the switches and the dark current of the photodiodes I_{dark} in eqn.3 . So to obtain sufficient resolution at very low light levels, the clock frequency T_{cl} and the number of bits M should be choosen in such a way that the following equation holds:

$$(T_{max} = 2^M \cdot T_{cl}) \quad > \quad \frac{kA_j}{I_{dark}}((V_d)^M - (V_{th})^M) \qquad [5]$$

Also at very high light levels the clock frequency determines the resolution. The comperator output is sampled only at the rising edges of the clock and therefore the junction voltage is monitored at discrete time intervals T_{cl}. The minimum conversion time can therefore be derived by substituting the maximum photocurrent I_{max} in eqn.3. So to obtain sufficient resolution at very low light levels the clock frequency T_{cl} should be choosen in such a way that:

$$T_{cl} < \frac{kA_j}{I_{max}}((V_d)^M - (V_{th})^M) \qquad [6]$$

Increasing the clock frequency will generally result an extra inaccuracy component, caused by extra charges induced by parasitic feedthrough of the switches, higher power consumption and higher demands on the speed of the comperator. Another approach is to improve the resolution by averaging over several measurement cycles.

5. Charge redistribution A-to-D converter

Another method for the read-out is the charge redistribution A-to-D converter shown in fig.6 If the voltage across the junctions of the photodiodes is smaller than the comparator threshold voltage, a charge-pump capacitor C_p, is switched in parallel with the photodiodes.

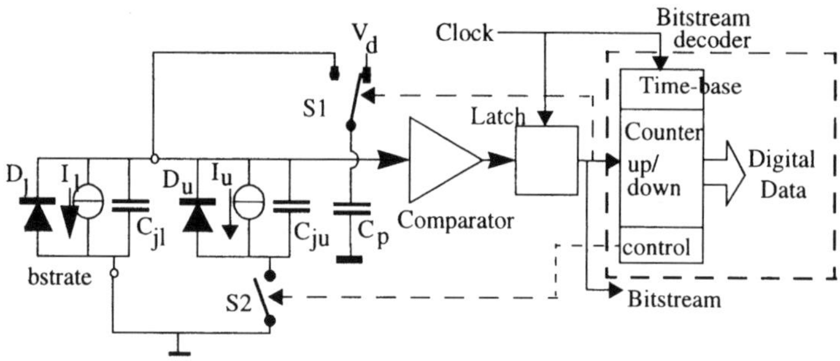

Fig.6 Charge redistribution A-to-D converter.

The capacitor has been charged to a reference voltage V_d, thereby introducing an additional charge $Q_p = C_pV_d$ in the system. This charge redistributes across the photodiodes causing the junction voltage to rise again. The charge transferred to the photodiodes equals: $Q_c = C_p(V_d - V_{th})$. This charge is drained by the photocurrent during the following clock periods ($\Delta Q = I_{ph}T_{cl}$ per period) until the comparator threshold voltage is reached again. Thus the voltage across the junction oscillates around the comperator threshold level. Equating the average transfer of charge by C_p and the loss of charge by the photocurrent I_{ph} over a fixed period of M clock cycles gives:

$$MT_{cl}I_{ph} = (V_d - V_{th})C_pN \qquad [7]$$

With:
- T_{cl} = clock period
- V_d = reference voltage
- V_{th} = comparator threshold voltage
- N = number of charge transfers

Where N is the number of clock cycles when charge is transferred from C_p to the photodiode.

From (7) follows:

$$N = \frac{MT_{cl}}{(V_d - V_{th})C_p} I_{ph} = \frac{MT_{cl}}{(V_d - V_{th})C_p} \cdot \frac{\lambda \eta_e A_J}{hc} \cdot P \qquad [8]$$

With:
P = optical power [W]
η_e = external quantum efficiency
λ = Wavelength
h = Planck's constant
c = speed of light

The number of charge transfer clock cycles is thus proportional to the photocurrent I_{ph} and the optical power P. The size of the charge pump capacitor can be used as an extra degree of freedom when, e.g. the circuit should be designed for a specific clock frequency T_{cl} and a specified size of the photodiodes.

6. Measurement results

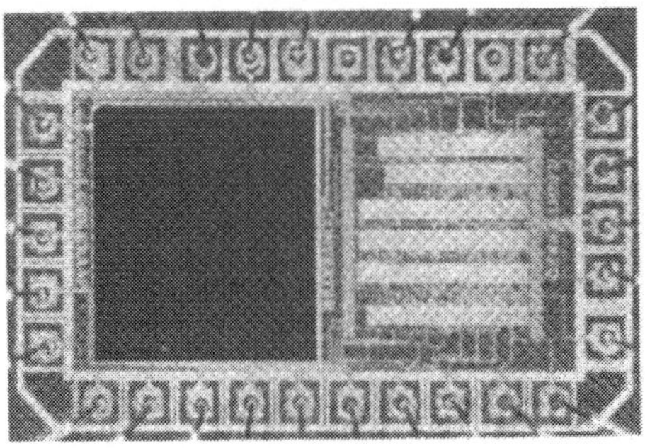

Fig. 7 Photodiode with read-out and bitstream conversion circuits

The microphotograph of the sensor and the read-out circuits for a digital output is shown in fig. 7. The photodiode measures 800 x 600 μm^2. The leakage current of the upper and the lower junction have been measured at room temperature as 50 pA/cm^2 and 200 pA/cm^2 respectively. The light intensity signal is derived from the sum of both photocurrents and is shown in fig. 8 a. The measured conversion time of the voltage switching A-to-D converter ranges from 0.2 Hz (dark) to about 20 kHz, giving a dynamic range of about 10^4. The ratio of the photocurrent of the upper diode and the sum of both diodes can be used as colour detection signal. Fig. 8 b shows the spectral response of the ratio of the discharge times (T1/T2) of the voltage switching A-to-D converter. The comparator switching frequency, which depends mainly on the photocurrent and the size of the charge pump capacitor, is generally much higher then in the voltage switching A-to-D converter. Therefore, the influence of the parasitic capacitances of the switches will be larger for the charge redistribution ADC.

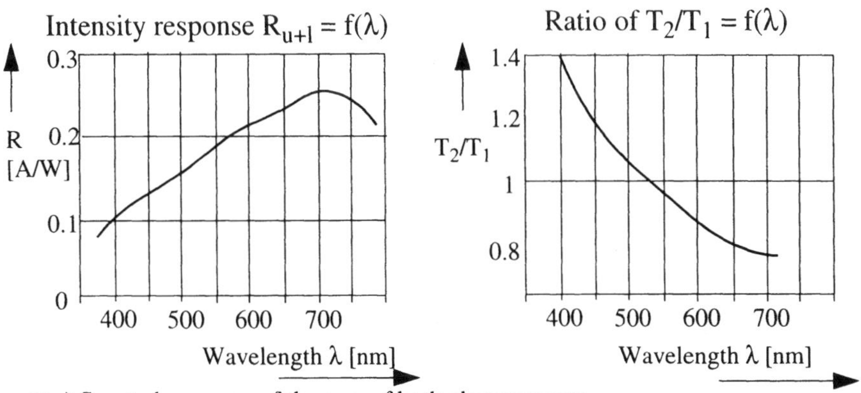

Fig. 8 a) Spectral response of the sum of both photocurrents.
b) Spectral response of the ratio of the upper- and sum of both photocurrents.

7. CONCLUSIONS

An optical sensor system for use as a intensity and colour detector with integrated read-out circuits has been realised in a CMOS standard process. Distinct peaks in the spectral response at 460nm for the upper and at 730nm for the lower junction have been measured. Simulations and measurement results have shown good results for the integrating mode read-out.
Two basic electronic circuits for on-chip A-to-D conversion have been developed, a voltage switching A-to-D converter and a charge redistribution A-to-D converter. A dynamic range of about 10^4 in the light intensity has been measured using both circuits. For colour sensor operation two conversion cycles are needed. The ratio of the conversion cycle times can be used as a wavelength dependent output signal in the range from 450-760 nm.
Both the sensor and the circuits should be optimized for the required application and the CMOS process available. Therefore, only quantitative results and a comparison have been given for both types of A-to-D converters. Important design parameters are the photodiode area and the parasitic capacitances of the MOS switches. The thickness of the oxide layer above the photodiodes has a large influence on the spectral sensitivity at short wavelengths.
By adding more integration cycles operations like offset- or dark-current compensation or better tuning of the spectral response can be implemented. A tri-colorimetric sensor can be realised by adding a third integration cycle if an extra PN junction with a good responsivity at 360 nm can be fabricated also. Very long integration times can be obtained by increasing the number of bits of the counters and by averaging over many cycles of the bitstream signal. To further develop the smart sensor concept, logic for an interface bus can be added.

ACKNOWLEDGEMENT

The authors would like to thank the IC processing section of the Delft Institute for Micro-Electronics and Submicrontechnology (DIMES) for the fabrication of the photodiodes and circuits.

REFERENCES:

[1] J.H. Huijsing,F.R. Riedijk and G. van der Horn, "Developments in integrated smart sensors", Sensors and actuators A, Vol. 43 (1994), pp. 276-288.
[2] R.F. Wolffenbuttel, "Silicon sensors and Circuits", Chapman & Hall, London, 1996, ISBN 0.412.70970.8, pp. 149-169.
[3] Texas Instruments, "Intelligent Opto Sensor Data Book", Section 3 Light-to-Frequency converters, pp. 3.1-3.21, 1995 Texas Instruments Technology Centre, 800 Pavillion Drive, Northampton,NN4 7YL, United Kingdom.
[4] G. P. Weckler, "Operation of p-n junction photodetectors in a photon flux integrating mode," IEEE Journal of Solid State Circuits, Vol. SC-2 (1967), pp. 65-70.
[5] S. G. Chamberlain, "Photosensivity and Scanning of Silicon Image Detector Arrays," IEEE Journal of Solid State Circuits, Vol. SC-4, No.6, December 1969, pp. .
[6] P.J.W. Noble, "Self-scanned silicon image detector arrays", IEEE Transactions on Electron Devices, vol. ED-15 (1968), pp.202-209.
[7] G. De Graaf and R.F. Wolffenbuttel, "Light-to-Frequency Converter using Integrating Mode Photodiodes". IEEE transactions on Instrumentation and Measurement 1997, Vol. 46 (1997), no. 4, pp.933-936.
[8] R.F. Wolffenbuttel and P.P.L. Regtien, "A Novel Approach to Solid-State Colour Sensing." Sensors and Actuators A, Vol.9 (1986), pp. 199-211.
[9] R.F. Wolffenbuttel, U.S.Patent no. 4749851,"Method and circuit for determining the wavelength of light", August 29, 1986

THERMAL ANALYSIS OF MULTI-CHIP-COMPOSED MICROSYSTEM

M.P. Brünner, M. Bartek and R.F. Wolffenbuttel
Delft University of Technology, Department of Electrical Engineering,
Laboratory for Electronic Instrumentation, Mekelweg 4, 2628 CD Delft,
The Netherlands, Phone: +31-15-2786287, Fax: +31-15-2785755,
E-mail: M.Bartek@et.tudelft.nl

0. Abstract

Thermal analysis of an instrumentation microsystem, based on MCM (Multi Chip Module) technology with an active silicon platform, is presented. Results of theoretical analysis (analytical solutions based on thermal-electrical analogy, numerical FEM-simulations) are verified by measurements on a real system. The glue bonding layer (used for die attachment) and the package-to-ambient interface have been identified as the dominant thermal resistors. The measured thermal response time (time constant < 1 s) and good temperature distribution of the MCM microsystem (no hot-spots) indicate that a single-point adaptive thermal management scheme is required and sufficient to increase functional availability of all subsystems, and to maintain reliability and accuracy of the entire microsystem.

Keywords: microsystem, thermal management, MCM, temperature measurement

1. Introduction

Microsystems have proven to be an optimum compromise between the aim for size reduction, enhanced functional performance and reliability [1]. The major disadvantage results from the processing complexity required to build on-chip microsystems [2, 3]. This fabrication compatibility constraint can be avoided by using a universally applicable active silicon platform, that contains all the infrastructural functions that are common to any measurements system and on which the necessary modules (sensors, actuators, microcontrollers) are flip-chip bonded. Such a microinstrumentation system, based on MCM (Multi Chip Module) technology (see Fig. 1), is a promising approach for complex measurements, while avoiding compatibility problems and maintaining dimensional advantages of on-chip smart sensors.

The active silicon platform contains already a basic set of functions, like a local sensor bus, power and thermal management and mounting sites for sensors, a microcontroller and other modules [4]. This flexible system ensures a reliable, and yet economically interesting class of microsystems, that can be used in a wide range of applications [1].

In such a system, thermal management is aggravated due to both volume reduction and integration of additional infrastructural functions in the platform. Compared to single chips, multichip modules have more complicated thermal profiles [5] due to a higher thermal resistance at the die interfaces, when compared to bulk silicon. Investigation was performed to find out whether thermal management is necessary, and if so, if measuring the temperature at a single spot on the silicon platform is sufficient for a stable and reliable thermal management.

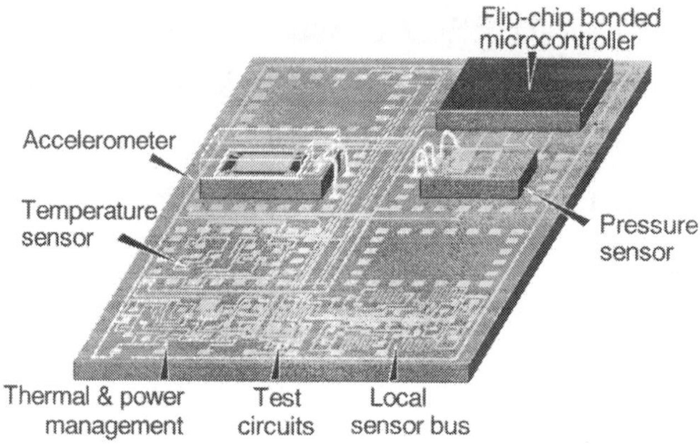

Fig. 1: The microinstrumentation system with active silicon platform, based on MCM (Multi Chip Module) technology.

2. Model description and theoretical analysis

The initial simplified analytical study on the platform was performed using thermal-electrical analogy. Heat generated from the attached die will flow via the flip-chip bonding and/or glue layer to the platform, leadframe and package (ceramic or plastic). The flip-chip/glue bonding interface is one of the major thermal resistors encountered in vertical direction. The silicon platform thermal resistivity in vertical direction is small, due to the high thermal conductivity of silicon, large area and small thickness. A low-cost plastic package will show a large thermal resistance.

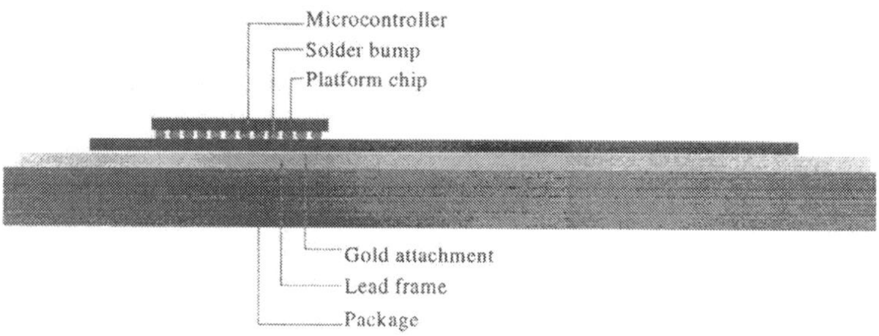

Fig. 2: Several layers encountered in the microinstrumentation system.

The variables can be split up in three groups; heat conduction, convection and radiation. A common equation in heat conduction is:

$$q = KA\frac{T_1 - T_2}{\Delta X} \quad (1)$$

which relates the heat flow q to the temperature difference over the object, by the thermal conductivity K, the flow area A and the flow path length ΔX. Thermal resistance can be then defined as:

$$R_{th} = \frac{\Delta X}{KA} \quad (2)$$

A differential equation that represents time-dependence can be written as:

$$\frac{\partial T_1}{\partial t} \times R_{th} \rho c V = T_1 - T_2 \tag{3}$$

which transforms in the Laplace domain to:

$$\frac{T_1(s)}{T_2(s)} = \frac{1}{1 + R_{th} \rho c V s} \tag{4}$$

This is a first order system, for which the response time increases with increasing resistance, specific heat, volume and density of the body. Equation (4) is the electrical analogy of a thermal circuit, and was used to model thermal behavior.

Heat convection is not so well defined for small structures as heat conduction, because of turbulent flows at sharp edges. It is possible to combine all complexities involved in convection in one parameter, the film coefficient h, which is defined as [6]:

$$h = \frac{-K_f \, \partial T / \partial l}{T_s - T_f} \tag{5}$$

in which K_f is the thermal conductivity of the flow, $\partial T/\partial l$ is the flow temperature gradient at the body surface and $T_s - T_f$ is the temperature difference between the body surface and the flow. Typical film coefficient values of air are 5-25 $Wm^{-2}K^{-1}$ for natural convection and 10-500 $Wm^{-2}K^{-1}$ for forced convection. Heat radiation can be disregarded because of relatively low temperatures.

The modeled system (Fig. 2) consists of a silicon strip carrier (10x2 mm^2), on which a silicon die (2x2 mm^2) is glued. From this structure a thermal circuit is built, by modeling each 2x2x0.5 mm block by a resistor, and modeling the glue layer as a single resistor as well as the whole package and the convection heat transport. The latter can be modeled as a resistor due to the quasi-static state and the constant area of the package. Heat is dissipated from the small die, which is modeled as a current source. Table 1 contains the values used in calculations. The values in the table are calculated from specifications of the glue used, and the ceramic DIL-40 package.

Table 1: Parameters used in theoretical analysis.

P_{diss}	0.5 W	R_{vert}	0.83 K/W	R_{conv}	9.5 K/W
R_{die}	0.83 K/W	R_{lat}	13.3 K/W		
R_{glue}	15 K/W	R_{pack}	13 K/W		

Results of the simulation are shown in Fig. 3. Note that the lateral temperature drop is very small, which is caused by the limitations of the model: the back of the platform is a single node. Important results are that the maximum temperature reached is still 42° C, with a highly thermal conductive ceramic package, and that a temperature drop of 8° C is present over the glue interface.

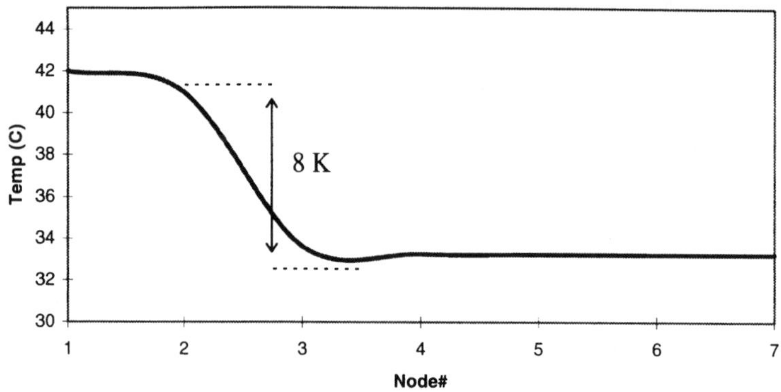

Fig 3. Expected temperature profile based on thermal-electric analogy model.

To obtain a more quantitative result, the ANSYS thermal analysis finite element simulation tool was used. The structure depicted in Fig. 2 is built, meshed and after applying the loads, the model is solved. To be able to plot the temperatures, a path is defined from the heated die, through the glue and over the platform, a typical result is plotted in Fig. 4. This FEM-model shows quite a similar vertical profile as the analytical model, but a better insight in the lateral heat distribution is obtained; a gradient of 0.5° C/mm is present over the platform. Note that the thickness and the area of the glue layer have a large influence on the thermal profile, and are difficult to forecast.

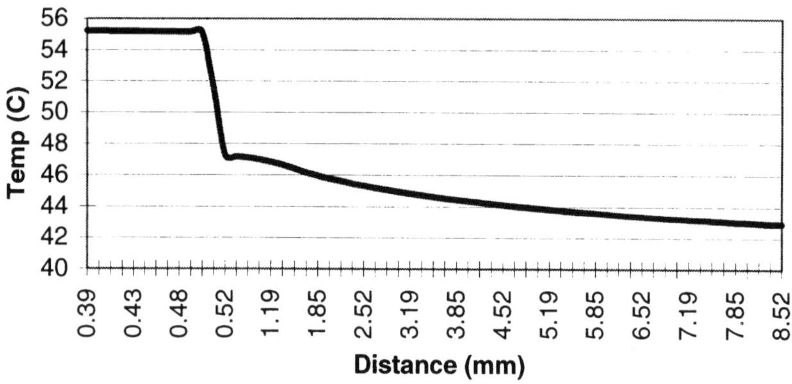

Fig. 4. A path-plot result, simulated with FEM.

3. Test structure

Analytical and FEM simulations were verified by measurements on a modelled microsystem (see Fig. 5). This was built up out of differently sized chips with a diffused meander heating resistor, and a diode array for the temperature measurement (Fig. 6). At the packaging stage, one or more smaller chips were bonded (using glue) onto a platform die, thus obtaining a similar structure as a real microinstrumentation platform. Before performing the actual measurements, each diode has been calibrated in the temperature range of 20°C to 95°C.

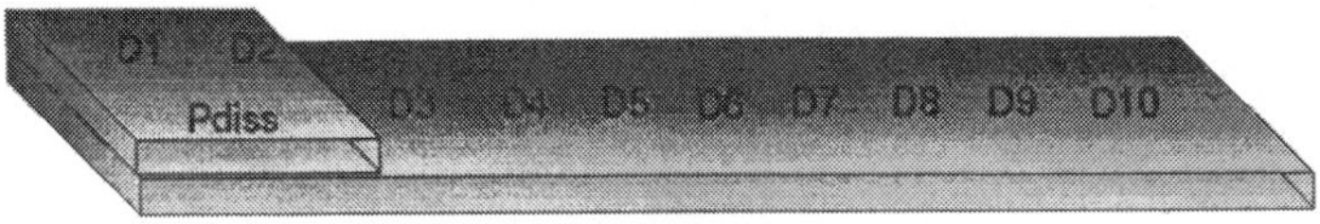

Fig. 5. The strip-like device and diode placement.

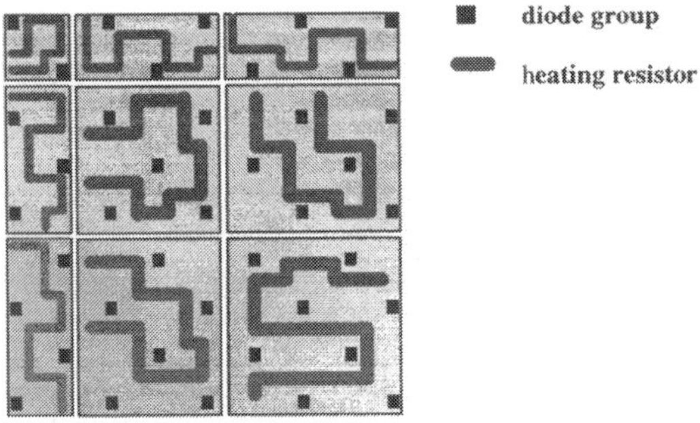

Fig. 6. A 10x10 mm^2 layout used to built up different test structures for thermal characterisation.

4. Measurements

All devices were packaged in DIL-40 ceramic packages, which has the consequence that a lot of heat (0.5 W) can be dissipated before reasonable rise in temperature is seen. A standard plastic package would cause a temperature offset in the structure, and hardly change the vertical and lateral temperature drops. Results from measurements on the strip-like device are presented in Fig 7.

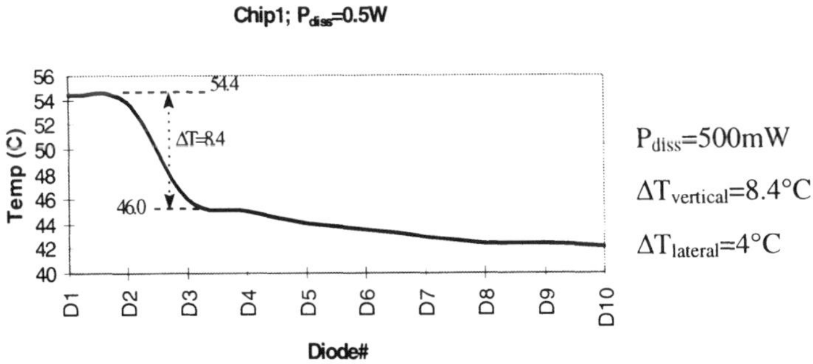

Fig. 7. Measured temperature profile.

For control of the thermal behaviour of the microinstrumentation system, the dynamics of the system is important. The measured dynamic behavior (Fig. 8) is dominated by several parameters, like the volume and thermal capacitance of both dies, the resistance of the glue layer, and the structure-to-ambient resistance. The graph shows two time-constants, one of the small die and one of the large platform die. These time-constants stay well below 1 s, and if a plastic package is used, much faster response is expected within the platform.

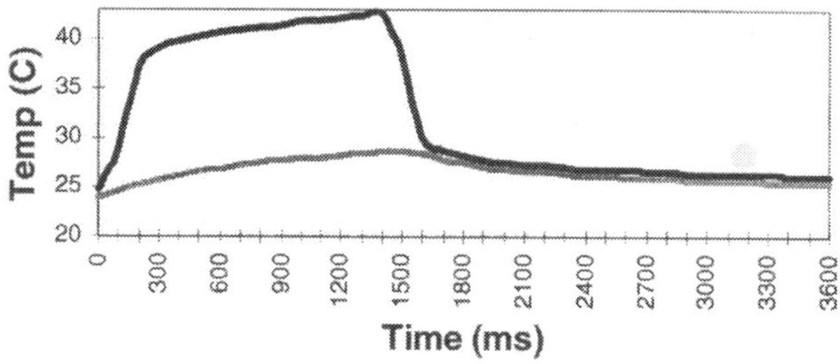

Fig. 8: Measured dynamic temperature response at the dissipating mounted die and at the platform.

5. Conclusions

Thermal analysis of an instrumentation microsystem, based on MCM (Multi Chip Module) technology with an active silicon platform, is presented. Results of theoretical analysis (analytical solutions based on thermal-electrical analogy, numerical FEM-simulations) are verified by measurements on a real system. The glue bonding layer (12...15 K/W) and the package-to-ambient interface (12...22 K/W) have been identified as the dominant thermal resistors. For anticipated power dissipation, thermal management is necessary, especially when a plastic packag is used. A fast response (time constant < 1 s) and temperature distribution of the MCM microsystem analyzed require an adequate adaptive thermal management scheme to increase functional availability of all subsystems, and to maintain reliability and accuracy of the entire microsystem. Best option is to assign priorities to system functions, and performing these only if temperature remains below pre-set maximum, e.g. an interrupt mask level depending on temperature. A single-spot measurement can be sufficient, because of fast response and small thermal differences within a platform.

Acknowledgments

The authors wish to thank the staff of Delft Institute of Microelectronics and Submicron Technology for technical assistance in fabrication of the devices. The work presented here has been supported in part by STW, project DEL 55.3733.

References

[1] H. Zinner, Microsystems - The European approach, Sensors and Actuators A 46-47, 1995, 1-7.

[2] H. Baltes, CMOS as sensor technology, Sensors and Actuators A 37-38, 1993, 51-56.

[3] R.F Wolffenbuttel, On-chip microsystems in silicon: opportunities and limitations, J. Micromech. Microeng. 6. 1996, pp. 138-143.

[4] A. Mason, N. Yazdi, K. Najafi, K. Wise, A low-power wireless microinstrumentation system for environmental monitoring, Transducers, 1995 pp. 107-110.

[5] Y. C. Lee, H. T. Ghaffari, J. M. Segelken, Internal thermal resistance of a multi-chip packaging design VLSI based systems, Proc. 38th Electron. Components Conf. (ECC), pp. 293-301, 1988.

[6] G. N. Ellison, Thermal computations for electronic equipment, Van Nostrand Reinold Company Inc. 1984.

IC POST-PROCESSING MODULE FOR DEPOSITION OF THIN METAL FILMS

D. Wijngaards, M. Bartek and R.F. Wolffenbuttel
Electronic Instrumentation Laboratory (DIMES), Delft University of Technology,
Faculty of Information Technology and Systems, Dept. of Electrical Engineering,
Mekelweg 4, 2628 CD Delft, The Netherlands
E-mail: D.D.L.Wijngaards@et.tudelft.nl

0. Abstract

The postprocessing module described in this paper enables the addition of a wide range of desired metal film, while maintaining compatibility with IC processing. When being able to combine any thin metal film with electronics, a new class of transducers becomes feasible with a performance often superior to existing silicon transducers. Although initially designed for transducer applications, the post-processing module is also capable of depositing metal film resistors for use as integrated passive components. Typical for the module are the flexibility of metal required, the flexibility of deposition order and the low temperature at which deposition takes place. Leaving the thermal budget of the electronics unaffected and the standard IC process intact, full compatibility with micro-electronic processing is assured. An example of the module is demonstrated in fabrication of low-TCR NiCr resistors.

Keywords: Thin metal film, process compatibility, post-processing, NiCr, lift-off etching

1. Introduction

Several approaches for integration of sensor and circuit can be found in literature. Hybrid and semi-hybrid technology are to be preferred when designing for low-production-volume dedicated sensor systems. In contradiction, dedicated processing is the right choice when considering batch-fabrication of fully integrated sensors. Compatible processing is in the middle of the two approaches, offering larger flexibility than dedicated processing, yet keeping the advantage of fully integrated sensors. This flexibility is available due to the use of mutually compatible modules [1], which can be applied on request, without interfering with standard IC processing. The module presented in this article is an example of such a compatible post-processing module, capable of depositing almost any thin metal film required.

2. Limitations of present smart Si sensor technology

On-chip integrated silicon (smart) transducers have the advantage of material compatibility that enables on-chip functional integration with micro-electronic circuits in silicon. This does not imply that co-fabrication of any integrated silicon sensing/actuating element with circuitry is possible or even desirable. Firstly, although silicon has many transduction effects [2,3], material properties of silicon are not optimal in certain applications: silicon is a non-piezoelectric and non-magnetisable material. Secondly, due to technological restrictions, a compromise has to be made between the actual transducer and the on-chip electronic circuit performance. Processing of transducer and circuitry is often incompatible due to high temperatures during fabrication, affecting the thermal budget of the circuitry [4]. Also, one has to keep in mind that cleanroom compatibility constraints are present, when depositing some metals and metal compounds. The major limitation of thin metal film technology, is the limited knowledge on thin metal film behaviour on a SiO_2 surface. This might lead to unexpected physical effects (diffusion, adhesion, step coverage, thermal stress) or chemical effects (silicidation).

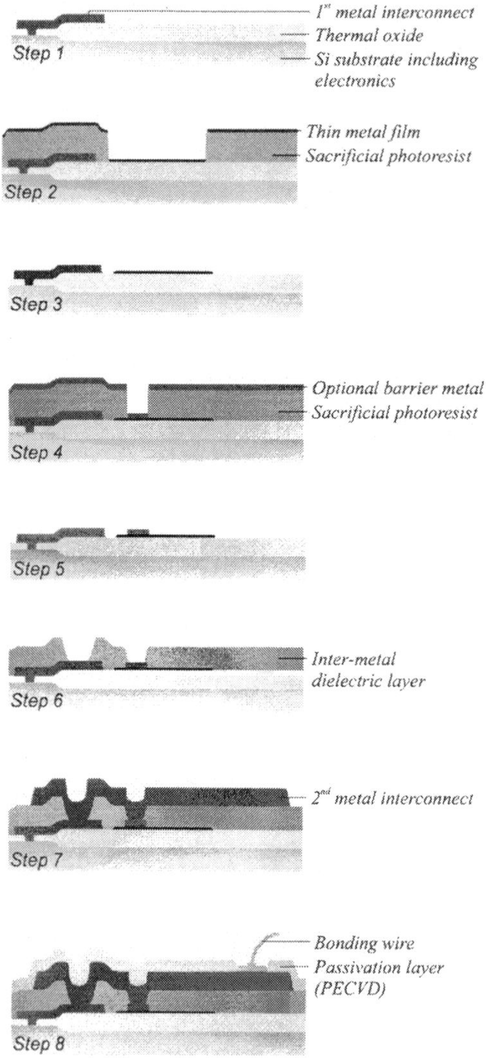

Fig. 1 Schematic of thin metal film deposition module (from steps 2 and further), to be applied after regular IC processing (as shown in step 1).

3. Advantages of thin film integration

The metal film post-processing module described in this paper allows deposition of metal layers on top of an integrated circuit aiming at on-chip integrated transduction elements, not achievable in standard silicon processing. Because the metal film is deposited using a post-processing module and it is deposited at very low temperatures (evaporation, sputtering), deposition does not interfere with the silicon processing. For this reason, no compromise between circuit performance and transducer performance has to be made, giving the module potential for the fabrication of smart transducers and on-chip integration of existing non-silicon (metal-film-based) sensing elements.

4. Metal film deposition module

A few disadvantages of integration of thin metal films with silicon circuitry have already been mentioned. These, together with some other constraints, serve as the initial design considerations for the module.

4.1. Initial design considerations

The first consideration involves the thermal budget during circuit processing. To make sure not to influence the circuitry, processing temperatures have to be low at all times. The second consideration is flexibility of metal choice. Aluminium is patterned by etching away all unwanted aluminium. For the module to be applicable for many different metals, this would require a large quantity of different etchants, making processing more difficult than desired. A better approach is the use of lift-off etching, where processing is dependent on the metal layer deposited. The third consideration is when to deposit the metal films. This can be done using either a pre- or post-processing module. Some metals or metal compounds do not withstand higher temperatures common in silicon processing, due to, for example, low eutectic temperatures. For this reason a post-processing module is more suitable, since all silicon processing steps have been performed previously to depositing the metal film. Regarding cleanroom compatibility, again a post-processing module is preferable, since the non-compatible metals can be deposited at dedicated lines, outside the cleanroom dedicated to standard processing, without having to re-enter this area.

4.2. Postprocessing module in detail

The deposition module is built up in such a way that full compatibility with standard IC processing is ensured and, therefore, it can be easily combined with any standard IC process as a simple additional post-processing step. Fig. 1 shows the fabrication sequence schematically. The actual post-processing module, from steps 2 through 5, is inserted in the IC fabrication process after the first metallisation has been completed. In Fig. 1 one mask is required to deposit the required metal film, and an extra mask is used to deposit an addition barrier metal layer. With both steps, lift-off etching technique is used to form the required layers. If required, processing then continues with PECVD oxide deposition, second metallisation and is completed by PECVD nitride passivation and alloying. A lift-off patterning scheme is preferred over the conventional etching, because it is independent on the actual metal applied and, therefore, the development of a dedicated etching process for every specific metal is not required.

The lift-off process works as follows. In step 2, first a layer of (sacrificial) photoresist is deposited and patterned. Thereafter the required thin metal film is deposited. Finally, in step 3, the sacrificial photoresist is etched away, removing all unwanted thin metal film. In a similar way the barrier metal layer is deposited. A barrier metal layer may be required when the 2nd metal interconnect and the thin metal film are not inert. The barrier metal layer serves as a contact region, assuring low contact resistance and a reliable contact between 2nd interconnect and thin film.

5. Example: Low-TCR NiCr resistors

The on-chip integration of stable low-TCR resistor networks has been investigated. Low-TCR metals are essential for fabrication of devices that require extreme stability. These resistors will be used for high-accuracy static D/A converters, operating on the principle of a weighted resistor network. Alloys for low-TCR resistors are Evanohm (75 wt% Ni, 20 wt% Cr, 2.5 wt% Al and 2.5 wt% Cu) and Karma (75 wt% Ni, 20 wt% Cr, 2.5 wt% Al and 2.5 wt% Fe) which have a reported TCR of less than $1 \cdot 10^{-5}$ K^{-1} over a large temperature range [5]. Presently however, work concentrates on NiCr 50/50 resistors, which have a reported TCR of $1.8 \cdot 10^{-5}$ K^{-1} [6,7]. The NiCr layers have been evaporated using electron beam heating of a NiCr 50/50 source at a pressure of $2 \cdot 10^{-3}$ Pa and a deposition rate of about 0.5 $nm \cdot s^{-1}$. As sacrificial layer for lift-off, 1.4 μm Hunt HPR504 positive photoresist has been used. The lift-off etching was performed in an acetone bath using ultrasonic agitation. Various resistance structures have been fabricated, with a thickness varying from 35 nm to 75 nm and a uniformity of about 5 %.

5.1. Matching

An average sheet resistance of 18.4 Ω/□, equivalent to a resistivity of 119.7 μΩ·cm at a thickness of 65 nm, is obtained. Matching between adjacent resistors, with a spacing of 150 μm, is within ±0.2 %.

5.2. Temperature coefficient of resistance (TCR)

As mentioned previously the main reason for deposition of thin NiCr films is the fabrication of low-TCR resistors. Unlike silicon bulk resistors, the resistance of NiCr resistors is almost independent of applied voltage and much less dependent of temperature. The best reported TCR is $18{\cdot}10^{-6}\,K^{-1}$. Thermal treatment for a duration of 24hrs at 150 °C is required to achieve stability [7]. As a second approach, thermal cycling has been applied, cycling from 25 °C to 150 °C for six times. Both yielded the same result, shown in Fig. 5. The first observation is that a TCR of $40{\cdot}10^{-6}\,K^{-1}$ has been achieved, a factor of two higher than reported in literature. This can be explained when regarding the type of substrate used. In [7] both steel and mylar were used as substrate, while this work applies to a silicon substrate, having a much larger coefficient of expansion. A second observation shows that there is a small hysteresis in the resistance when cycling. Though research on this phenomena has not been completed, ideas indicate that the type of glue used, was not proper for such temperature variations. Special low-stress glues are available which could help eliminating the hysteresis.

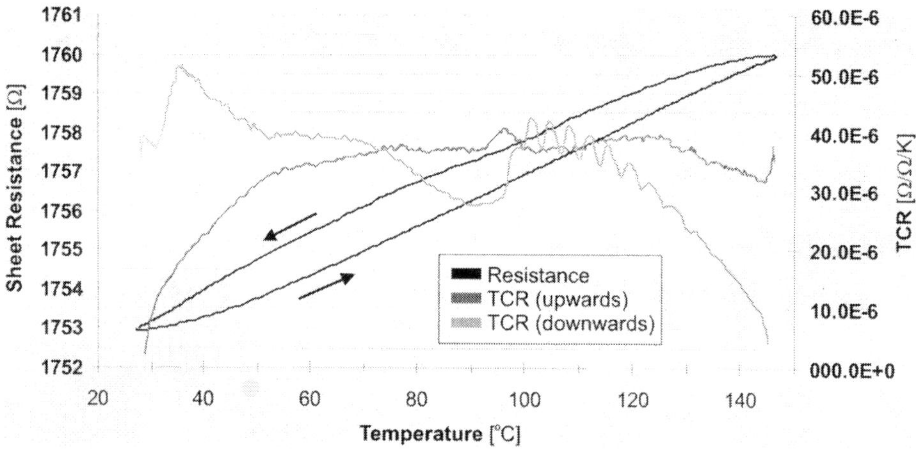

Fig. 2 Graph showing variation of resistance and TCR when temperature cycling is applied, after application of temperature stabilisation treatment.

5.3 Yield improvement

Though the results obtained are promising, two practical limitations had to be overcome. Firstly, peeling of the photoresist was observed and secondly, improper lift-off had to be dealt with.

Peeling of photoresist - It was found that excessive overheating before and during deposition can cause this peeling of the photoresist from the underlying silicon dioxide layer. A very thin layer of metal still deposits on regions initially masked, causing short-circuits. The only solution is to carefully control the thermal treatment of the photoresist.

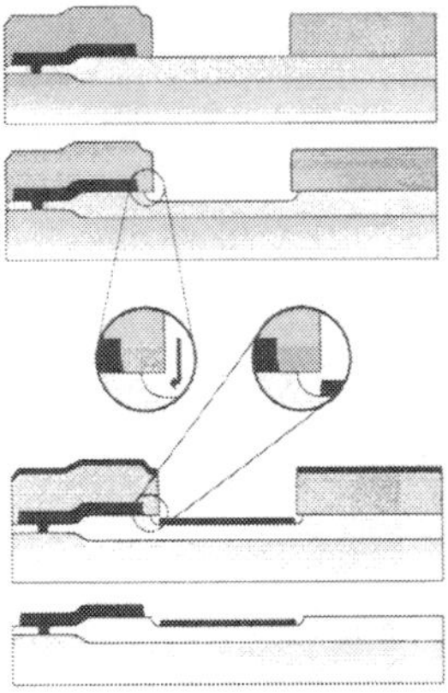

Fig. 3 Double-layer resist scheme

Improper lift-off - Experiments have shown that for thicker films (> 200 nm) the photoresist used (with a positive wall slope) is not sufficient. Fig. 4a shows a test structure, from which it is visible that the redundant thin metal film has been removed rather poorly, resulting in greater mismatch. When high quality is required, and for thicker metal layers (> 150 nm), dedicated single layer photoresist (with negative wall slopes) or multi-layer resist schemes [8] are needed. Using a double-layer resist scheme as shown in Fig. 3, the results, as depicted in Fig. 4b, were obtained. Besides improved deposition accuracy, planarity can also be improved, when the thickness of sacrificial oxide and metal film are tuned to be equivalent.

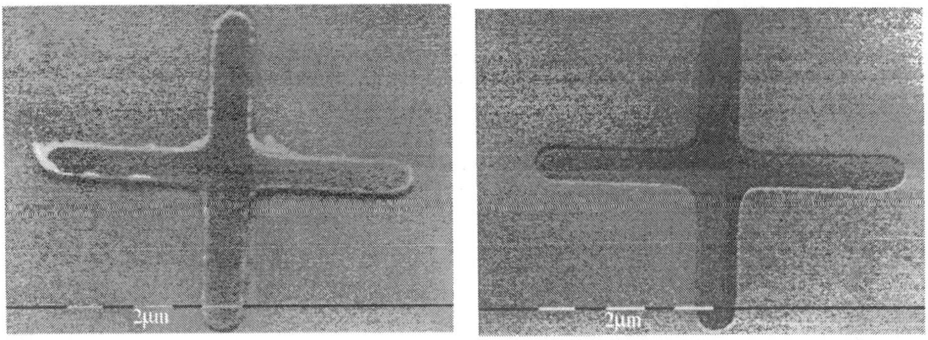

Fig. 4 (a) Thin metal film deposition, using standard photoresist. (b) Thin metal film deposition using a double-layer resist scheme. Both deposition and planarity are improved.

6. Conclusions and outlook

The metal film post-processing module presented is fully compatible with standard IC processing and has high flexibility with respect to the deposited materials. Four important characteristics contribute to this result. Firstly full cleanroom compatibility can be maintained, since deposition of the metal films can take place outside the cleanroom. Secondly the thermal budget of the circuitry is not significantly influenced, since deposition and patterning take place at low temperatures. Thirdly, due to lift-off etching using photoresist as sacrificial layer, a great flexibility in choice of metal is achieved. A fourth characteristic, is the flexibility of deposition order. It is very easy to add or stack additionally required layers, for example adhesive layers [9], or barrier metal layers. These four characteristics enable the co-fabrication of on-chip integrated metal-film-based transducers, not achievable in standard silicon processing, with on-chip smart systems in silicon. The similarity between deposition of metal films for transduction purposes and deposition of metal films for integrated passive components, makes the postprocessing module also applicable for the latter application area. Extensive overheating of the photoresist layer can lead to peeling of the photoresist used, and therefore, to undesired deposition of metal on initially masked regions. Initial experiments showed that when sputtering or/and thicker films (> 150 nm) are used, dedicated single layer photeresist (with negative wall slopes) or multi-layer resist schemes are required for a successful lift-off process. Future research will focus on the optimisation of photoresist schemes and adhesion properties of metal films or metal film compounds. Future applications include reference resistors, low-TCR strain gauges, thermocouples [10], microrelays [11] and fabrication of free-standing metallic microstructures [12].

1. P.M. Sarro, Sensor technology strategy in silicon, Sensors and Actuators, A31 (1992) 138 - 143.
2. S. Middelhoek and S.A. Audet, Silicon sensors, Delft University Press, Delft, The Netherlands, 1994.
3. J. Schubert, Physikalische Effekte, Physik Verlag, Weinhein, 1984.
4. R.F. Wolffenbuttel, Silicon sensors and circuits: On-chip compatibility, Chapman&Hall, London, UK, 1996.
5. C.J. Smithells, Resistivity and temperature coefficient of pure metal and alloys, Metals reference book 5th edn., Butterworth Scientific Publications, London, UK, 1979, pp. 1035-1039.
6. F. Hegner, The industrial production of high quality nickel-chromium resistors with controlled temperature coefficient of resistance, Thin solid films, vol. 57 (1979) 359 - 362.
7. A. García-Alonso, J. García, E. Castaño, I. Obieta and F.J. Gracia, Strain sensitivity and temperature influence on sputtered thin films for piezoresistive sensors, Sensors and Actuators A, Vol. 37 - 38, 1993, pp. 284-789.
8. L.F. Thompson et al., Introduction to Microlithography - 2nd ed., American Chemical Society, Washington DC, 1994.
9. J.A. Foerster and R.F. Wolffenbuttel, High-temperature accurate thermal sensing using integrated platinum resistors on silicon, Proceedings IMTC/95, Boston, April 1995, pp. 538-540.
10. E. Castaño, E. Revuelto, M.C. Martín, A. García-Alonso and F.J. Gracia, Metallic thin-film thermocouple for thermoelectric microgenerators, Sensors and Actuators A, 60 (1997) 65-67.
11. P.M. Zavracky, S. Majumder and N.E. McGruer, Micromechanical switches fabricated using nickel surface micromachining, IEEE J. Micromechanical Systems, vol. 6 no. 1 (1997) 3 - 9.
12. M. Bartek and R.F. Wolffenbuttel, Dry-release of metal structures in oxygen plasma: process characterisation, Proc. MME'97 Micromechanics Europe, September 1-2, 1997, Southampton, UK, pp. 71-74.

A FAST AND ACCURATE TEMPERATURE-CONTROL SYSTEM USING SMART TEMPERATURE SENSORS.

Paul R. van der Meer[1], Gerard C.M. Meijer[1], Michiel J. Vellekoop[2], Harry M.M. Kerkvliet[1], Ton J.J. van den Boom[3].
Delft University of Technology/DIMES, Faculty of Information Technology and Systems, Electronics Research Laboratory[1], Electronic Instrumentation Laboratory[2], Control Engineering[3], Mekelweg 4, 2628 CD Delft, The Netherlands,
Phone: +31 15 2781715, Fax: +31 15 2785922,
E-mail: P.R.vanderMeer@et.tudelft.nl

0. Abstract

This paper deals with the thermal and temperature-control aspects of a gas sensor based on Surface-Acoustic-Wave (SAW) devices in a dual delay-line configuration. Application of smart temperature sensors combined with a micro-controller in a temperature-control system provides for a high-performance low-cost A/D conversion. An anti-windup control mechanism has been implemented to solve the problem of actuator saturation.

Keywords: SAW devices, smart sensors, anti-windup temperature control.

1. Introduction

Surface-Acoustic-Wave (SAW) delay-line oscillators are very suited to measure concentrations of specific gaseous chemical compounds (dopings) in a carrier gas [1]. Adsorption and resorption processes of doping gasses in a chemical sensitive film, applied to the SAW delay line, influence the phase velocity through mass and conductivity changes of this layer. Unfortunately, SAW devices are burdened with temperature cross-sensitivity, because temperature variations also change the phase velocity [2, page 130]. A better performance is obtained by the addition of a reference delay line. However, the two delay lines will never match exactly, thus causing the difference in phase velocity between the two delay lines to be temperature dependent. This phenomenon reduces the accuracy of the chemical sensor substantially, hence temperature stabilization is a necessity.

In addition to temperature stabilization, adjustment of the set-point temperature within a predefined interval is also desirable. This enables the user to choose the optimal set point for a specific application. Features such as selectivity, sensitivity (Hz/ppm), oscillator instability, chemical response time and the type of chemical interface used play an important role in choosing the optimal set-point temperature [1]. To enable accurate temperature control and stabilization, a new encapsulation for the gas sensor has been designed. This gas-tight encapsulation combines good thermal insulation of the gas sensor from its environment with a short response time.

The temperature sensor applied in the control system is a smart temperature sensor based on the asynchronous conversion mechanism [3, 5]. The output signal of this sensor is a duty-cycle modulated square-wave voltage. The pros and cons of this sensor are here discussed.

For the controller design many aspects have been taken into account, such as accuracy, time response, settling time, overshoot and costs. The most important aspects of this control system are the accuracy of the temperature stabilization and low costs. To achieve these aims, the interrelations between the time constants of the process, the properties of the temperature sensor, the costs, the accuracy and speed will be discussed.

2. Design criteria

In the design, the following operational conditions have been assumed:

Operational conditions

- The gas flow over the SAW device has to be constant in the range 0 l/h to 10 l/h.
- The environment temperature has to be constant in the range 0 °C to 30 °C.

The following functional demands have to be met:

- The operational temperature of the SAW device should be adjustable between 40 °C and 120 °C.
- Temperature variations should be less than ± 0.01 °C around the set point. This will limit the temperature-dependent part of the frequency difference to approximately 3 Hz. With this resolution, it is possible to measure concentrations of a few ppm.
- The encapsulation has to be gas tight and able to withstand temperatures up to 150 °C.
- At the end of the sensor life-cycle, which is limited by the ftalocyanine layer, as many as possible of the expensive parts have to be reusable.
- The digital controller has to be implemented in firmware.

3. Thermal model of the gas-sensor and its encapsulation

The SAW chemo-sensor is based upon a dual delay-line oscillator [1]. Each SAW delay line consists of two aluminum Inter-Digital Transducers (IDT) fabricated on a quartz substrate (12 mm × 10 mm × 0.5 mm). The chemo-sensor has to be insulated thermally from its environment, otherwise it is not possible to stabilize the set-point temperature of this device within ± 0.01 °C. The delay-line oscillator operates at approximately 80 MHz, hence electrical shielding is required to limit ElectroMagnetic Emission (EME). The IDT is a capacitive device and, therefore, susceptible to electromagnetic fields (EMS). The shielding also reduces the EMS. To heat the quartz die only,

(a)  (b) 

Figurel. (a) Bottom side of the quartz die (12×10×0.5mm) containing an aluminum heater and a smart temperature sensor. (b) Top side of the quartz die containing the IDTs.

a resistive meander-shaped heater is etched into an aluminum layer (0.6 μm thick) sputtered on the back of the die. An aluminum-free spot in the middle of the heater is reserved for the smart temperature sensor, which is glued to the quartz substrate (figure 1a). This enables direct measurement of the die temperature. The absolute measurement error is, amongst other things, determined by the thermal resistance between the temperature sensor and the quartz substrate, therefore, a heat-conductive adhesive is applied [2, page 51]. The quartz substrate is mounted on a circular Duroid printed circuit board (PCB), by using a silicone adhesive. The temperature sensor and the aluminum heater are wire bonded on gold-plated PCB wires through a hole in the center (figure 2b). Duroid laminate is applied, because it is composed of inert glass micro-fiber embedded in a matrix of polytetrafluor-ethylene (PTFE). PTFE is exceptional among plastic materials because of its chemical inertness, insolubility, non flammability and non toxicity. It can also withstand temperatures of up to 280 °C, exhibits no outgassing and has a low thermal conductivity (0.26 $Wm^{-1}K^{-1}$). These are important features, because the quartz substrate can reach temperatures of up to 120 °C. A tin-plated steel cylindrical box containing one radial gas inlet and two lateral gas outlets are soldered gas tight on top of the PCB. The box and the ground plane of the PCB provide for the necessary electrical shielding and also reflect the heat radiation, thus keeping a large part of this heat inside. The complete box is placed into a PTFE encapsulation, which provides for the required thermal insulation. Disadvantages of the PTFE are that it is expensive and more difficult to process, because

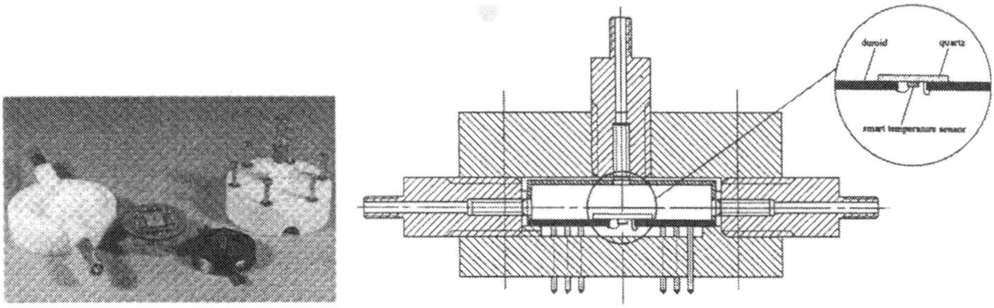

(a) (b)

Figure 2. (a) The sensor encapsulation. (b) Cross-section of the sensor encapsulation.

it is quite plastic. Because the life-time of the sensor is limited by the ftalocyanine layer, it is economical to re-use the PTFE encapsulation. After the bottom and the top are screwed together, the PTFE gas connectors are screwed onto the gas inlet and outlets with PTFE tape between, to form gas-tight connections. Figures 2a and 2b show the encapsulation.

Controller design starts with thermal modeling. In this case, the process under control contains the chemo-sensor and its encapsulation. Figure 3 shows the time responses of the process temperature θ, in °C, to power steps of two Watts for different air-flow velocities and a constant flow temperature. A higher air-flow velocity is responsible for a lower final temperature, because of the heat drain on the flow. A second-order transfer function of the thermal process has been derived by fitting the function:

$$\tilde{\theta}(t) = K_1\left(1 - e^{-\frac{t}{\tau_1}}\right) + K_2\left(1 - e^{-\frac{t}{\tau_2}}\right) \tag{1}$$

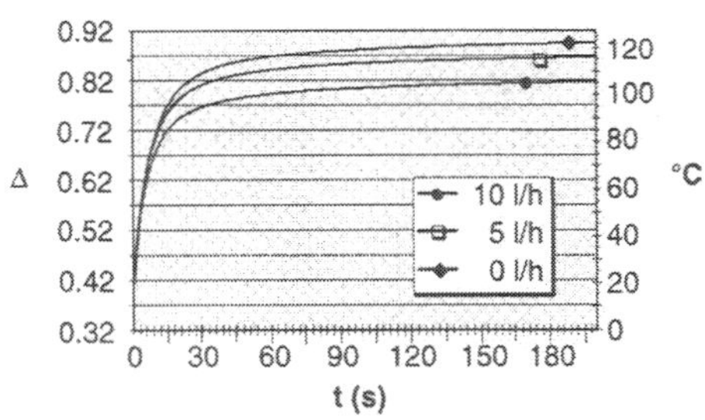

Figure 3. Time responses to power steps of 2 Watts for different air-flow velocities and a constant flow temperature: $\theta_{flow} = \theta_{amb}$. (Δ = duty-cycle, see equation 6).

onto the curves of figure 3. The parameters K_1, K_2, τ_1 and τ_2 are determined by applying, e.g., the least-squares criterion. The least-squares criterion minimizes:

$$\int_0^t \left(\tilde{\theta}(t)-\theta(t)\right)^2 dt \tag{2}$$

in which $\tilde{\theta}(t)$ is the approximation of the measured curve $\theta(t)$. The Laplace transform of equation (1) consists of the two Watts power step and the transfer function $G_p(s)$ of the thermal process:

$$\frac{2}{s}\cdot\frac{\left(\frac{K_1+K_2}{2}\right)\left(\frac{K_1\tau_2+K_2\tau_1}{K_1+K_2}s+1\right)}{(\tau_1 s+1)(\tau_2 s+1)} \equiv \frac{2}{s}\cdot G_p(s) \tag{3}$$

For a flow of 5 lh^{-1}, $G_p(s)$ was found to be:

$$G_p(s)=\frac{48(35s+1)}{308s^2+51s+1} \tag{4}$$

The thermal model of the whole sensor can be represented by an electrical equivalent circuit [2, chapter 2], shown in figure 4.

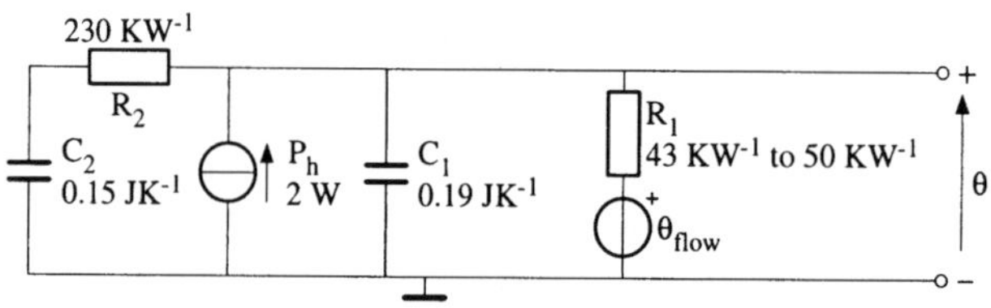

Figure 4. The thermal model of the whole sensor represented by its electrical equivalent circuit. R_1 varies in the range 50 KW^{-1} to 43 KW^{-1} for flows in the range 0 lh^{-1} to 10 lh^{-1}.

This electrical equivalent is obtained by physical insight and has the following transfer function:

$$G_p(s)=\frac{R_1(R_2C_2s+1)}{R_1R_2C_1C_2s^2+(R_1C_1+R_2C_2+R_1C_2)s+1} \tag{5}$$

which is identical to equation (4). In this model, R_1 (KW^{-1}) and C_1 (JK^{-1}) represent the thermal resistance to the flow and the thermal capacitance of the quartz substrate, respectively. The values of R_2 an C_2 were obtained by the curve-fitting process and belong, in regard to the values, most probably to the silicone adhesive between the quartz and the Duroid PCB. The values of R_1, R_2, C_1 and C_2 are obtained by comparing equations (4) and (5). The heat flow to the environment can be neglected because of the high thermal resistances of the PTFE encapsulation. The value of R_1 varies in the range 50 KW^{-1} to 43 KW^{-1} for flow velocities in the range 0 lh^{-1} to 10 lh^{-1}. The power (heat) dissipated in the heater is represented by the current source P_h. The voltage source θ_{flow} represents the changes in the flow temperature. The temperature measured by the temperature sensor is represented by θ. The environment temperature is used as a reference (ground).

4. The smart temperature sensor

For the temperature sensor we applied the SMT 160-30 [4,5]. The *duty cycle* (Δ) of its square-wave output signal is directly proportional to the temperature (equation 6) in the -45 to +130 °C range [5]:

$$\Delta = 0.32 + 0.0047\theta \tag{6}$$

where θ (in °C) represents the temperature. A low-cost Analog-to-Digital conversion is realized by connecting this sensor to a micro-controller with a "capture function", e.g., the Intel 87C51FB(FX). The micro-controller captures the content of a free-running counter on each rising and falling edge of the output signal of the sensor. Thus a digital representation is obtained in two steps. In the first step, the analogue temperature signal is converted into a duty cycle by the temperature sensor. The conversion of the duty cycle in the second step is performed by the micro-controller.

The *duty cycle* contains the information which is present in each whole period. However, since also the frequency is temperature dependent, the data speed is temperature dependent too. Since for discrete time systems it holds:

$$z = e^{sT} \tag{7}$$

where T is the sampling time. This means that the time response characteristics are a function not only of s, but also of T. Variations in the sampling rate change the z-plane pole locations, which determine the transient response of the feedback control system. A constant measurement time solves this problem. This is accomplished by varying the number of periods, N, so that the product of N and the period of the temperature sensor's output signal, t_p, is as close as possible to a predefined value of the total measurement time, t_m. However, a constant measurement time causes the total noise to be temperature dependent, hence the minimal noise requirements have to be satisfied for the highest output frequency of the temperature sensor, because the sampling noise is inversely proportional to the square root of the period t_p.

Precautions have to be taken to minimize the amplitude of the variations in the heater signal, because they introduce cross-talk in the SAW device. Therefore a 12-bit digital-to-analog converter is implemented to provide for the heater signal, instead of pulse-width modulation, which could be implemented much more cheaply by using a timer in the micro-controller. For the same reason, the noise of the temperature sensor has also to be significantly suppressed. In other applications, this noise is less critical, because it is filtered by the thermal process, which possesses a low-pass characteristic, usually combined with a very small bandwidth.

5. The digital control system

The response time of the thermal process shown in figure 3 is minimal for the maximum flow velocity. In this situation, the smallest time constant of the process (τ_l) equals the product of C_l and the minimal value of R_l, hence:

$$\tau_{1,\min} = C_1 \times R_1 = 0.19\ JK^{-1} \times 43\ KW^{-1} \approx 8 \sec \tag{8}$$

As a rule of thumb, the sampling time (T) is taken at least ten times smaller than the smallest time constant of the process under control. Note that at a sampling rate of $\omega = 0.1\omega_{sample}$, the zero-order hold introduces a phase lag of 18°, which is an appreciable amount. The maximum allowable total noise level of the temperature sensor dictates the minimal measurement time. Although these two demands could conflict with each other, they do not, because of the big time constant τ_I. The time required to perform filter calculations in the micro-controller introduce a time delay in the control loop. A time delay (t_0) is represented by $e^{-j\omega t_0}$ in the frequency domain. The phase lag caused by this time delay equals:

$$\phi(\omega) = -\omega \cdot t_0 \tag{9}$$

Hence, the time delay decreases the phase margin, which has a destabilizing effect.

To comply with all of the above-mentioned requirements and to meet the Nyquist sampling criterion, a sampling time of T = 0.1 seconds and a temperature measurement time of 0.09 seconds were chosen. The block diagram of the control system is shown in figure 5. $R^*(s)$ is the set-point temperature. $I(s)$ is the temperature θ averaged over 0.09 seconds by the micro-controller. The transfer function 0.0047/(0.09s+1) represents the temperature measurement and averaging process. $G_p(s)$ is the transfer function of the process under control. The zero-order A/D and D/A conversions are modeled by $(1-e^{-Ts})/s$. D_1, D_2 and the limiter represent a non-linear anti-windup controller [6].

Now, the controller action is explained. The controller was initially developed by designing a PI controller, with transfer function :

$$K_p\left(\frac{s\tau+1}{s}\right) \equiv 0.03\left(\frac{9s+1}{s}\right) \tag{10}$$

in the s-domain, such that the feedback system exhibits no overshoot on a step input. The reason for avoiding overshoot is that the warming-up process is much faster than the cooling-down process. The block diagram for warming up is given in figure 5. An active cooling facility is not available

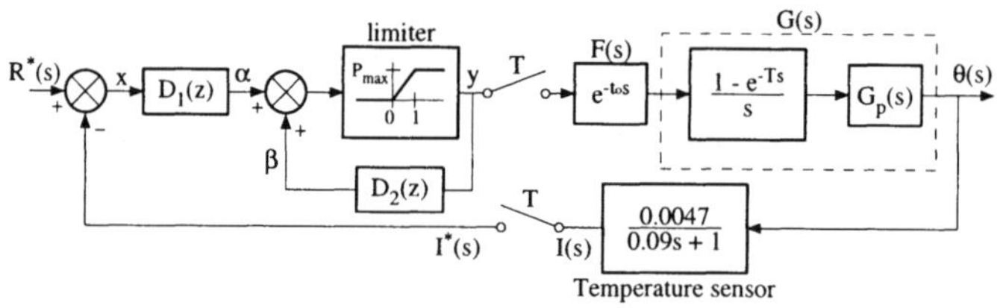

Figure 5. The block diagram of the temperature-control system for warming up.

(P_{min} = 0) and the power delivered to the heater is finite (P_{max} = 2.65 Watts). To model this actuator saturation a limiter is added to the control scheme. Outside the linear range, the PI-controller output does not drive the process thus causing the controller states to be updated wrongly. This effect is called controller windup. Therefore the PI controller is divided into:

$$D_1(s) = \frac{K_p \tau_a (\tau s+1)}{P_{\max}(\tau_a s+1)} \equiv \frac{0.006(9s+1)}{2.65(0.2s+1)} \tag{11}$$

$$D_2(s) = \frac{1}{P_{\max}(\tau_a s+1)} \equiv \frac{1}{2.65(0.2s+1)} \tag{12}$$

The transfer function of the anti-windup controller is, in the linear range, identical to the transfer function of the PI controller. Outside the linear range, the anti-windup circuit bounds the transient, which could be large when the system returns to the linear regime, and could lead to instability. The continuous-time transfer functions $D_1(s)$ and $D_2(s)$ are transformed into discrete-time transfer functions:

$$D_1(z) = \frac{\alpha(z)}{x(z)} = Z\left[D_1(s)\left(\frac{1-e^{-sT}}{s}\right)\right] \tag{13}$$

leading to the difference equations:

$$\alpha(n) = 0.1x(n) + Q(n-1) \tag{14}$$

$$Q(n-1) = 0.606531\alpha(n-1) - 1.346612x(n-1) \tag{15}$$

Because $D_2(s)$ and the limiter form a mathematical loop, $D_2(s)$ is transformed directly by using the Euler approximation for numerical integration:

$$\beta(n) = 0.606531\beta(n-1) + 0.148479y(n-1) \tag{16}$$

If the time delay t_0 becomes significant in comparison to the sampling time T, the modified z-transform has to be applied in the above transformations to account for the non synchronous sampling process [7, subsection 4.7].

To obtain a constant measurement time and a minimal total time delay, the input and output data rates of the micro-controller are synchronized by two software timers.

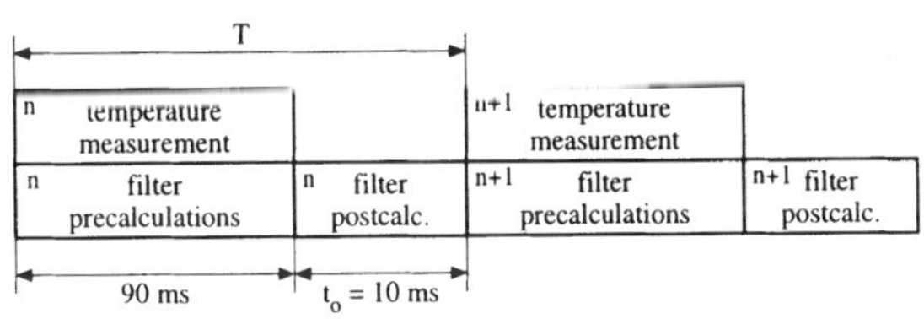

Figure 6. Timing diagram of the synchronization process.

The timing diagram is shown in figure 6. The parameters Q(n-1) and β(n), equations (15) and (16) respectively, are calculated in the precalculation time. α(n) is calculated in the postcalculation time by using equation (14).

6. Experimental results

The temperature-control system has been tested under various circumstances. As an example, figure 7a shows the measured time responses for different set-points and a flow velocity of 10 lh^{-1}. It has been concluded that the set point is reached within 90 seconds. For a flow velocity of 10 lh^{-1}, temperature fluctuations remain within ± 0.01 °C around a 120 °C set point. It has been found that even when the flow velocity varies ± 0.1 lh^{-1} at a frequency of 2 Hz, the temperature deviation from the set point remains within ± 0.02 °C. Figure 7b shows the effects of flow steps on the system response. The peak at 75 seconds and the dip at 135 seconds are caused by switching off and on a flow of 10 lh^{-1} respectively. It can be concluded from this plot that the system is quit robust.

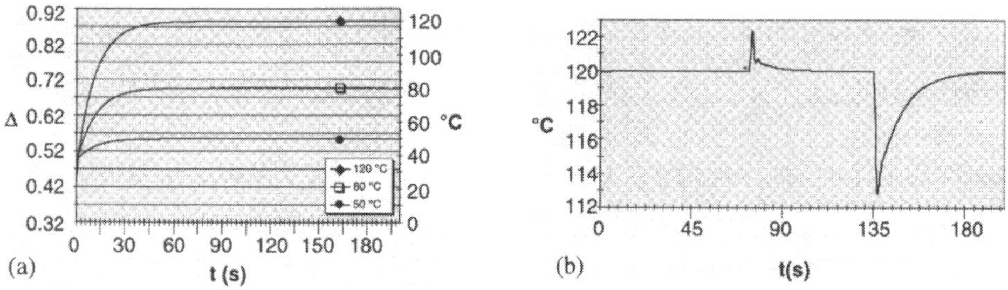

Figure 7. (a) Reaching different set points for a flow velocity of 10 lh^{-1}. (b) Time responses to flow steps of 10lh^{-1}.

7. Conclusions

It can be concluded that the resolution and accuracy of SAW devices can be significantly improved by applying temperature stabilization. SAW devices have to be shielded electrically and thermally from their environment, in order to minimize disturbances.

Application of the SMT 160-30 smart temperature sensor combined with a micro-controller in a temperature-control system provides for a low-cost A/D conversion. The output frequency of the smart temperature sensor changes with temperature, which means that the sample rate is temperature dependent, hence the z-plane pole locations are also temperature dependent. The use of a constant measurement time solves this problem.

The process under control determines the maximal sampling time, whereas the noise suppression requirements determine the minimal sampling time (temperature measurement time). The filter calculations in the micro-controller introduce a delay time in the control loop, thereby decreasing the phase margin. Long delays with respect to the sampling rate introduce non synchronous sampled systems, which may lead to instability.

The controller is designed in such a way that no overshoot occurs, because the warming-up process is much faster than the cooling-down process. To deal with actuator saturation, an anti-windup control mechanism has been implemented in a micro-controller, yielding a temperature-control system that satisfies the specifications.

References

1. Nieuwenhuizen M.S., Venema A., *Surface acoustic wave chemical sensors*, Sensors and Materials 5 (1989) 261.
2. Meijer G.C.M. and Herwaarden A.H., 1994 *Thermal Sensors* (Institute of Physics Publishing Bristol and Philadelphia).
3. Goes van der M.L. and Meijer G.C.M., 1996 *Sigma-Delta versus Oscillator-based A/D Converters in Low-Cost Accurate Sensor Systems.* (Conf. Proc. Instrum. and Meas. Imtc/96).
4. Meijer G.C.M., Gelder van R., Nooder V., Drecht van J. and Kerkvliet H.M.M., 1989 *A three-terminal integrated temperature transducer with micro-computer interfacing*, Sensors and Actuators, 195-206.
5. Smartec, 1991 *Specification sheet SMT 160-30*, Breda, Netherlands.
6. Mayuresh V. Kothare, Peter J. Campo, Manfred Morari and Carl N. Nett, 1994 *A Unified Framework for the Study of Anti-windup Designs*, Automatica, Vol. 30. No. 12. pp 1869 -1883.
7. Phillips C.L. en Nagle H.T., 1990 *Digital Control Systems and Design* (Prentice-Hall INC).

AN ACTUATION PRINCIPLE: THE ELECTROLYSIS OF WATER

Cristina Neagu, Henri Jansen, J.G.E. Gardeniers and Miko Elwenspoek

MESA Research Institute, University of Twente, P.O.Box 217, 7500AE Enschede
Phone: 053-4892805/2751; Fax: 053-4894434; E-mail: C.Neagu@el.utwente.nl

0. Abstract

In this paper the theory of water electrolysis in a closed electrochemical cell, that contains two electrodes, an electrolyte and a pressure sensor is described. From the leakage and electrochemical experiments done with this cell it is possible to obtain information about the applicability of the electrochemical principle in a closed cavity, the choice of the electrodes and electrolyte, and different types of leakages. Further, tests done with a control system, that consists of an electronic feedback system together with the electrochemical macro cell and the pressure sensor are presented. A value of the pressure is set and the regulator will actuate the electrochemical cell in such a way to get the desired pressure.

Keywords: electrolysis

1. Introduction

In the field of MEMS there is an ever increasing need for actuators able to deliver a relatively large amount of mechanical power, i.e. they should create sufficient displacement together with a big force or pressure. When a big force is needed, generally piezoactuators are used, however, this type of actuators suffer (except for piezoelectric bimorphs) from a small displacement. In contrast, electrostatically driven comb actuators are found when the displacement should be high, but, now the force is relatively small.

To overcome this limitation, in 1996 the authors introduced a new actuation principle in the field of MEMS characterised by a large displacement while maintaining a high pressure: the electrochemical (EC) actuator [1, 2]. In such an actuator, a miniaturised gas sealed cavity filled with an electrolyte and two electrodes is connected to an external power supply to generate gas from the electrolyte by way of electrolysis. The produced gas is used to deflect a membrane, which in turn can be used to perform mechanical work.

The paper will start with a short description of the reaction kinetics underlying the electrolysis and vaporisation of water. After this, the design of an EC cell and experiments will be discussed. To stabilise generated gas pressure at a desired value, as well as the deflection of the membrane, a feedback control system has been incorporated as discussed in the last section.

2. Theory

2.1 Electrolysis of water

The electrolysis of water is a well-known principle to produce oxygen and hydrogen gas from water. In figure 1a, a schematic of a sealed electrochemical cell is shown. The cell is partly filled with pure water and has two electrodes which are connected with an external power supply. At a certain critical voltage between both electrodes, the electrodes start to produce hydrogen gas at the negatively biased electrode and oxygen gas at the other one. The amount of gas evolved per unit time is directly related to the current passing the water.

In water, there is always a certain percentage found as ionic species; H^+ and OH^- represented by the equilibrium equation:

$$H_2O(l) \Leftrightarrow H^+(aq) + OH^-(aq) \tag{1}$$

At the negative electrode, the H^+ ions will gain an electron thereby forming hydrogen. At the positive electrode, the OH^- ions will deliver an electron and finally creates oxygen gas:

positive electrode: $4OH^- - 4e^- \rightarrow 2H_2O + O_2$ (2a)

negative electrode: $4H^+ + 4e^- \rightarrow 2H_2$ (2b)

In case of acidic or basic water, the reactions which occur at the electrode interface are slightly different. Anyhow, the overall reaction kinetics, i.e., presuming no additional side reactions is:

$$2H_2O \Rightarrow (4e^-) \Rightarrow O_2 + 2H_2 \quad (3)$$

So, the production of 3 gas molecules out of two water molecules needs 4 electrons. The complete electrolysis of 1 mole of water molecules is, therefore, accompanied by 2 moles of electrons. The charge of 1 mole of electrons is known as Faraday's constant, $F = 96400$ [C]. For the electrolysis of water, the standard reaction enthalpy ΔH^0 is, $\Delta H^o = 285830$ [J/mol] and the free energy can be calculated as:

$$\Delta A^o = \Delta H^o - TR\Delta n - T\Delta S^o = 233147 \text{ [J/mol]} \quad (4)$$

In our case of a closed cell we use ΔA instead of ΔG, that is used for an open cell. The voltage necessary for an electron to overcome the Helmholtz energy barrier is, thus: $\Delta V_a^o = \Delta A^o/n \cdot F =$ 1.21 volt, defined as the standard free voltage.

To start a reaction, it is normally necessary to overcome an (extra) energy barrier; the activation energy E_{act}. The number of molecules able to overcome this barrier is controlling the reaction rate, ρ, and it is given by the statistical Maxwell-Boltzman relation which has an exponential behaviour: $\rho \sim \rho_o \cdot \exp(-E_{act}/RT)$. So, the activation energy is depicting the speed with which a reaction occurs.

We are now able to write down the energy balance during the electrolysis of water. Assume that 5 V is applied to the sealed EC cell, fig. 1a. The energy, ΔE_{elec}, supplied by the voltage source, ΔU, per mole of water is $\Delta E_{elec} = \Delta U \cdot nF = 5 \cdot 2 \cdot 96400 = 964$ kJ/mol. This energy equals the reaction free energy plus the IR drop ($E_{IR} = IR \cdot nF$; i.e. losses causing Joule heating) plus the overpotential, η, per electron ($E_\eta = \eta \cdot nF$), fig. 1b. The IR-drop is caused by the low electrical conductance of the water. It can be improved by making the electrolyte acid or alkaline, or by adding a support electrolyte which delivers extra charge carriers. The overpotential is responsible for the typical exponential behaviour of the current flowing through the solution as given by the so-called Bultler-Volmer equation. The activation energy can be lowered by using electrodes with a catalytic action such as platinum. This action can be described as [3]:

$$H_2 + 2Pt \Rightarrow 2Pt\text{-}H \quad (5a)$$

$$O_2 + 2Pt \Rightarrow 2Pt\text{-}O \quad (5b)$$

$$2Pt\text{-}H + Pt\text{-}O \Rightarrow 3Pt + H_2O \quad (5c)$$

As can be observed in fig. 1b; the balance of energy per mole of reaction events is:

$$\Delta E_{elec} = \Delta A + E_{IR} + E_\eta \quad (6)$$

All the energies involved in this equation of balance, except for ΔA, will leave the reaction as heat: $T\Delta S$. In first instance, this heat will be used for the vaporisation of the water

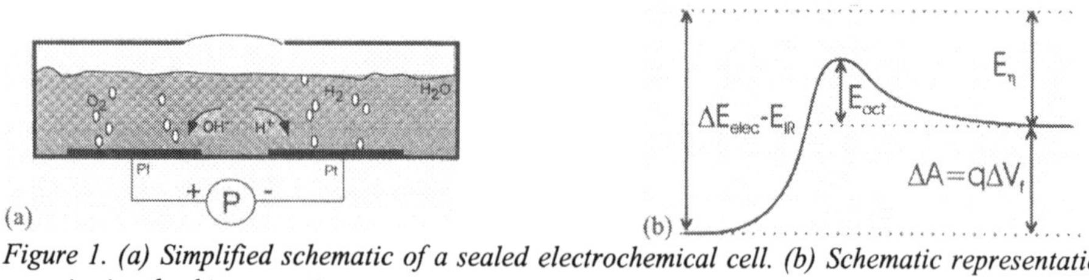

Figure 1. (a) Simplified schematic of a sealed electrochemical cell. (b) Schematic representation of energies involved in a reaction.

and, thus, increasing the pressure inside the cell. However, after a while all this heat will be lost due to heat transport to the surroundings and the pressure will decrease again. Of course, heat will be exchanged with the surroundings. So, in fact we have to take the enthalpy instead of Helmholtz as the energy involved in the electrolysis process. The efficiency, ε, of an electrolysis cell is thus:

$$\varepsilon = \Delta H/\Delta E_{elec} = \Delta H/(\Delta A + E_{IR} + E_{\eta}) \qquad (7)$$

For water electrolysis, the value of ΔH = 285.830 J/mol and ΔA = 233.147 J/mol. Hence, under ideal reversible conditions, ε = 120%. This means that the electrolysis would take place with 120% efficiency in respect to the electrical energy source. Thus, heat would have to flow into the cell from the surroundings. When the value of the denominator in eq. 7 is 1.48 nF, i.e. electrolysis is performed at an over-potential of 0.25 Volt, the electrolyser would perform at 100% efficiency. Under these conditions, the cell neither heats nor cools. V_{th}= ΔH/nF is denoted as the thermoneutral potential. In practice, the IR drop may itself be around 0.25 V. Anyway, the overpotential η should be kept low in order to maximise the efficiency and to minimise the production of heat. Of course, the lower the overpotential the slower the reaction will proceed and we would be forced to make a compromise. A way to increase the current without increasing the overpotential is to increase the contact area between the electrodes and the liquid.

2.2 Vaporisation of water

Another way to increase the pressure inside a closed volume is by increasing the temperature. In case of liquids, the temperature dependency of the pressure is normally much higher due to the production of vapour gas. In case of water, the overall reaction is:

$$H_2O(l) \Leftrightarrow H_2O(g) \qquad (8)$$

ΔH^0 = 44004 J/mol, the standard enthalpy of vaporisation. The vapour pressure is increasing exponentially with the temperature as can be found in most books considering thermodynamics. In case of a constant volume the relation is:

$$P_v(T) = P^o \exp\left(\frac{-\Delta A^o}{RT}\right) = P^o \exp\left(\frac{R\Delta n + \Delta S}{R}\right)\exp\left(\frac{-\Delta H^o}{RT}\right) \qquad (9)$$

The sensitivity of the vapour pressure with respect to the temperature can be found by differentiating $P_v(T)$ with respect to T. We arrive at:

$$\frac{dP_v}{P_v} = \frac{\Delta H^o}{RT}\cdot\frac{dT}{T} \qquad (10)$$

3. Experimental

As a start, it is necessary to close the macro cell hermetically from the surroundings to harness the gases produced. In principle this is not difficult, however, we would like to have a cell which can be easily modified with respect to the electrode materials and electrolytes. The final version did incorporate some basic ultra high vacuum components from the vacuum technology. In figure 2, an art impression of a basic macro cell is shown. A stainless steel blank flange was machined (a hole was drilled) to be able to mount the pressure sensor. The pressure sensor was connected by way of a Viton 'O' ring and Teflon tape to ensure a leak-free montage. The electrodes were inserted inside standard K40 "window-glasses" that has a stainless steel support-ring. Holes, 1mm in diameter, were drilled into the glass through which the electrodes were inserted. The residual gaps were filled with Torrseal to prevent leakage. After filling the cell with the electrolyte, all the parts were assembled and closely tight by way of a flange clamp.

During long-term measurements (exceeding many days) pressure and temperature fluctuations in the atmosphere could alter our monitored signal and validity of the conclusions drawn drastically. When the macro cell is closed under atmospheric pressure and is placed inside a vacuum surrounding, only fluctuations of the base pressure caused by the vacuum pump will disturb the signal. The set-up of the macro cell inside the vacuum chamber is given in figure 3. The pressure sensor to measure the gas evolution due to electrolysis is a miniature vented gauge transducer of Data Instruments (MM10013). The sensor is temperature compensated in stainless steel housing and able to measure pressures up to 100PSI~ 7bar. It uses piezo-resistive elements driven by a stabilised 5 V supply from an external voltage source (Voltcraft TNG35). To minimise the ripple from this source, a voltage regulator is utilised (LM7805). The voltage monitor of the pressure sensor is a programmable unit from Hewlett Packard (HP34401a) with an IEEE488 (GPIB) port to enable data exchange. This data from the pressure sensor is recorded with the use of HP-Vee software.

Now the cell has been placed inside a vacuum chamber, the problem arises how to connect the electrical leads from the pressure sensor to the units outside the vacuum chamber. The standard electrical feed-through flange from the vacuum technology is not appropriate because temperature differences between the two leads of the sensor signal at the flange interface would cause thermo-voltages much too large to be neglected. Therefore, again a modified blank flange is used. The leads are directly led through the flange without introducing an extra electrical contact to avoid thermo-voltages. The gaps left are filled with Torrseal to prevent leakage.

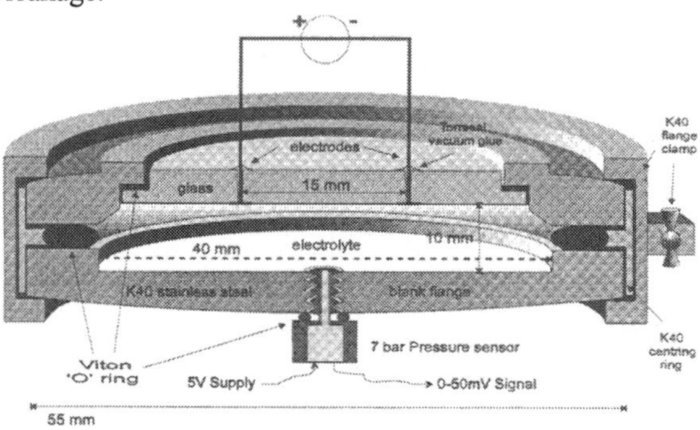

Figure 2. Basic set-up of the vacuum sealed macro cell with standard UHV components.

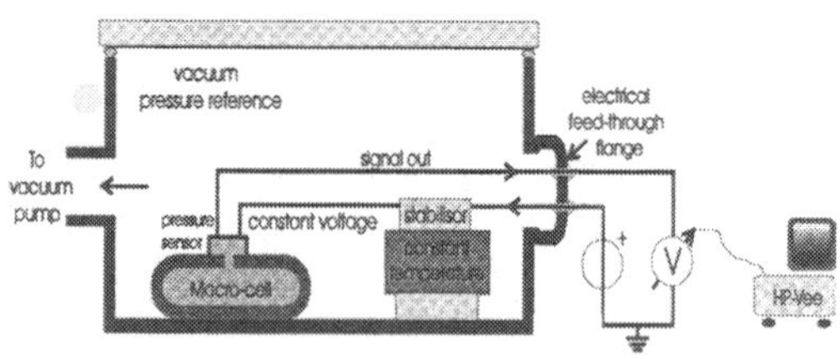

Figure 3. Basic experimental set-up of the macro-cell.

4. Results

Gold: In the first experiment, the macro cell has two gold electrodes and the result is given in figure 4. The plot shows three peaks; the first peak is caused by a current through the cell of 500μA for 200sec at 3.8V, the second one by 100μA for 1000sec at 3.0V, and the third peak is caused by 20μA for 5000sec at 2.5V. So, the total charge transported to the cell was constant for all peaks. Clearly, the increase in pressure is identical for all the three peaks, as it

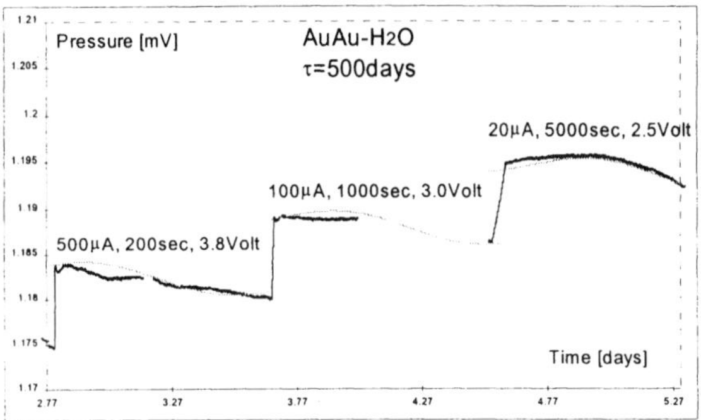

Figure 4. Behaviour of the EC cell with gold electrodes, the 'lifetime' of the open circuit state is > 500 days.

should because the current of electrons controls the flow of reaction gases out of the electrolysis process. The drop in pressure after switching the electrolysis current of is a very slow process with a response time better than 500 days. This means that the leakage of the cell or the pressure drop caused by other processes is minimal.

Platinum: The next experiment is using two platinum electrodes (figure 5a). Again three electrolysis sequences have been performed; the first peak is caused by 2mA for 50sec, the second by 1mA for 100sec, and the third is 0.5mA for 200sec. It can be seen that the signal is reaching a plateau after a sharp initial peak directly after switching of the electrolysis current. The sharp peak is thought to be caused by the relatively large over voltage necessary to sustain such a high current. This voltage will heat the liquid and vaporise some water. After switching the current off, the cell is loosing its heat quickly to the surroundings and, therefore, the water vapour will condense back to liquid water. It is found that the height of the plateau is directly related to the amount of electrons driven through the cell, i.e. the flow is proportional to the current times the time the cell is fed with this current. To show the influence of the catalytic action of the platinum, a current of 0.5mA was driven through the cell for a very long time of 90 minutes (fig. 5b). It is observed that the pressure drops quickly to its value back with a response time of 3.5 days. This drop is thought to be caused by the catalytic action of the platinum.

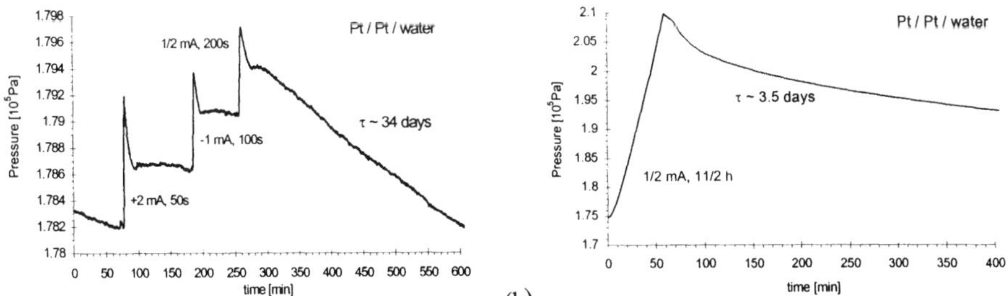

(a) (b)

Figure 5. The EC cell with platinum electrodes. The lifetime of open circuit state is very short (~3.5 days). a) The behaviour for short current pulses and b) for a long pulse.

Silver: The last material which has been tested for the electrolysis of water is silver. In fig. 6 the result is shown. As can be observed in fig. 6a, the increase of pressure is again directly related to the charge which has been driven through the cell. A remarkable difference, however, is that there is no sharp peak is observed directly after the current is switched off. This difference is probably caused by a leakage of the cell: This leakage will cool the water inside the cell very quickly due to the heat needed for the vaporisation (ΔH). After the cell was cleaned and assembled again, the characteristic sharp peak appeared. The decay time was approximately 34 days. Moreover, after disassembling the cell, it was found that the electrodes were completely eroded and a black substance, probably silver oxide, was found as an emulsion in the water.

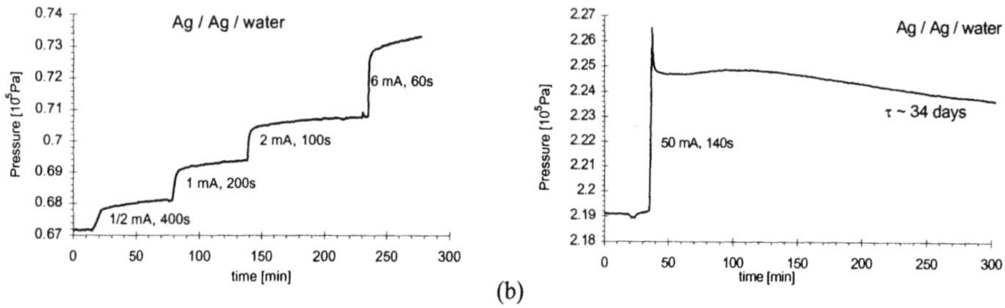

(a) (b)

Figure 6. The EC cell with silver electrodes. The lifetime of the open circuit state is intermediate (34 days). (a) The behaviour for short current pulses and (b) for a long pulse.

5. The feedback control system

A constant cell pressure can be achieved by incorporating a feedback controller, together with the sensor and actuator forming a regulating system. In fig. 7a, such a total regulated system is shown and in fig. 7b the electronic circuit of feedback controller is given in detail. It consists of three sections; a 100 times amplifier for the sensor signal (OP27), an integrator build around a comparator to pre-set the cell pressure (CA3140), and a current amplifier to feed the electrolysis sequence (BD139). After amplification, the signal is fed to the CA3140 where it is compared to the requested pressure. This is achieved by a trim-potentiometer "Pressure Set" which is able to maintain a pressure inside the cell up to 7bar. The signal is integrated to filter out any noise. The integration time is controlled by a replaceable capacitor, in this case 10nF. The capacitor across the base emittor of the current amplifier BD139 is meant to prevent parasitic oscillations.

The result of the feedback controller upon the pressure inside the cell is shown in fig. 8. Clearly, the circuit is stabilising the cell pressure much better as would be possible without the controller.

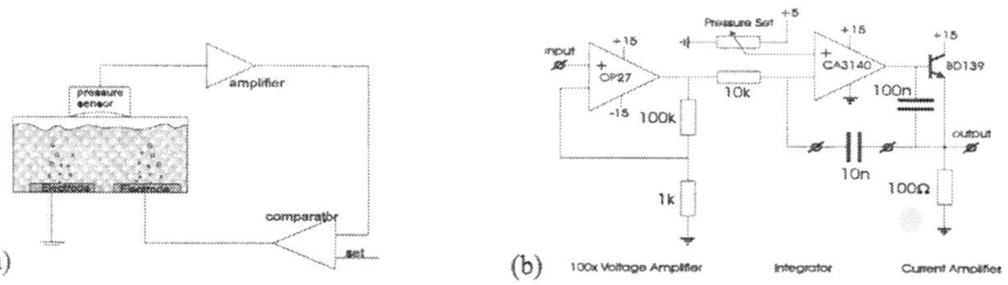

(a) (b)

Figure 7. (a) Schematic of the feedback controller to stabilise the cell pressure. (b) Electronic circuit of the feedback controller to settle the cell pressure.

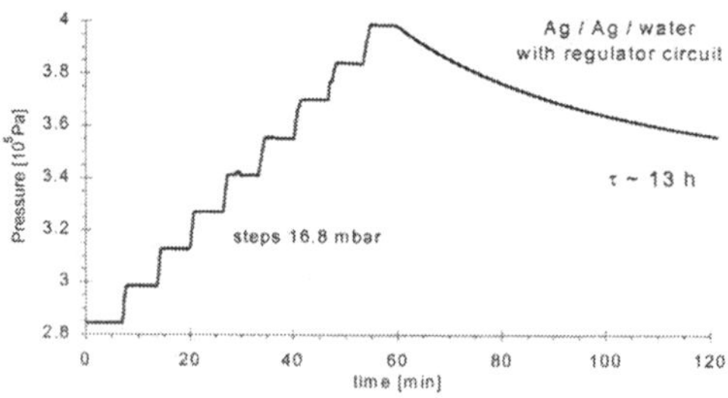

Figure 8. Output of the cell pressure with feedback controller

6. Conclusions

In this paper the results of an electrochemical cell have been presented. It is found that the pressure drops relatively fast (a few days) back to its initial value when platinum is used. The reason would be that platinum has a low so-called activation level to make the transition of hydrogen with oxygen gas into water possible. Due this characteristic, the non-equilibrium state, i.e. hydrogen and oxygen gas, is difficult to sustain. It is proven experimentally that the decay of the pressure after switching the cell off (the so-called passive state of the cell) is indeed improving by the use of gold electrodes. However, a disadvantage of this higher activation energy level is the higher so-called over-potential necessary to drive the cell from one passive state (e.g. low cell pressure) into another passive state (e.g. high cell pressure). This higher overpotential will give rise to the production of heat which will be used to convert water into water vapour. After the current to the cell is switched off, this vapour will condense until thermal equilibrium is reached again. This thermal decay effect is a relatively fast process with respect to the electrolyse back reaction and, therefore, an extra and sharp peak on top of the step caused by the electrolysis is found.

With respect to the usability of an electrolysis process as an actuator, it can be stated that the effect of vaporisation of water is the main bottleneck. Rapid changes in temperature will cause large water vapour pressure changes. These changes are compensated for by the electrolysis process with the help of a feedback control circuit. This circuit has been tested in this paper with very good results. However, the circuit was in fact meant to regulate the cell pressure and not the vapour pressure! So, under certain circumstances it is possible that the EC cell will be more busy with controlling the vapour pressure of the cell than with regulating the cell pressure.

When the efficiency of the electrolysis process and the additional vapour pressure peak is not a problem, it can be said that the EC actuator is able to create and maintain a very high pressure inside a closed cell. Pressures beyond 7 bar have been reached without any difficulties. This maximum is only restricted by the limitation of the pressure sensor itself. Due to this characteristic, the EC cell can open possibilities far beyond the reach of other types of actuators, such as the thermal actuator, piezo actuator, or magnetic actuator.

To create a more efficiently operating cell, we may study equation 7. A first approach would be to minimise the IR drop, E_{IR}, by increasing the conductance of the solution. Secondly, E_η can be minimised by using a catalytic material like platinum or increasing the active surface of the electrode. This can be achieved by creating very rough surfaces, i.e. so-called black platinum. However, to avoid a fast recombination of the gases towards water, the activation energy should be high (i.e. gold instead of platinum electrodes).

References

[1] C. Neagu, J.G.E. Gardeniers, M. Elwenspoek and J.J. Kelly, An Electrochemical Microactuator: Principle and First Results, *J. Microelectromech. Syst.* 5(1), March, 1996, p.2-9.

[2] C. Neagu, J.G.E. Gardeniers, M. Elwenspoek and J.H.J. Fluitman, An electrochemical actuated microvalve, *Actuator '96*, Bremen 19-21 June, 1996.

[3] E. Gileadi, E. Kirowa-Eisner, J. Penciner, *Interfacial Electrochemistry. An experimental approach,* Addison-Wesley, 1975.

A MIXER BASED ON DIFFUSION-MIXING DESIGNED FOR THE MAFIAS-PROJECT

T.T. Veenstra, T.S.J. Lammerink, A. van den Berg, M.C. Elwenspoek
MESA Research Institute, University of Twente
P.O. Box 217, 7500 AE Enschede, The Netherlands
E-mail: t.t.veenstra@el.utwente.nl

0. Abstract

The MAFIAS project (Micro Ammonia Flow Injection Analysis System) aims at the development of a monolithic micromachined flow injection analysis system for the detection of ammonia. From the applied chemistry, design restrictions for the micromechanical ammonia detection system are deduced. These restrictions involve the temperature control of the system at 37° C. Also the time available for the mixing of the compounds used in Berthelot's method is restricted to less than 20 seconds. This last restriction calls for a mixer since the compounds itself would take at least three times as long to mix. A new type of mixer based upon diffusion is proposed. Theory for deriving the exact geometry for the mixer is deduced. With this mixer, the mixing time can be reduced to 2 seconds.

Keywords: Ammonia, Phenol, Mixing, Diffusion

1. Introduction

The MAFIAS project uses Berthelot's reaction for the measurement of ammonia. Berthelot's reaction mechanism transforms ammonia into indophenol-blue. The concentration of indophenol-blue can be measured by its feature that it has an absorption-spectrum with a peak at 625 nm. Beer's law states that the light absorption at this peak will be a measure for the concentration of indophenol-blue. The layout of the system is shown in figure 1.

Since the detection system will be monolithic, all incorporated fluidical components such as pumps, mixers, detection cell and channels have to be designed to be processable with each other. Therefore components with a lot of processsteps are not favored, since these components might seriously reduce the overall yield.

For the mixer component, design restrictions will be derived. These restrictions are the result from the various specifications which the system has to meet and from the chemistry to be used. The complete range of specifications that has been set up, includes the following:

- 1 sample throughput/minute.
- Total system volume < 1 cm^3
- Use of Berthelot's method

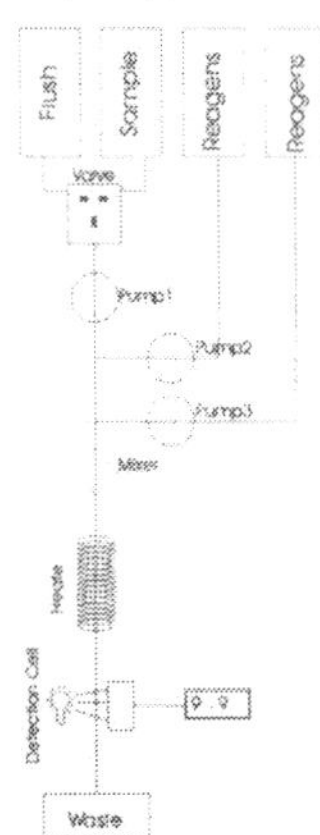

Fig.1: Layout of the MAFIAS-system

2. Theory

For the detection of ammonia, chemistry is invoked to convert the ammonia molecules into a compound that has a large absorption coefficient. The so-called Berthelot-method will be used for this conversion. In 1859 Berthelot [1] wrote a letter to the 'Repertoire de chimi appliquee', a periodical on chemistry. He stated, that he found the blue coloration of ammonia mixed with phenol and warm hypochlorite to be of the exact same nature as the blue coloration of aniline under comparable circumstances. Berthelot therefore claimed that very probably the reagents he used were polluted with aniline.

Though his conclusion was not right, he was the first to notice the blue coloration formed by ammonia, phenol and hypochlorite. In the 1960's Bolleter [2] proposed a reaction mechanism for the formation of indophenol-blue from ammonia, phenol and hypochlorite. In the 1980's [4] proved that Bolleter was right. Also some work was done on the reaction kinetics of Berthelot's. From this it is concluded, that Berthelot's reaction can be divided into two steps, from which one is very fast and the other very slow. Optimizing the first -fast- reactionstep can be done by controlling the pH of the solution at pH=11. In the region around this pH, the reaction intermediate will be purest.

The second step involves the reaction of phenol with the intermediate product monochloramine which results in the formation of indophenol-blue. This is a very slow reaction step taking about 40 seconds to complete at 37°C. Higher temperatures cannot be allowed, since airbubbles will appear [5].

Since the complete procedure has to be completed within one minute, this 40 seconds will take up most of the time. For all other functions only 20 seconds are left. Functions to be completed are the pulling in of the sample, mixing with 2 reagents and measuring the absorption. The arising problem is the mixing of the second reagent with the sampled liquid. At microscale all flows will be laminar. Therefore mixing will have to be due to diffusion. The speed of diffusion depends on the square of the distance x that has to be crossed and the *diffusionconstant* $\boldsymbol{D}$ of the diffusing compound:

$$t = \frac{\pi\langle x\rangle^2}{4D} \qquad \textbf{(1)}$$

For phenol $\boldsymbol{D} = 0.89\times10^{-9}$ m^2sec^{-1}, which means that it takes about 50 seconds for phenol to cross (half) a channel of 300 μm. It is clear, that these 50 seconds are too long. Therefore a mixer is needed in order to reduce the time needed for the mixing of phenol into water. The only way reducing this mixing-time is to reduce the distance over which phenol has to diffuse.

For the mixing of different fluids, the following idea for the perfect mixer is proposed (fig.2). It consists of a channel of which the cross-section changes from square to low and broad to high and narrow back to square.

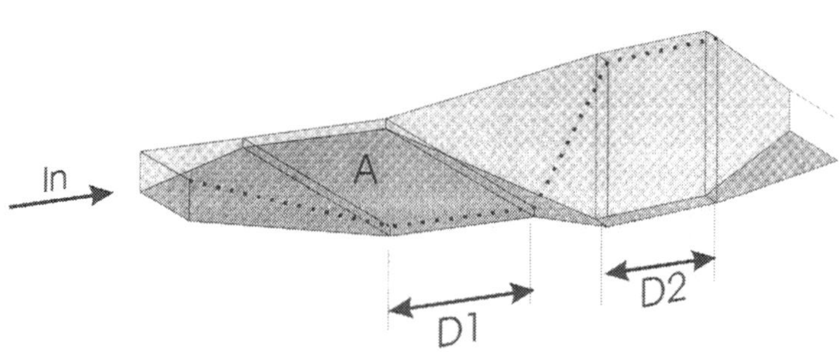

Fig. 2 : 3D-impression of the mixer/channel

This layout is based upon the idea, that since all flows are laminar, the transition from square to rectangular will not change the order of the layers in the fluid. In the low part of the channel (A), diffusion results in mixing in the vertical dimension whilst in section B mixing in horizontal direction will be accomplished.

For mixing-purposes which do not need complete randomization of all molecules, this complex channel structure can be simplified. In that case the part of the mixer designated *A* in figure 2 can be left out, which reduces the mixer to a narrowing in the channel. An even further simplification -inspired by technology restrictions- leaves the height of the narrowed part of the channel equal to the height of the channel at the inlet.

The structure depicted in figure2 therefore reduces to a structure as in figure 3.

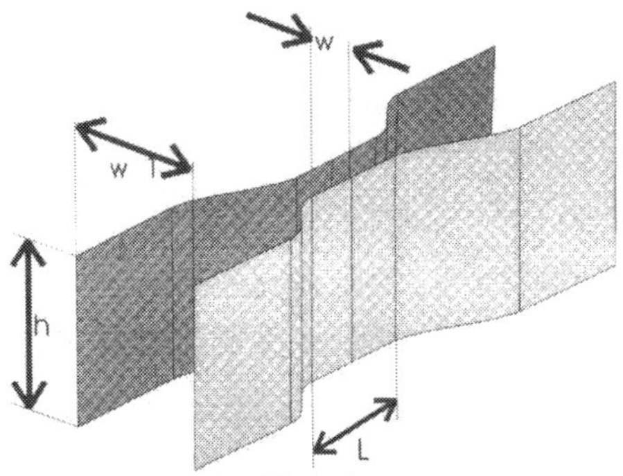

Fig. 3: simplified mixer structure.

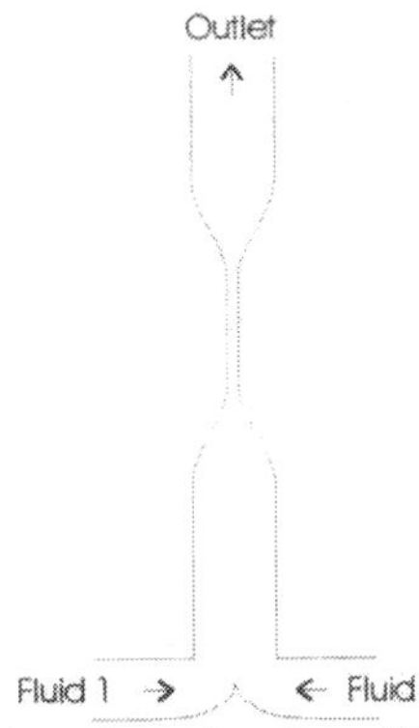

Fig.4: injection scheme for mixer of figure 3.

Using the injection scheme as shown in figure 4, the distance to diffuse over is reduced from $w1$ to $w2$. The relative time-reduction for the mixing will be equal to the square of $(w2/w1)$. The 50 seconds for mixing in a 300μm wide channel ($w1$) will be reduced to 2 seconds for a 60μm channel ($w2$).

In formula 1 [6], the time it takes for the fluids to get mixed over a distance <x> is given. This time should equal the time that both fluids are present in the small part of the channel. This time during which the fluids are in the small part of the channel is the volume of the channel divided by the flowrate through the system:

$$t = \frac{h \cdot W_2 \cdot L}{\Phi_{system}} \tag{2}$$

These two formulas should be combined to result in some design parameters for the mixer-to-be. In order to make this coupling, the average diffusion distance (<x>) should be expressed in the width of the channel (w_2). An estimation can be made through the following reasoning.

The particles that are on the side of the channel have to diffuse over the entire width of the channel (figure 5). Particles that are at the interface of the both fluids have to diffuse for about half the channelwidth.

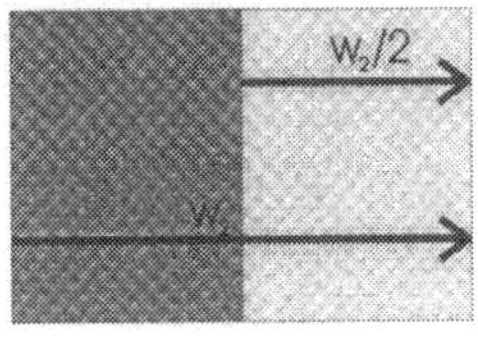

Fig 5: diffusing distances across the channel

From Fig 5 it can be seen that the average distance one fluid has to come across is ($w_2/2$ + 2)/2=3/4 w_2; three quarters of the channelwidth:

$$\langle x \rangle = \frac{3 \cdot W_2}{4} \tag{3}$$

The formulas 2-4 result in the following formula for the length *L* of the mixing part of the channel:

$$L = \frac{9\pi}{64D} \cdot \frac{\Phi_{system}}{h} \cdot W_2 \tag{4}$$

Using this length as the length of the mixing part of the channel, all compounds with a diffusionconstant larger as the one used will be mixed very efficiently over the whole channelwidth.

3. Discussion

Next to the proposed mixer, other mixers could be used. All other mixers however are more complex and need more process-steps. Miyake's [6](figure 6) mixer probably is the best alternative. His mixer consists of a sieve-like structure through which the reagent is injected in the sample liquid which flows over the sieve. In this way a lot of plumes of the reagent are inserted in the sample liquid. Biggest advantage of Miyake's mixer over the proposed one is the fact that the mixing process is *initiated* in the mixerstructure and carried out downwards the stream, whilst the proposed mixer mixes efficiently in the mixer itself and not outside the mixing-structure. This illustrates even more the need for a carefull design of the mixer that is proposed in this paper.

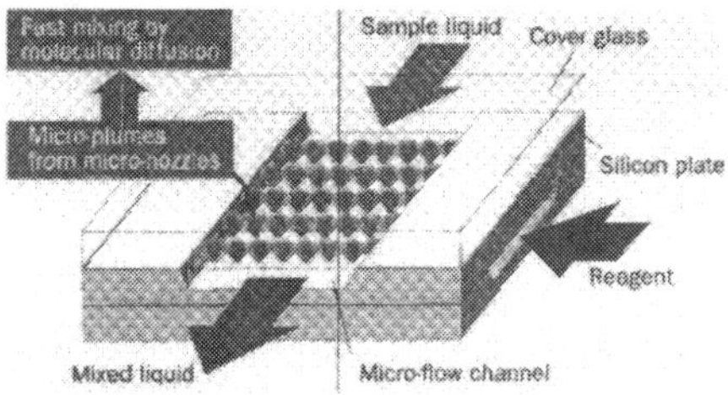

Fig.6: Mixer as realised by Miyake.

4. Conclusions

The proposed mixer mixes within 2 seconds the necessary compounds for completing Berthelot's reaction. These 2 seconds are more than sufficient for the MAFIAS-system to ensure a completed reaction within one minute.

The proposed mixer is very easily integrated with existing micro-fluidical components, since channels are needed anyway to couple the different fluidical components. Etching these necessary channels, the mixer can be created at the same time without the need for extra process-steps. This way, the overall production yield will increase.

The proposed mixer is only efficient for compounds with a diffusionconstant larger as the diffusionconstant it is designed for. Compounds with a larger diffusionconstant will not be mixed completely.

References

1. Berthelot, Repertoire de chimi appliquee, 1, 284, 1859
2. Bolleter, Anal. Chem. 33, 1961
3. P.L.Searle, The Berthelot or Indophenol Reaction and its use in the analytical Chemistry of nitrogen - A review, Analyst 109, 549-568, 1984
4. R.G.Harfmann and S.R.Crouch, Kinetic study of Berthelot reaction steps in the absence and presence of coupling reagents, Talanta, 36, 261-269, 1989
5. F.J.Krug, Zone trapping in flow injection analysis, Analytica Chimica Acta 151, 39-48, 1983
6. P.W.Atkins, Physical Chemistry, 4th edition, Ch 25, Oxford University Press, 1990
7. R.Miyake, Micro mixer with fast diffusion, Proc. of MEMS'93, 1993

LOW POWER SIGMA-DELTA ANALOG-TO-DIGITAL CONVERTER INTERFACE FOR BIOMEDICAL MEASUREMENT SYSTEM

Miodrag Djurica, Johan H. Huijsing

Electronic Instrumentation Laboratory
Delft University of Technology
P.O.Box 5031, 2600 GA Delft
e-mail: M.Djurica@ET.TUDelft.NL

0. Abstract

A dedicated CMOS Sigma-Delta modulator for use in a biomedical measurement system is presented. Measurement system is developed at the Medical Physics Laboratory of the University of Amsterdam. It is portable and battery operated. System consists of 32 measurement active electrodes, analog-to-digital converters, and transmitter. Every channel having its own dedicated analog-to-digital converter (ADC). In this way, time skew of measured biopotentials is avoided. Large number of ADCs in a battery powered system emphasizes Low Power Low Voltage operation. ADC was designed at Electronic Instrumentation Laboratory, TU Delft.

Keywords: interface electronics, Sigma-Delta, Analog-to-digital converter, Low power, biomedical measurement system

1. Introduction

Trend in modern electronic systems for personal use is to go towards portable equipment. Everybody got used to the idea of carrying walkman or a mobile phone. They make our life more comfortable, and due to the their small size, they are just another accessory we are carrying. Similar idea of having a portable equipment lies behind portable universally applicable biomedical measurement system. In case of patients whose biopotentials (ECG, for example) have to be monitored, it is important that he/she behaves as naturally as possible. And that means that the measurement system has to be as small as possible, in order not to interfere with patients everyday habits. At University of Amsterdam's Medical Physics Department a multichannel biomedical measurement system is developed. In Fig.1 a wireless biomedical system is shown. It consists of a 32 active electrodes, a box with analog-to-digital converters and a transmitter, and a monitoring unit (PC). System is used for 'body mapping', when a map of body's biopotentials is taken. Miniaturisation of the complete portable instrument necessitates battery powered operation, and therefore low-power operation of the components comprising it.

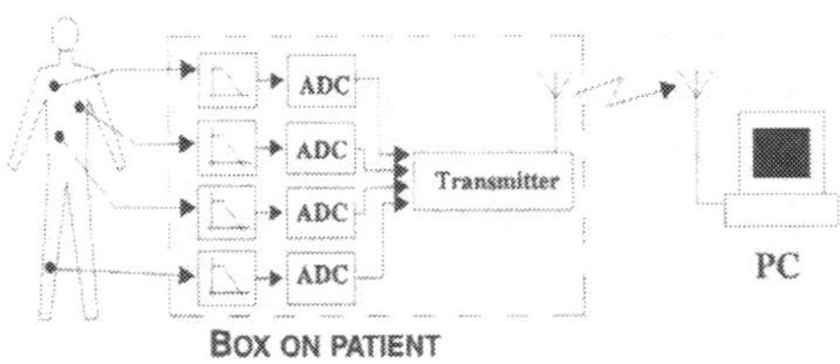

Figure 1. Wireless biomedical measurement system

2. About specifications of Analog-to-digital converter

On a complete measurement system level, current consumption of the ADCs was set to be 100mA maximally. Since there are 32 channels, average current consumption per channel (anti-aliasing filter plus ADC) is 3.125mA. For a voltage supply of 3V, power of an ADC can be maximally 9.375mW.

Another of the design requirements is the amplitude of the input signal of the ADC. Due to the configuration of the system, signal from the active electrode is transferred via wire to the ADC. To minimize decrease in the signal-to-noise ratio (SNR) during transfer of the signal, output signal of the active electrode is made rail-to-rail. This signal ($3V_{pp}$) is the input signal of the ADC.

Number of bits was determined by the need for universal application of the measurement system. Level of measured biopotentials depends on their type. In EEG measurements signals are in order of 10uV, for ECG in order of 1mV and for EMG in order of 10mV. To be able to measure all of these signals without the need to adjust gain of the amplifiers used in active electrode, 16 bit converter was chosen.

2.1 Design

Dedicated ADC was designed. It is second-order continuous-time Sigma-Delta Converter (Fig.5) . Sampling frequency is 16ksamples/s. Due to the requirement for the rail-to-rail input and 16 bit linearity, designed Sigma-Delta modulator uses resistors as a linear V-I converters. Amplifiers used in integrators are two stage fully differential class-AB Miller compensated circuits. Comparator is implemented with return-to-zero control (RTZ) circuit, to avoid clustering and signal dependent delay time of the comparator.

Due to the part of the digital circuitry that is also part of the Sigma-Delta modulator, there is a lot of digital noise passing through the substrate. To avoid its influence, complete circuit is fully differential as to improve its immunity to substrate and/or power supply interference.

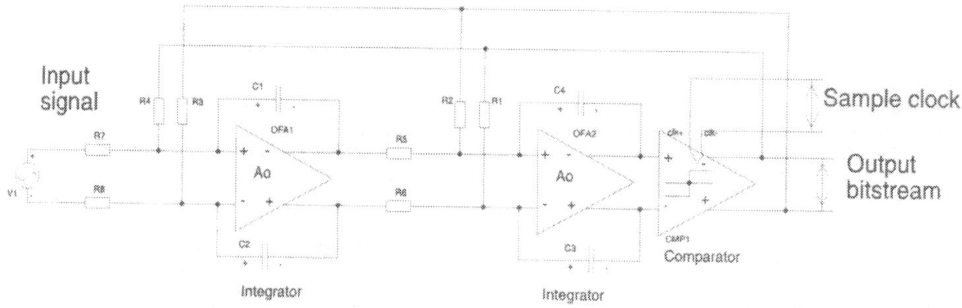

Figure 2. Block diagram of the Sigma-Delta modulator

Sigma-Delta ADCs are most suitable for high accuracy, low power conversions. They achieve high accuracy based on a use one-bit quantizer and large oversampling (much higher than Nyquist rate). Minimal oversampling ratio is always calculated on basis of the ideal circuit, where the only noise source present in the system is quantization noise. So, total amount of the noise in the signal's range of frequencies is:

$$\overline{\sigma_{qua}^{2}} = \frac{\pi^{4}}{5}\left(\frac{1}{OSR}\right)^{5}\frac{q^{2}}{12} \qquad (EQ\ 1)$$

OSR is a oversampling ratio, and *q* is amplitude of the (assumed) uniformly distributed error signal of the 1-bit quantizer (comparator). From (EQ 1) follows that by increasing *OSR* amount of the quantization noise inside signal's bandwidth decreases.

For 16 bit Sigma-Delta ADC, minimal OSR is 153. But in this design, oversampling rate is 512, to make more room for other noise sources (1/f noise, thermal noise, clock jitter).

Since this design is continuous-time, precise timing is important since it defines amount of charge that flows into integrating capacitors. This amount of charge represents reference value with whom the input signal is compared. To avoid signal dependent delay time of the comparator, comparator is implemented with return-to-zero control (RTZ) circuit. Now, the output signal from the comparator is disabled for a controlled period of time. During that time, comparator will make a decision. So, when RTZ is deactivated, there is a stable signal at the output of the comparator block (comparator with RTZ). In this way delay time of the comparator is controlled. Therefore, absolute amount of charge sent to integrating capacitors is always the same. RTZ is implemented with the duty cycle of 3/4.

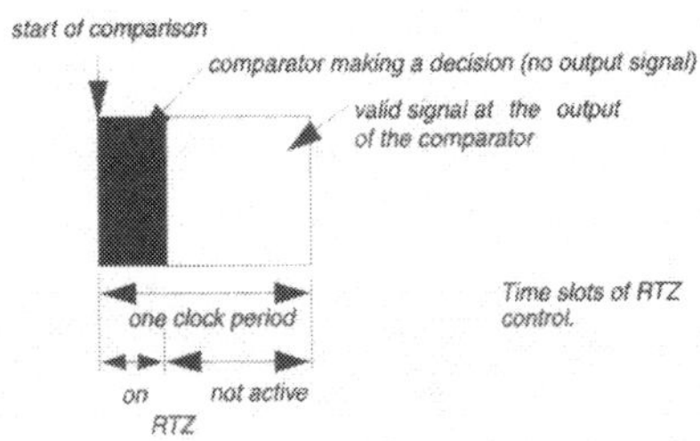

Figure 3. RTZ control time slots

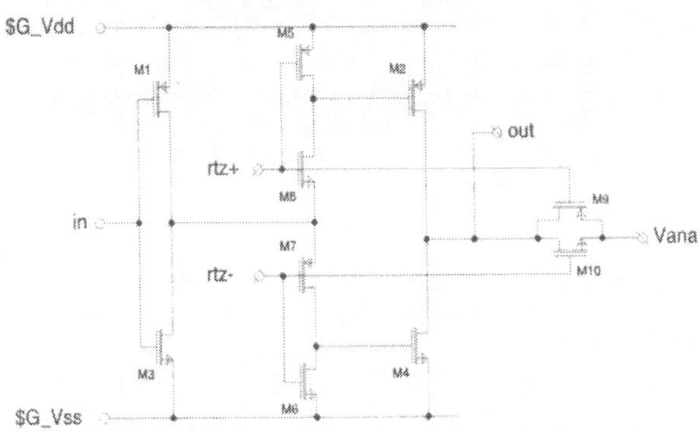

Figure 4. 3-state circuit

Signal at the output of the comparator is used as a feedback voltage. This voltage can have one of three values digital '0' (or Vss), digital '1' (or Vdd) and analog reference voltage (half a way between Vdd and Vss). Analo reference voltage is introduced to make sure that the parasitic capacitors (associated with integrated resistors) a discharged.

Output voltage swing of integrators is large enough to accomodate 4-5 clock periods integration with the constar reference current. This requirement comes from the need to avoid clipping of the amplifiers, and as a result dete rioration of the performance of the converter.

$$C_{integrator} R_{feedback} = 4T_{sample} \qquad (EQ\ 2$$

Specifications for the bandwidth of the amplifier are result of MATLAB simulations of the stability of the circui and its noise-shaping performance. According to the simulation results, for a unity gain frequency of the amplifie that is about two times larger than the sampling rate of the converter, decrease in performance is not significan The more bandwidth we need, the more current is needed to realize it.

3. Implementation

Circuit was implemented in 1.6umCMOS process at DIMES (Delft Institute for Microelectronics and Submicro technologies). Process is standard with 1 poly, 2 metal. Size of the circuit is 3.1x1.8 mm2.

Figure 5. Photograph of the implemented analog part

According to the simulation results a total power consumption of 0.9mA (at 3V voltage supply) was expected Measurements of the IC (analog part and decimation filter) are in progress, but before full results, a current con sumption of 1.4mA for analog part, and 1.7mA for digital part was observed.

Acknowledgements

This project is sponsored by Dutch Technology foundation (STW), project AGN 44.3416.

Literature

[1] R. van de Plassche, "Integrated analog-to-digital and digital-to-analog converters", Kluwer academic publish

rs, 1994.
2] E.J. van der Zwan, E.C. Dijkmans, "A 0.2 mW CMOS SD Modulator for Speech Coding with 80 dB Dynamic :ange", IEEE Journal of Solid-State Circuits, vol.31, pp. 1873-1880.
3] R. Hogervorst, "Design of Low-Voltage Low-Power CMOS Operational Amplifier Cells", Delft University 'ress, 1996.
4] R.G.H. Eschauzier, "Wide Bandwidth Low Power Operational Amplifiers", Delft University Press, 1994.
5] E.A. Vittoz, "The Design of High-Performance Analog Circuits on Digital CMOS Chips", IEEE Journal of olid-State Circuits, vol. SC-20, pp. 657-665.

DEVELOPMENT OF CRYOGENIC CURRENT COMPARATORS WITH DC SQUID READOUT FOR THE CALIBRATION OF ELECTRICAL STANDARDS

L.Vargas, G.C.S.Brons, B.Bouverat, J.Flokstra and H.Rogalla
Low Temperature Division, University of Twente,
PO Box 217, 7500 AE Enschede, The Netherlands.

G.Rietveld
Nederlands Meetinstituut, Delft, The Netherlands

C.Rillo, J. Sesé and A.Camón
University of Zaragoza, Zaragoza,Spain

W.Jaszczuk and H. Altenburg
Fachhochschule Münster, Steinfurt, Germany

0. Abstract

For the realization of the electrical quantum metrology triangle (V-A-Ω) a device to amplify very small currents with high precision is needed. The cryogenic current comparator (CCC) is by far the best instrument to do this. In order to make a very current sensitive CCC for calibration of electrical standards, we have developed optimum dc Superconducting QUantum Interference Devices (SQUIDs). The design, fabrication and characterisation of these devices is presented. The measurements concern the flux-to-voltage transfer and the noise properties, especially the input current noise. The optimisation of the flux transformer circuit that links the CCC with the SQUID will be treated. In addition, typical fabrication aspects of the CCC as the wires and tube assembly, the shields and the support system will be addressed.

Keywords: Current standard, Current comparator, Resistance, SQUID, metrology

1. Introduction

Electrical standards for voltage and resistance [1,2] are presently based on physical quantum effects, with values determined by frequency and constants of nature. The voltage standard is related to the small quantised voltage steps produced across a Josephson junction when it is irradiated with photons with a certain frequency. The steps are proportional to the frequency and the proportionality constant is determined by the natural constants e (electron charge) and h (Planck's constant). In an analogous way, the resistance standard is based on the Quantum Hall effect of a two-dimensional electron gas at low temperatures (~1K) and high magnetic fields (~10T). The standard resistance can be expressed in terms of h/e^2.
A possible realisation of a quantum current standard is based on the clocking of single electrons in synchronism with an applied frequency through a single electron tunnelling device (SET) at very low temperatures (~10mK) [3] . With switching frequencies of 10 MHz, currents in the pA range can be realised.
The availability of the three quantum electrical standards will enable the realization of the electrical metrologic triangle. That is to say, a SET current, amplified to the μA level, will

feed a Quantum Hall Resistance (~12kΩ) and the voltage (~10mV) will be compared to the Josephson voltage (Fig. 1). No errors within the uncertainty of the quantum standards (10^{-9}) should appear.

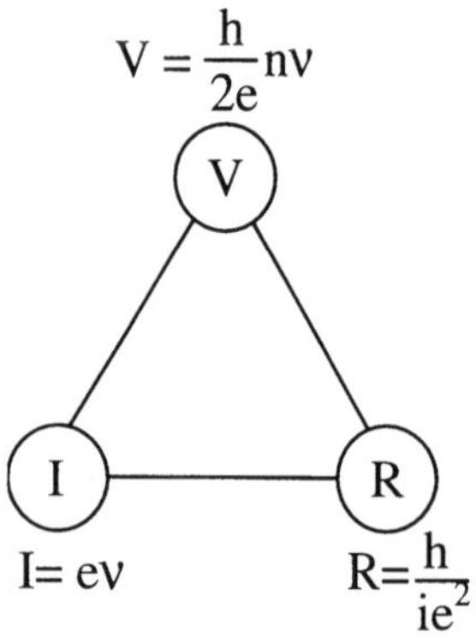

Fig. 1. Electrical Metrologic Triangle: e and h are the elementary charge and Planck's constant respectively. n is an integer number, i=2 and ν is the frequency.

A Cryogenic Current Comparator (CCC) is by far the best device to amplify the small SET current. The CCC is also the best system to compare very precisely currents in the resistance standard. Nowadays, the CCC is widely used in national metrology institutes as a measuring instrument for calibrating resistance standards used in industry.

In this paper the development of a CCC for a current standard is described. Specific dc SQUIDs are needed for the readout of this CCC, and the design, fabrication and test of these devices is presented.

2. Cryogenic Current Comparator

A CCC is a three superconducting winding transformer (two primaries and one secondary). The primary windings, with N_1 and N_2 turns, pass through the interior of a superconducting hollow cylinder and are fed by currents I_1 and I_2 whose ratio has to be precisely established (Fig. 2a). There exists a Meissner current in the inner surface of the cylinder that returns on itself via the external surface, I_e, and equals the magnetomotive force unbalance: $I_e = N_1I_1 - N_2I_2$. The secondary, typically a few turns of superconducting wire, also called sensing coil, connected to the input coil of a SQUID magnetometer [4], is used to sense the magnetic flux generated by I_e. The sensing coil and the input coil of the SQUID form a superconducting flux transformer. Ideally, the magnetic flux seen by the SQUID should be proportional to I_e. For this to be true, direct coupling between the ratio windings and the sensing coil has to be avoided by magnetic shielding. This is implemented by using an overlapped tube structure like a snake swallowing its tail (Fig. 2b) [5,6]. The SQUID controls the value of I_2 by feedback, such that I_e becomes zero. The dc ratio of the primary currents is then given by the inverse of their turn number ratio ($I_1/I_2=N_2/N_1$).

The current of the SET device is known with large precision (the frequency can be measured with an accuracy of 10^{-13}). This current (I_1) needs a large amplification and thus a CCC with a large ratio N_1/N_2 is needed (not larger than 10^5 for practical considerations). A current $I_2 = N_1/N_2 I_1$ is then known at the μA level with the same relative precision as the SET current. The current resolution of the CCC-SQUID system should be small enough in order not to introduce deviations from this scaling factor.

The ultimate current resolution of a CCC has been investigated [7]. The key parts are the coupling between the superconducting tube and the SQUID, and the input current noise of the

SQUID. It has been found that due to the "image effect" the inductances of the sensing coil and the overlapping tube do change [8]. When this effect is considered coupling improvements of up to one order of magnitude can be obtained. Using wire sensing coils the optimum (maximum) coupling is always smaller than the ideal due to the unavoidable gap between the superconducting filaments of the wire and the overlapping tube. Ideal coupling has been proposed to occur in a new configuration without sensing coil in which the SQUID is directly connected to the superconducting overlapped tube.

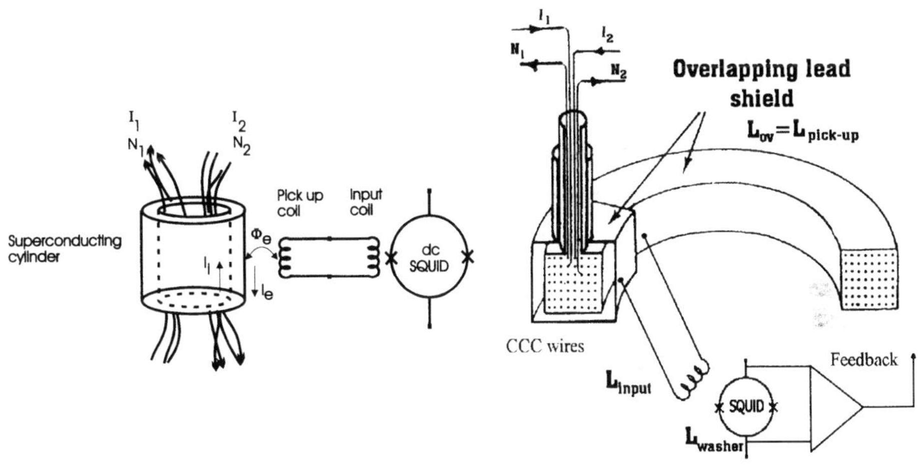

(a) (b)

Fig. 2: (a) Superconducting cylinder with N_1 wires carrying a current I_1 and N_2 carrying I_2. The current I_e that flows on the outer surface of the cylinder is equal to the I_i flowing on the inner surface ; (b) Overlapped tube configuration for the CCC using the toroid as pick-up coil of the SQUID

Our CCC will have 27 separate windings of Nb-Ti monofilament wire with 65 μm diameter. The windings comprise 1,1,2,2,4,4,10,20,...4000 and 10000 turns in order to have a self-calibration system. A total of 30000 turns will have an approximate rectangular cross section of 40 mm^2. A toroidal lead shield will cover the coil having a two and a half turns of overlap with Kapton insulation between turns. The complete toroid will be mounted on a Celotex holder again shielded by a layer of lead. The SQUID mounted on a SQUID module will be mechanically fixed to this holder and it will be electrically connected to the toroid.

The CCC-SQUID holder will be placed in a cryostat with a diameter of 160 mm (Fig.3). A cryoperm μ-metal cylinder with a coaxial, closed lead cylinder inside are used for further magnetic shielding. A small stainless steel tube, for filling liquid helium, will be mounted outside the shield. The inside diameter of μ-metal-Pb can will be about 133 mm. The Celotex holder will just fit in this cylinder. Due to the shielding effect of the Pb-shield of the Celotex holder, the diameter of the CCC will be around 100 mm which means an effective toroid inductance of approximately 100 nH.

A very important issue in the design of the support of the system is the avoidance of mechanical vibrations. The SQUID-CCC support should be rigid and not move with respect to the superconducting shields.

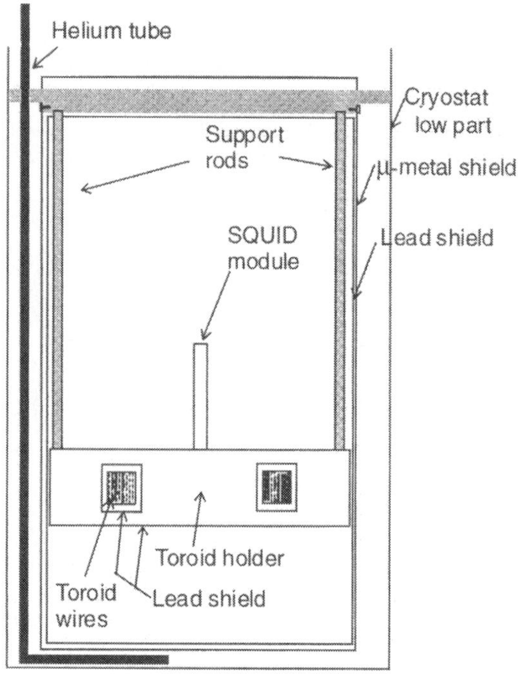

Fig.3: Lower part of the cryostat where the CCC will be placed together with the SQUID

3. Optimum dc SQUIDs for the CCC

SQUIDs are the most sensitive sensors of magnetic flux available. They combine two physical phenomena, flux quantization and Josephson tunnelling [4]. A dc SQUID consists of two Josephson junctions connected in parallel in a superconducting loop of inductance L_{washer}. Each junction has a critical current I_0, a capacitance C and a resistive shunt R_s. Junctions are resistively shunted to eliminate hysteresis on the current-voltage characteristics. The critical current of the SQUID, $I_c=2I_0$, can be varied by sweeping the values of the externally applied magnetic flux. When the SQUID is biased with a constant current larger than two times the critical current of each junction, the voltage across the SQUID oscillates with period $\phi_0=h/2e$ (flux quantum) when the external flux is increased.

Usually the magnetic flux to be measured is not applied directly to the SQUID because, due to the small SQUID dimensions, it would not be very effective. In practice a flux transformer is used as shown in Fig.2. This transformer consists of a closed loop of superconducting wire with a part of inductance $L_{pick\text{-}up}$ called pick-up coil coupled to the external field and another coil of inductance L_{input} tightly coupled to the SQUID called input coil. The most effective coupling of the input coil to the SQUID is obtained by a planar multiturn superconducting coil located on top of a thin film dc SQUID washer [9].

The dc SQUIDs we made were fabricated with our Nb/Al technology using photolithography techniques [10]. In the design chosen for our devices the SQUID ring is patterned in the shape of a washer of self-inductance L_{washer}. The energy resolution of the SQUID is directly proportional to this self-inductance so the smaller the L_{washer}, the better the energy resolution

is, in general. The junctions, two 4×4 μm^2 Nb/Al/AlO_x/Al/Nb sputtered junctions, were placed on the outside part of the washer shunted respectively by 2 Ω Pd thin film resistors.
Two coils were integrated on top of the washer, one with n turns called input coil and a second coil called modulation coil which is used to couple a feedback and a flux modulation signal to the SQUID. The input coil inductance of a SQUID in the optimised configuration should be equal to the pick-up coil inductance, which in our case is the inductance of the overlapping tube. The latter inductance has been calculated to be about 100 nH for the geometrical parameters of the overlapping tube. A series of low T_c dc SQUIDs with different input coil inductances (20, 40, 80 and 160nH) was designed in order to match the overlapping tube inductance.
Looking for the best energy resolution several washer inductances were considered for each input coil inductance resulting in the design and fabrication of the following SQUIDs:

n = number of turns of input coil	L_{input}=20 nH	L_{input}=40 nH	L_{input}=80 nH	L_{input}=160 nH
L_{washer}= 100 pH	14	20	#	#
L_{washer} = 150 pH	11	16	23	32
L_{washer} = 200 pH	10	14	20	28
L_{washer} = 250 pH	9	13	18	25

Table 1: Relation of the different fabricated SQUIDs. Those ones marked with # could not be realised with the present design

The SQUIDs have been electrically characterised by measuring the current-to-voltage curves, the voltage-to-flux and flux noise versus frequency graphs. The I-V curve of a SQUID is used to determine the SQUID critical current and shunt resistor values. From these quantities information can be obtained about the material properties of the SQUID.
Based on the fact that the critical current varies periodically with the applied flux the washer inductances could be experimentally obtained. These experimental values fitted very well with the designed ones.
V-ϕ_{appl} measurements are used to obtain the optimal settings of the bias current. In this case the voltage modulation amplitude is maximal. It is possible to determine the mutual inductance counting the number of ϕ_0 created on the SQUID when a current flows through the coil. So for V-ϕ_{appl} measurements we bias the SQUID at a constant value, I_b, and sweep a current in the input coil or in the feedback coil. The experimental mutual inductances between the SQUID and the input coil, M_{input}, and the SQUID and the feedback coil, M_{fb}, were determined then in this way. These values for the M_{input} are in general smaller than the theoretical ones which is usually the case. Values for the mutual inductance between feedback coil and SQUID washer were quite good and according to the designed ones.
Flux noise design values were calculated based on a 2 Ω shunt resistor and the theoretical washer inductances. Values ranged from 2 $\mu\phi_0/\sqrt{Hz}$ for the 100 pH SQUID to 5 $\mu\phi_0/\sqrt{Hz}$ for the SQUID with 250 pH. Measured values are between 4.0 to 6.3 $\mu\phi_0/\sqrt{Hz}$, higher than predicted. Most likely this is due to the voltage noise of the read-out electronics used for the measurements. The amplifier that was used has an input voltage noise of about 1 nV/$\sqrt{Hz}$. The 1:25 step-up transformer at room temperature that is used at the input of the amplifier, would in principle give 40 pV/$\sqrt{Hz}$. Due to thermal noise of the resistance of the transformer an effective noise of 0.2 nV/$Hz^{1/2}$ is obtained that leads to an equivalent flux noise of about 4 $\mu\phi_0/\sqrt{Hz}$, in the order of the noise we measured. A solution might be to place the transformer close to the SQUID at 4.2 K.

The flux noise can be converted to a current noise using the experimental values of the M_{input}. Based on the measured noise data we find values ranging from 6 pA/√Hz to 15 pA/√Hz. We think that the current noise values will be a factor of two better when a cold step-up transformer is used.

4. Conclusion

A cryogenic current comparator has been designed for amplification of the currents produced by a single electron tunnelling device . This CCC with a total number of 30000 turns and an inductance of approximately 100 nH will be made with a novel configuration in which the toroid is used as pick-up coil of the SQUID. Optimum dc SQUIDs with inductance values around the value of the toroid inductance have been fabricated. The dc SQUIDs, made with a technology based on Nb/Al/AlO_x/Al/Nb sputtered junctions, have been characterised and values for the current sensitivity ranging from 6 pA/√Hz to 15 pA/√Hz have been obtained. These values are most likely determined by the noise in the electronics. The use of a step-up transformer at 4.2K would help to overcome the problem.
In this CCC system other parts are at this moment being designed and fabricated. Once all the different parts are ready, they will be assembled and fixed to make the first calibrations.

References

1. Kautz R., “On a proposed Josephson-effect voltage standard at zero current bias”, Appl.Phys.Lett., 36 (1980) 386-8
2. A.Hartland, “Quantum Hall Effect and Resistance Standards”, Metrologia 29, (1992) 175-190
3. L.J.Geerligs, V.F.Anderegg, P.A.A.Holweg, J.E.Mooij, H.Pothier, D.Esteve, C. Urbina, and M.H.Devoret, “Frequency-Locked Turnstile Device for Single Electrons”, Phys.Rev.Lett. 64, (1990) 2691
4. C.D.Tesche and J.Clarke, “DC SQUID: noise and optimisation”, J.Low Temp.Phys.29, (1977) 301-331
5. D.B.Sullivan and R.F. Dziuba,”Low temperature direct current comparators”, Rev.Sci. Instrum. 45 (1974) 517
6. A.Hartland, “Development of a cryogenic current comparator for the measurement of small currents”, International Conference on electromagnetic measurements BEMC 18 (1993) 1-4
7. J. Sesé, F. Lera, A. Camón, C.Rillo, “Calculation of effective inductances of superconducting devices. Application to the Cryogenic Current Comparator”, submitted to IEEE Trans. Appl. Supercond.
8. J.Sesé, A.Camón, C.Rillo, G. Rietveld, “Ultimate current resolution of a cryogenic current comparator”, submitted to Applied Physics Letters
9. J.M.Jaycox and M.B. Ketchen, “Planar coupling scheme for ultra low noise dc SQUID”,IEEE Trans. Magn. MAG-17 (1981) 400
10. D.J.Adelerhof, M.E.Bijlsma, P.B.M.Fransen, T.Weiman, J.Flokstra and H.Rogalla, ”Fabrication of Nb/Al,AlOx/Al/Nb Josephson tunnel junctions using reactive ion etching in SF_6”, Physica C 209 (1993) 477-485

SQUID READOUT OF CRYOGENIC PARTICLE DETECTORS

A.W. Hamster, M.J. van Duuren, G.C.S. Brons, J. Flokstra and H. Rogalla
Low Temperature Division, University of Twente,
P.O.box 217, 7500 AE Enschede, The Netherlands
A.W.Hamster@tn.utwente.nl

0. Abstract

Cryogenic radiation and particle detectors are widely investigated nowadays. Their high intrinsic energy sensitivity makes them suitable for a wide range of applications. In order to take full advantage of the sensitivity of these newly developed cryogenic detectors, in general current sensors are needed with a white noise level of the order of 10^{-12} A/√Hz and a bandwidth larger than 1 MHz. We applied a Double Relaxation Oscillation SQUID (DROS) as a low-noise current preamplifier for measuring the current-voltage characteristics of a superconducting Nb/Al-AlO_x/Al/Nb Josephson tunnel junction. A white current noise level smaller than 2 pA/√Hz was obtained. Thanks to the large flux-to-voltage transfer of a DROS (typically 1 mV/ϕ_0), no modulation techniques are required, allowing simple room-temperature flux locked loop electronics. Since the Josephson junction was voltage biased, no magnetic field had to be applied to suppress the Josephson supercurrent, which eliminates distortion of the subgap I-V characteristics by Fiske resonances. The measurements show that a DROS is a serious alternative for the high resolution readout of an X ray spectrometer based on a superconducting tunnel junction.

Keywords : superconductivity, SQUID, superconducting tunnel junction, particle detector

1. Introduction

Present state of the art particle and radiation detectors, like gas proportional counters, dispersive techniques using Bragg single crystals or grating spectrometers and semiconductor-based detectors all suffer from either a lack in sensitivity, a small quantum efficiency or a narrow energy band of operation. A detector performing well in all three aspects might be obtained using superconducting tunnel junctions (STJs). Theoretically, STJ-based cryogenic sensors [1] are well suited for a wide scope of applications requiring an energy sensitivity of $\Delta E/E \leq 1$ ‰ within an energy band ranging from 0.1 to 10 keV. Exemplary fields of application are in astrophysics for detection of optical, UV and X-ray radiation and in condensed matter research for Energy Dispersive X-ray analysis (EDX).
For operation as a particle or radiation detector, an STJ is voltage biased in the subgap, $V < V_g$, of its *I-V* curve (see fig. 1). Absorption of an incident photon in either the top or the bottom electrode of the STJ results in a rise of the subgap current, which lasts no longer than about 5 μs. The magnitude of the current pulse, typically of the order of 1 μA, is a measure for the energy deposited by the absorption of the photon. To exploit the full potential of the STJ-based spectrometer a current sensor with a bandwidth larger than 1 MHz and an input current noise of the order of 10^{-12} A/√Hz [2] should be applied as preamplifier. Moreover, the

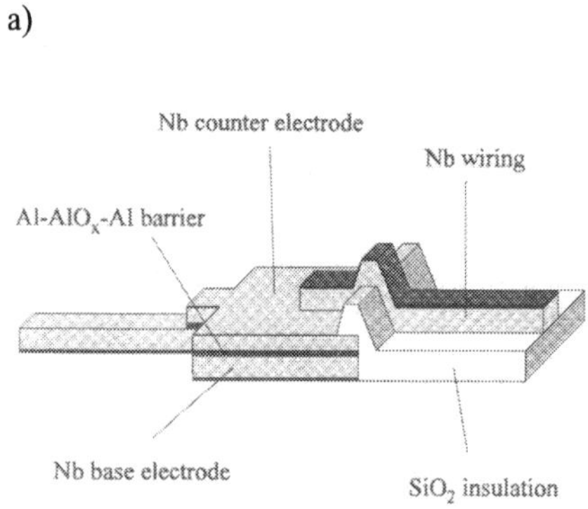

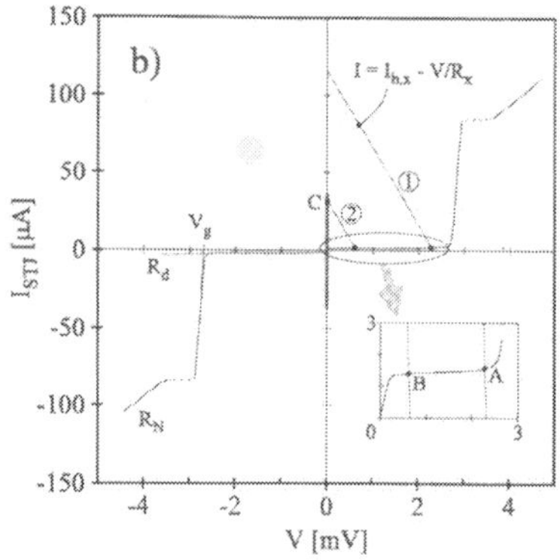

Figure 1. *a) Schematic view of a superconducting Nb/Al-AlOx/Al/Nb Josephson tunnel junction. b) The current voltage characteristic of the 10 x 10 μm² Josephson tunnel junction as measured by current biasing with standard electronics. A magnetic field was applied to suppress the critical current while recording the subgap.*

intrinsic resistance of the sensor should be as small as possible in order to obtain optimum voltage biasing of the STJ.

In principle, by applying a Superconducting QUantum Interference Device (SQUID) one can fulfill these demands. At present, SQUIDs are the most sensitive magnetic flux sensors available. The relative simplicity with which numerous physical quantities, amongst which current and voltage, can be transformed into a magnetic flux makes SQUIDs of interest for a broad range of applications far beyond the scope of direct magnetic field sensing. However, in most common dc SQUID systems, the bandwidth is limited to about 100 kHz. This restricted bandwidth is caused by the traditional dc SQUID readout scheme which is based on flux modulation, step-up transformers between SQUID and pre-amplifier and lock-in detection of the SQUID output voltage. Such a complex readout scheme is necessary because of the small flux-to-voltage transfer of standard dc SQUIDs. Direct voltage readout with a room-temperature amplifier would substantially decrease the sensitivity because of the noise of the pre-amplifier. Alternative SQUID types, such as SQUIDs with Additional Positive Feedback [3], series arrays of dc SQUIDs [2,4] or Double Relaxation Oscillation SQUIDs (DROSs) [5] have much larger flux-to-voltage transfer coefficients and therefore do not need modulated readout schemes, which allows straightforward room temperature voltage readout with a large signal bandwidth.

In the present work we present our preliminary measurements performed with an STJ coupled to the input coil of a gradiometric DROS. With this setup we recorded the subgap *I-V* characteristic of a voltage biased STJ in the temperature range between 4.2 K and approximately 1.6 K in order to demonstrate the feasibility of DROSs as current readout sensors for STJ-based X-ray spectrometers.

2. Experimental setup

The experimental setup is depicted in fig. 2. A 10 x 10 μm² superconducting tunnel junction was placed in series with the input coil of a gradiometric DROS. The main parameters of the DROS are listed in table I; detailed information about the DROS and its readout electronics has been published by van Duuren *et al.* [6] Since so far only low

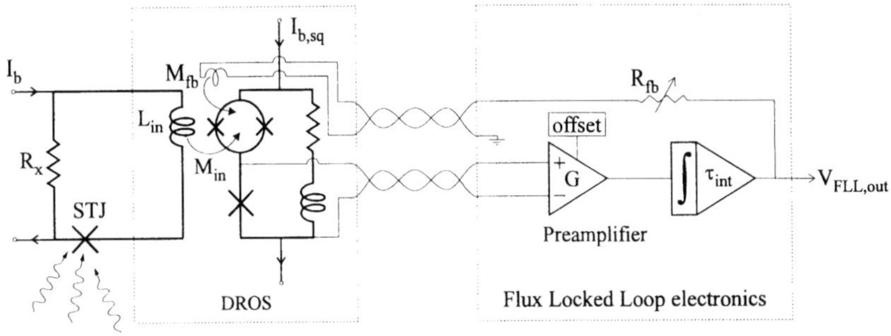

Figure 2. *Experimental setup for readout of superconducting tunnel junction with a Double Relaxation Oscillation SQUID.*

frequency *I-V* measurements have been performed, our current flux locked loop electronics with a bandwidth of only 100 kHz sufficed; for future X-ray spectroscopy experiments faster electronics is currently being developed.

The high quality STJ was fabricated on a thermally oxidized 2" silicon wafer, using standard Nb/Al technology. First, a Nb/Al-AlO_x/Al/Nb trilayer was deposited by dc magnetron sputtering and *in situ* oxidation of the first Al layer for one hour at an O_2 pressure of 10 mbar. The junction area of 10 x 10 μm^2 was defined by reactive ion etching in an SF_6 plasma. Insulation of the bottom electrode and the barrier was obtained with an RF magnetron sputtered SiO_2 layer. Finally, the top electrode of the junction was connected with a dc magnetron sputtered Nb wiring layer. The junction's current-voltage characteristics, recorded at 4.2 K with standard electronics, is shown in fig. 1b. At 4.2 K, the critical current of the junction I_0 was 37 μA. The gap voltage V_g was 2.8 mV and the normal resistance R_N was 42 Ω. Hence the critical current density of the junction J_c was 37 A/cm^2. The quality parameter of interest for application of STJs as radiation and particle detectors is the ratio between the dynamic resistance in the subgap, $R_d = \partial V/\partial I$ at $V = 2$ mV, and the normal resistance R_N. The R_d/R_N ratio for this junction was about 10^4.

Voltage biasing was obtained using a shunt resistor $R_x = 1.5\ \Omega$, satifying the condition $R_x \ll R_d$. The principle of the voltage biasing circuit is illustrated in the *I-V* graph of fig. 1b.

Signal SQUID		
L_{sq}	(signal SQUID inductance)	~ 550 pH
2 x 25 turns integrated input coil		
L_{in}	(input coil inductance)	150 nH
M_{in}	(coupling input coil ↔ SQUID)	6.7 nH (k ≈ 0.75)
f_{RO}	(relaxation frequency)	~1 GHz
$\sqrt{S_\Phi}$	(white flux noise)	5 .. 6 $\mu\Phi_0/\sqrt{Hz}$
$\sqrt{S_I}$	(=$\sqrt{S_\Phi}/M_{in}$, current noise)	< 2 pA/$\sqrt{Hz}$
ε	(=$S_\Phi/2L_{sq}$, energy resolution)	~ 200 h
$\partial V/\partial\Phi$	(flux-to-voltage transfer)	1 mV/Φ_0 typically

Table I. *Main parameters of the DROS.*

The intersections of the loadline of R_x and the I-V characteristic of the junction represent stable operation points. For bias currents $I_{b,x} > I_o$ (loadline 1 in fig. 2), only one stable point ("A" in the inset) exists, and the junction is effectively voltage biased at a voltage equal to $V_{bias} = I_{b,x} \times R_x$. For bias currents $I_{b,x} < I_0$ (loadline 2), two stable points exist: one in the subgap of the junction, and one in the superconducting state (points "B" resp. "C"). In this case the junction can switch from the voltage state to the superconducting state. As a consequence, stable voltage biasing in the subgap is obtained only for voltages $V_{bias} > V_{min} = I_o \cdot R_x$ in which case $I_{b,x} > I_0$.
It is not possible to bias the junction at a voltage below its minimum return voltage V_{return}, typically some hundreds of microvolts, even for very small bias resistors R_x. At voltages below V_{return} the junction will switch to the superconducting state. If $I_0 < I_{b,x} < V_{return} / R_x$, the bias circuit will act like a relaxation oscillator [7]. Because of these relaxation oscillations, the current through the junction oscillates between I_0 and (almost) zero, while the junction voltage oscillates between V_g and zero. The frequency of these oscillations is mainly determined by the time constant L_{in}/R_x. In the present setup, this time constant amounts to about 100 ns, implying a relaxation frequency of around 10 MHz.

3. Measurements

Experimentally, the relaxation oscillation effect at low voltages as described in section 3 could be observed in two ways: first, at a voltage between 100 and 200 mV, the $I_{b,x}$-V characteristic showed a discontinuity, marking the transition from the oscillating mode

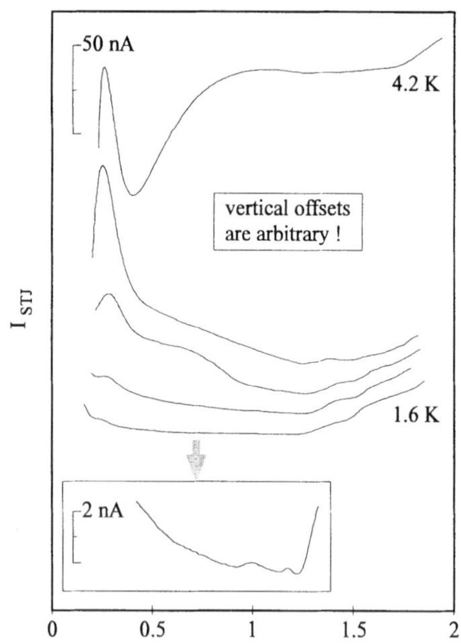

Figure 3. *Current-voltage characteristics of the Josephson junction at several temperatures with the presented setup. The temperature was varied by controlling the pressure above the helium bath. No magnetic field was applied. The inset shows an enlargement of the 1.6 K trace.*

to the stable mode. Second, at bias voltages between zero and ~ 150 mV, the DROS's *V-Φ* modulation disappeared completely, since the relaxation oscillations in the bias circuit sweep the input flux of the DROS over a range of more than 100 Φ_0 at a frequency of about 10 MHz, which wipes out the *V-Φ* modulation. At bias voltages $V_{bias} > V_{return}$, typically a few hundreds of microvolts, no relaxation oscillations can occur, and stable voltage biasing of the STJ is possible.

The subgap *I-V* characteristics of the STJ, measured in a helium-4 bath cryostat at different temperatures, is shown in fig. 3. Since a SQUID system in flux locked loop mode only measures flux changes, and not absolute values, the vertical offsets in the graph are arbitrary and different for each trace. An absolute zero could not be obtained since as soon as V_{bias} drops below either V_{return} or V_{min} the SQUID loses its flux locking.

Thanks to the bias circuit, no magnetic field was needed to suppress the Josephson current of the STJ, and hence the subgap structures are not obscured by Fiske resonances. As the junction is voltage biased, also the parts of the *I-V* trace with negative conductance $\partial I/\partial V$ could be measured. The peak at 260 μV in the 4.2 K trace is caused by the difference of 0.26 meV between the gap energy in the top and the bottom electrode of the STJ. The sensitivity of the DROS is demonstrated in the inset of fig. 3, which shows an enlargement of a part of the 1.6 K trace. Small structures of less than 1 nA are revealed, and the sharp rise of the subgap current at a voltage corresponding to half the gap voltage is evident [8].

4. Conclusions

We realized a measurement system for characterization of the subgap *I-V* characteristic of tunnel junctions based on a DROS. After upgrading of the present flux locked loop electronics to a larger bandwidth, the system is suitable for application in the field of particle detection. The sensitivity of the DROS, 2 pA/√Hz, will make readout of particle detectors with a spectral resolution of $\Delta E/E < 1$ ‰ attainable [2]. Further optimization of the input circuitry of the DROS for the readout of particle detectors or the introduction of an intermediate transformer could significantly improve the sensitivity.

References

1. N.E. Booth and D.J. Goldie, *Supercond. Sci. and Technol.*, **9** (1996) 493-516.
2. M. Frank, C.A. Mears, S.E. Labov, F. Azgui, M.A. Lindeman, L.J. Hiller, H. Netel and A. Barfknecht, *Nucl. Instr. and Meth. in Phys. Res.*, **A370** (1996) 41-43.
3. D. Drung, H. Matz and H. Koch, *Rev. Sci. Instrum.*, **66** (1995) 3008-3015.
4. R.P. Welty and J.M. Martinis, *IEEE Trans. Appl. Supercond.*, **3** (1993) 2605-2608.
5. D.J. Adelerhof, J. Kawai, G. Uehara and H. Kado, *Appl. Phys. Lett.*, **65** (1994) 2606-2608.
6. M.J. van Duuren, G.C.S. Brons, D.J. Adelerhof, J. Flokstra and H. Rogalla, *J. Appl. Phys.*, **82** (1997) 3598-3606.
7. M.J. van Duuren, D.J. Adelerhof, G.C.S. Brons, J. Kawai, G. Uehara, H. Kado, J. Flokstra and H. Rogalla, *J. Appl. Phys.*, **80** (1996) 4164-4173.
8. T.M. Klapwijk, G.E. Blonder and M. Tinkham, *Physica B*, **109&110** (1982) 1657-1664.

LOW CREEP AND HYSTERESIS LOAD CELL BASED ON A FORCE TO LIQUID PRESSURE TRANSFORMATION

Robert A.F. Zwijze, Remco J. Wiegerink, Theo S. J. Lammerink and Miko Elwenspoek

Mesa Research Institute, University of Twente
P.O. Box 217, 7500 AE Enschede, The Netherlands

0. Abstract

Important problems in load cells are creep and hysteresis. Expensive high grade steels are used in order to reduce these effects. In this paper a load cell design based on a force to liquid pressure transformation is presented. The design is insensitive to hysteresis and creep and can be made at very low costs. Analytical, numerical and experimental results are in very close agreement with each other.

Keywords: load cell, force measurement, creep, hysteresis

1. Introduction

A load cell is a sensor for the measurement of forces and masses. Load cells are for example used in weighing bridges for lorries, cars and trailers. Also in industries where bulk material is worked up, it is necessary to measure masses as accurate as possible.

Most current load cells are made of steel. The performance of these load cells is limited by hysteresis and creep even when expensive high grade steels are used. Hysteresis means that the output is a function of the loading history. Creep means that for a fixed load the output changes in time. In order to increase stability and to decrease the costs, a new type of load cell is presented which is nearly free of hysteresis and creep and which has a very simple design. The load cell we are aiming at has the following specifications:

- maximum load: 10000 N (1000 kg)
- accuracy of full scale: 0.03 % = 0.3 kg
- temperature range: -10 up to 50 °C
- production costs: less than US $ 75
- calibration: once in two years

The load cell discussed in this paper transforms the force into a fluid pressure. This pressure is measured by a commercial silicon pressure (pressure range: 0-8.8 bar; price: US $ 25) sensor which is because of its mono cristallinity not sensitive to creep and hysteresis. In [2], a fluid pressure load cell was introduced as a new kind of load cell. It consist of a piston under which the fluid is locked up by a seal. Characteristic for this kind of load cells is the high sensitivity. However, the load cell presented in [2] is still rather sensitive to hysteresis. Another disadvantage is that it needs a standard Teflon seal which requires a proper surface finish of the mating steel parts. The load cell discussed in this paper does not have these drawbacks.

In section 2 the concept of the force to liquid pressure transformation is explained. Section 3 considers the modeling of the load cell. The realization and experiments are treated in section 4. Finally, in section 5 some conclusions and recommendations are drawn.

2. Concept of force to liquid pressure transformation

In most current load cells the force F is applied to a steel bar which will be deformed (see figure 1a). The deformation is sensed by strain gages. This is a force to strain transformation and the strain ε of this type of load cell can be expressed as

$$\varepsilon = C\frac{F}{E}, \tag{1}$$

where C is a constant and E the elastic modulus of the bar. As the output signal depends on the elastic modulus, this sensor is sensitive to hysteresis and creep.

Another option is to use a force to liquid pressure transformation. The most simple example of this kind of load cells is given in figure 1b. The force to pressure transformation is given by

$$P = \frac{F}{A}, \tag{2}$$

where A is the area of the piston. The pressure is sensed by a silicon pressure sensor which is because of its mono cristallinity independent of hysteresis and creep. It follows from (2) that the signal is independent of an elastic modulus and, therefore, insensitive to hysteresis and creep. The drawback of this kind of load cell is that a seal has to be used which imposes demands on the surface finish of the piston and bucket. Besides, the friction of the seal may be a new source of hysteresis. In order to eliminate these drawbacks the load cell presented in this paper is designed.

Figure 2 shows a schematic drawing of this load cell which is axial symmetric. A liquid is locked up between the steel membrane and the steel bucket. The force F is applied to the steel boss which is attached to the steel membrane. A silicon pressure sensor measures the difference between the fluid pressure P_{fluid} and the air pressure P_{air}. It is seen that no seal is needed. In the next section it will be shown that this load cell has a same kind of relation as in (2). The only difference is that the area for calculating the pressure is some effective area (A_{eff}) which only depends on the ratio of radii r_1 and r_2:

$$P_s = P_{fluid} - P_{air} = \frac{F}{A_{eff}}. \tag{3}$$

In the next section this load cell is modeled and the function A_{eff} is determined.

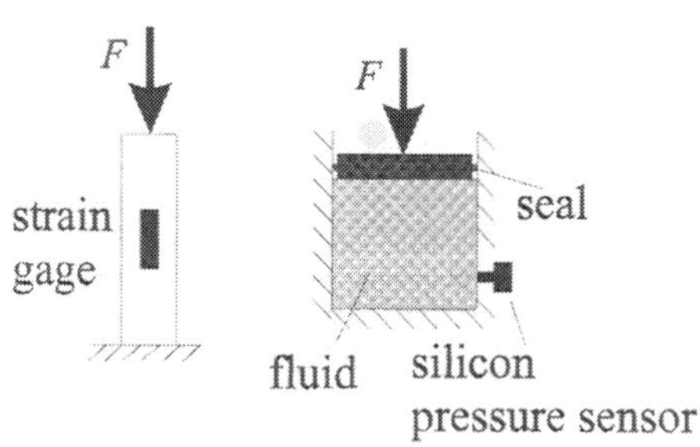

figure 1: (a) conventional steel load cell. (b) fluid compressed by a piston.

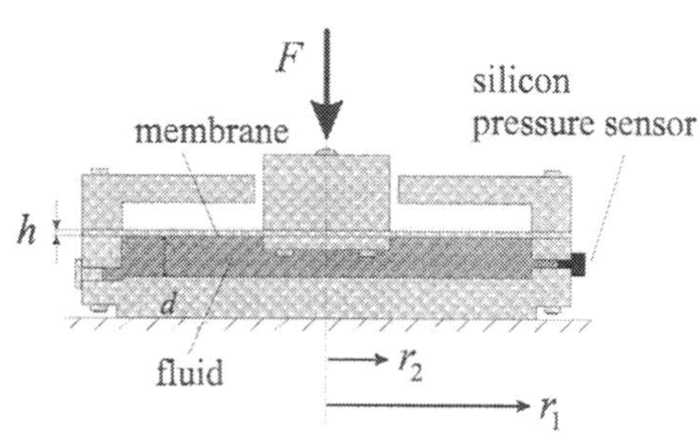

figure 2: Layout load cell.

3. Model

In this section the load cell is modeled mathematically. In §3.1 it is assumed that the water is incompressible. In this paragraph the pressure-force relation, the deformation profile of the membrane and the stresses in the membrane are calculated. Then, in §3.2 the compressibility of the water is taken into account. In §3.3 the temperature dependence of the load cell is considered.

3.1 Incompressible fluid

In figure 3 the deformation of the membrane and the deflection of the boss for some load F is shown. The pressure-force relation and deflection profile of the membrane are derived in the next way:

- Use classical plate elasticity theory [1].
- The deflection profile $w(r)$ of the membrane has to satisfy some boundary conditions:

$$w(r=r_1)=\frac{dw}{dr}(r=r_1)=\frac{dw}{dr}(r=r_2)=0 \tag{4}$$

- Incompressibility of the fluid is satisfied if the following condition is fulfilled:

$$\pi r_2^2 w(r=r_2)+\int_{r=r_2}^{r_1} 2\pi r w(r)dr=0 \tag{5}$$

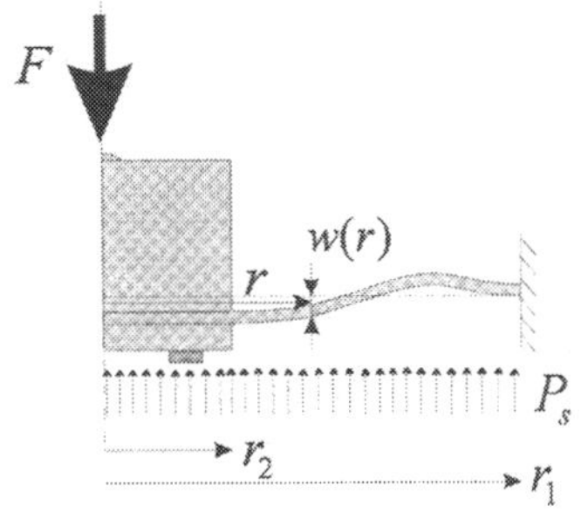

figure 3: Deformation of membrane and boss.

The general expression for the deflection $w(r)$ of the membrane is given by

$$w(r)=\frac{F(\nu^2-1)}{\pi E h^3}\lambda(\frac{r_1}{r_2},r_1,r), \tag{6}$$

where $\lambda(r_1/r_2,r_1,r)$ is a rather complex function. An impression of the deformation of the membrane is given in figure 4.

The pressure-force relation is given by:

$$P_s=\frac{F}{A_{eff}},\quad A_{eff}=\frac{\pi r_2^2}{\pi}\frac{\left(1-4s^2+6s^4-4s^6+s^8\right)}{\left(3.82\ln(s)\left[s^2-s^4\right]+0.95\left[1-s^2-s^4+s^6\right]\right)},\quad s=\frac{r_1}{r_2}. \tag{7}$$

From (7) it follows that there is no dependence on the elastic modulus of the steel membrane so that it is concluded that the load cell is insensitive to hysteresis and creep. A_{eff} is the effective area which was discussed in section 2. An impression of the effective area in

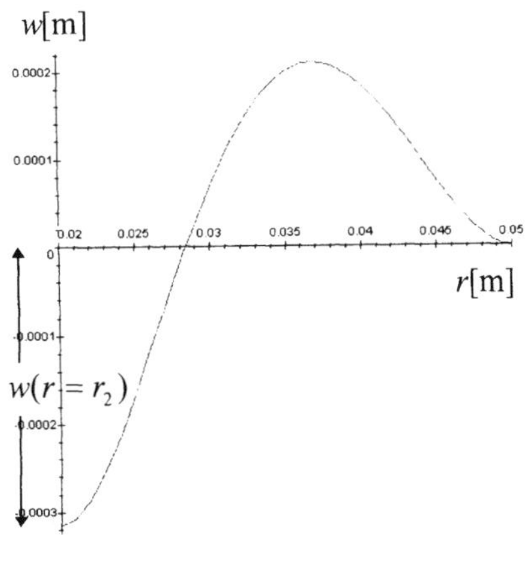

figure 4: Deformation of the membrane ($F=10000\ N$, $E=210\ Gpa$, $\nu=0.3$ $h=1\ mm$, $r_1=5\ cm$, $r_2=2\ cm$, $d=1\ cm$).

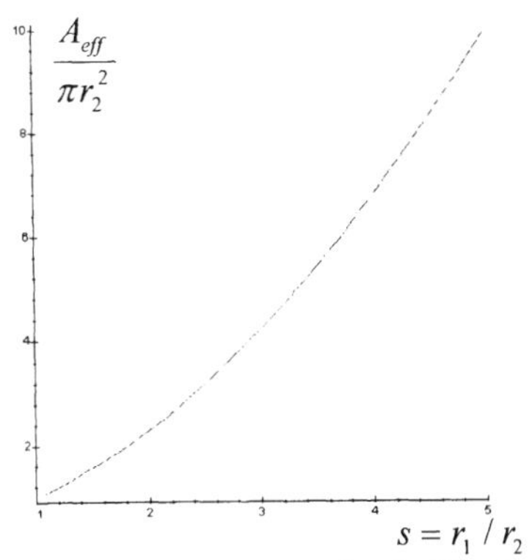

figure 5: Effective area divided by the area of the boss as a function of the ratio r_1/r_2.

comparison to the area of the boss and to the total area of the bucket (=area of boss + membrane) are respectively shown in figure 5 and figure 6. From these plots, it is concluded that the effective area lies somewhere between the area of the boss and the bucket.

As the membrane is rather thin it must be checked whether the critical stress is not exceeded. In the membrane three stresses are present: radial, tangential and shear stress acting in the direction of force F (σ_r, σ_t and τ respectively). The first two are calculated from their corresponding moments [1] and they have a maximum on both sides of the membrane. As all three stresses occur at the same place in the membrane, the Von Mises stress criterion is used for comparison to the maximum allowable stress. It is given by

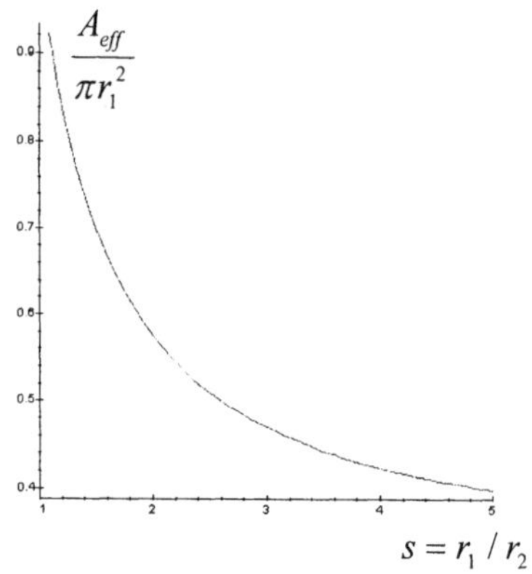

figure 6: Effective area divided by the area of the bucket as a function of the ratio r_1 / r_2.

$$\sigma_v = \sqrt{\frac{1}{2}\left[(\sigma_r - \sigma_t)^2 + \sigma_t^2 + \sigma_r^2 + 6\tau^2\right]}. \tag{8}$$

A plot of all the stresses is given in figure 7. The ultimate stress of steel is about 1 to 1.5 Gpa, so it should just be possible to bear a load of 10000 N. As the shear stress τ is very small in comparison to $\sigma_{r,\max}$ and $\sigma_{t,\max}$ this stress can be neglected. Besides, it is seen that the maximum stresses occur at $r = r_2$. Therefore, the maximum Von Mises stress in the membrane is approximately given by

$$\sigma_{v,\max} = \frac{3F\sqrt{(1-\nu-\nu^2)}}{4\pi h^2}\gamma(s), \qquad \gamma(s) = \frac{\left(4\ln(s)s^4 - 5s^4 + 8\ln(s)s^2 + 4s^2 + 1\right)}{(1-2s^2+s^4)}. \tag{9}$$

The function $\gamma(s)$ is drawn in figure 8. It is concluded that the maximum stress strongly depends on the thickness h of the membrane. The dependence on the ratio r_1 / r_2 is less strong.

For the parameters given in the caption of figure 4 the pressure sensitivity is calculated from (7) giving $\partial P_s / \partial F = 250.03$ Pa / N.

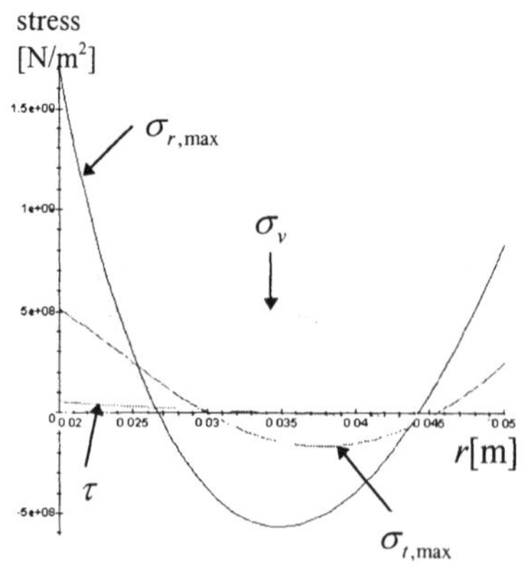

figure 7: Radial, tangential, shear and Von Mises stresses in the steel membrane.

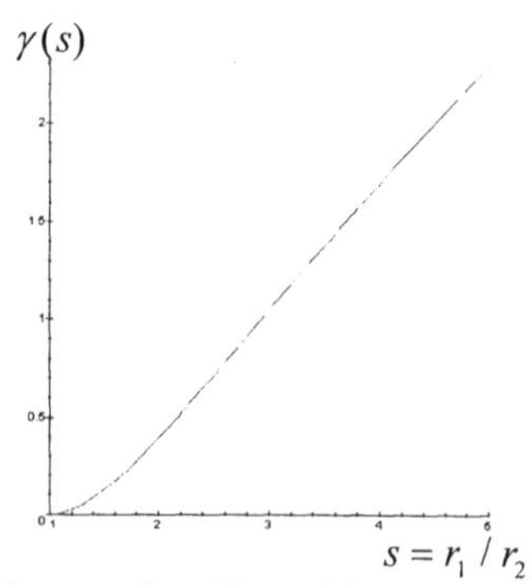

figure 8: Von Mises stress parameter $\gamma(s)$.

3.2 Compressible fluid

By including compressibility of the fluid, equation (5) is changed to

$$\pi r_2^2 w(r = r_2) + \int_{r=r_2}^{r_1} 2\pi r w dr = \Delta V = V_0 \frac{P_s}{E_v}, \tag{10}$$

where E_v is the bulk compressibility modulus, V_0 the initial volume and ΔV the decrease in volume of the compressed fluid. For water $E_v = 2.24 \text{ GN/m}^2$. The effective area is calculated in the same way as done in §3.1 giving

$$A_{eff} = \pi r_2^2 \frac{\left\{1 - 3s^2 + 3s^4 - s^6 + \frac{16 V_0 E h^3}{\pi E_v r_1^6 (v^2 - 1)} s^6)\right\}}{3.82 \ln(s) s^2 + 0.95\left[1 - s^4\right]}. \tag{11}$$

From this equation it is concluded that hysteresis and creep are present, because of Youngs modulus. Therefore, the term $\frac{16 V_0 E h^3}{\pi E_v r_1^6 (v^2 - 1)}$ must be chosen as small as possible. The most simple tool for reducing this term is by giving a very small value to the initial volume of the fluid. For the parameters given in the caption of figure 4 the pressure-sensitivity equals $\partial P_s / \partial F = 248.92$ [Pa / N]. So it is concluded that the condition of incompressibility can be maintained.

The load cell is also analyzed in the finite element program Ansys. For the (compressible) fluid, FLUID79 elements are taken. The deformation profile and displacements in the direction of the force are shown in figure 9. The deformation profile appears to be the same as was obtained with analytical formulas. The pressure sensitivity equals $\partial P_s / \partial F = 247.05$ [Pa / N] which is about the same as the analytical result.

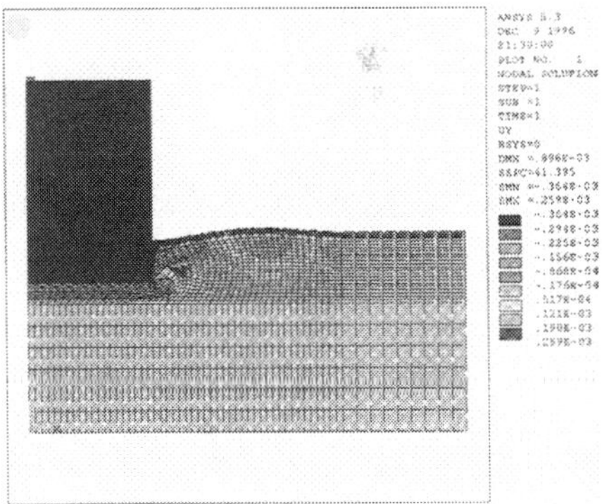

figure 9: Displacements in direction of applied force.

3.3 Dependence on temperature

The effective area on which the fluid presses, changes linearly with temperature:

$$\frac{dA_{eff}}{dT} = 2\alpha_{steel} A_{eff}. \tag{12}$$

T is the temperature and α_{steel} the temperature expansion coefficient of steel. However, due to large differences in temperature expansion coefficients of the steel and fluid, this effect is overruled. Practical values are $\alpha_{Steel} = 12 \cdot 10^{-6}$ °C^{-1} and for water at 20 °C, $\alpha_{water} = 207 \cdot 10^{-6}$ °C^{-1}. For these parameters Ansys calculates a temperature sensitivity of $\partial P_s / \partial T = 4849$ Pa / C°. If $\alpha_{fluid} = \alpha_{steel}$ is taken, then it follows from Ansys that $\partial P_s / \partial T = 0.54 \cdot 10^{-7}$ Pa / C° which is very low so that it is concluded that the temperature coefficients of the steel and fluid should match as close as possible.

The temperature sensitivity can also be decreased by reducing the volume of the fluid. Decreasing this volume by a factor 10 gives a decrease in temperature sensitivity by about a factor 10.

4. Realization and experiments

The load cell is tested by using a commercial Honeywell pressure sensor. The repeatability and hysteresis of this sensor is less than 0.15 %. For the fluid, water is used. Hysteresis of the load cell is tested by loading, then unloading and hereafter loading the load cell with some weight. The output of the Wheatstone bridge of the pressure sensor is shown in figure 10. The load cell behaves linearly and the pressure sensitivity equals $\partial P_s / \partial F = 249$ [Pa / N] which is in very close agreement with the analytical and numerical results. The sensor could not be loaded above 400 kg, because the pressure sensor is not able to withstand a higher pressure. The hysteresis as a percentage of the full scale maximum load (that is 1000 kg) is less than 0.009 % which is far more accurate than the 0.03 % that was specified in section 1. For the calculation of the full scale output of the pressure sensor linear behavior of the pressure sensor is assumed which is correct, because its linearity error is less than 0.25 % of full scale.

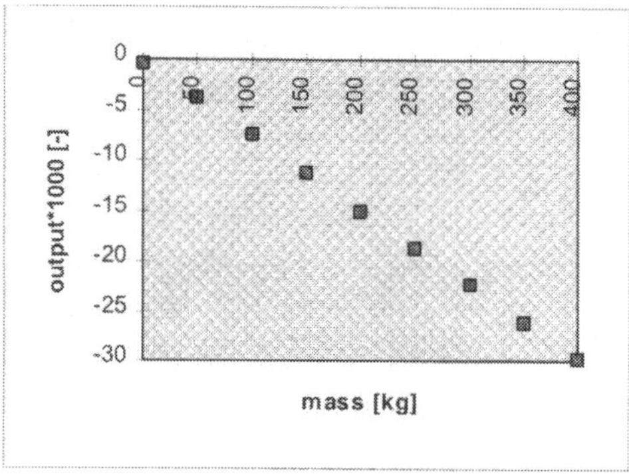

figure 10: Hysteresis in the load cell.

So far, it was not possible to test creep, because the pressure sensor and the force to fluid pressure transformation are rather sensitive to temperature variations.

5. Conclusions and recommendations

It is proved that the force-pressure relation of the designed load cell is independent of the elastic modulus and therefore is independent of hysteresis and creep. The analytical, numerical and experimental results are in very close agreement with each other. It is shown by experiments that the hysteresis of the load cell is less than 0.009 %. As the hysteresis of the pressure sensor can be greater than 0.009 % the hysteresis in the force to liquid pressure transformation might even be smaller. The dependence of the fluid pressure on temperature variations can be eliminated by choosing a very small initial volume of the water and by matching the temperature expansion coefficients of the fluid and steel.

At the moment the load cell is being made completely of silicon. Calculations have shown that this is indeed possible. The advantage is that load cells can be made in large amounts and at very low costs.

In the future creep must be investigated experimentally. In addition, a pressure sensor has to be used which is insensitive to temperature variations and which has an accuracy better than 0.03 %. Fluids must be found with temperature expansion coefficients which equal the coefficient of the steel as close as possible.

6. References

[1] S.P. Timoshenko and S.W. Woinowsky-Krieger, Theory of plates and shells, McGraw-Hill International Editions, Singapore, 1959.

[2] H. Kazerooni, M.S. Evans and J. Jones, Hydrostatic force sensor for robotic applications, Journal of dynamic systems, measurement and control, 119(Issue 1), 1997, 115-119.

STABILITY AND APPLICATIONS OF ELECTROCHEMICALLY-ACTIVE SELF-ASSEMBLED MONOLAYERS

M.W.J. Beulen; F.C.J.M. van Veggel; D.N. Reinhoudt
MESA Research Institute, Supramolecular Chemistry and Technology
Chemical Technology, University of Twente,
P.O. Box 217, 7500 AE Enschede, The Netherlands
E-mail: Beulen@ct.utwente.nl

0. Abstract

The electrochemical stability of monolayers is important for future applications. A potential window for self-assembled monolayers in aqueous inert electrolyte was obtained using Cyclic Voltammetry, Electrochemical Impedance Spectroscopy and Heterogeneous Electron Transfer. Also electrochemical techniques can be used for sensing applications. We investigated mixed monolayers of ferrocenethiol and an acid alkylthiol. Deprotonation of the acid by an alkaline electrolyte influences the other part of the mixed monolayer. The formal potential of ferrocene is shifted about 150 mV cathodically by the negative charge in the carboxylate part of the layer.

Keywords: monolayer, electrochemistry

1. The Electrochemical stability of self-assembled monolayers.

The use of self-assembled monolayers of sulfur containing adsorbates on gold in electrochemical applications requires monolayers that are electrochemically stable [1,2,3]. Therefore, monolayers of the five adsorbates in Figure 1 were prepared and investigated for their stability towards an electrical field. The adsorbates differ in type of attachment to the gold (thiol vs. sulfide) and number of attachments points (1-4). Two parameters were used to follow the change in quality of the monolayer: the resistance and the capability to block an external redox-couple (HET). Figures of the quality of the monolayer vs. the applied potential for the decylthiol **1** and cavitand **5** are depicted in figure 2. The gray window indicates the potential area where the decylthiol and cavitand monolayers are electrochemically stable. This windows ranges from -0.9 to +0.4 V_{MSE}. The decylthiol monolayer is a little bit more stable at the cathodic side of the window. Figure 3 shows the electrochemical stability for three differently benzene derivatives with two sulfide units for the attachment to gold. The quality of these three monolayers is of an order of magnitude lower than the decylthiol and cavitand monolayer (more current is going through the monolayer in the HET and the resistance of the monolayer is smaller). The potential window for these monolayers is a little bit smaller: +0.4 to only -0.8 V_{MSE}. Furthermore there are no big differences in stability between the three differently substituted benzenes.

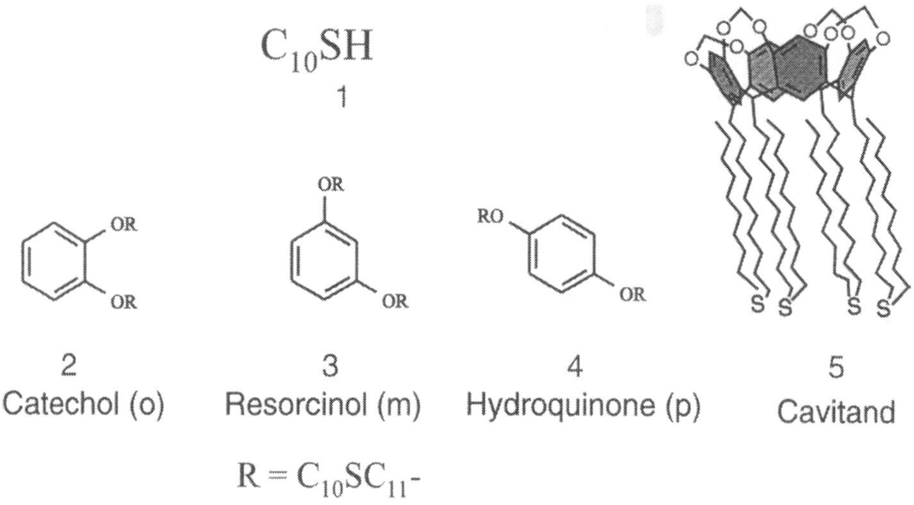

Figure 1. Adsorbates used for the electrochemical stability measurements.

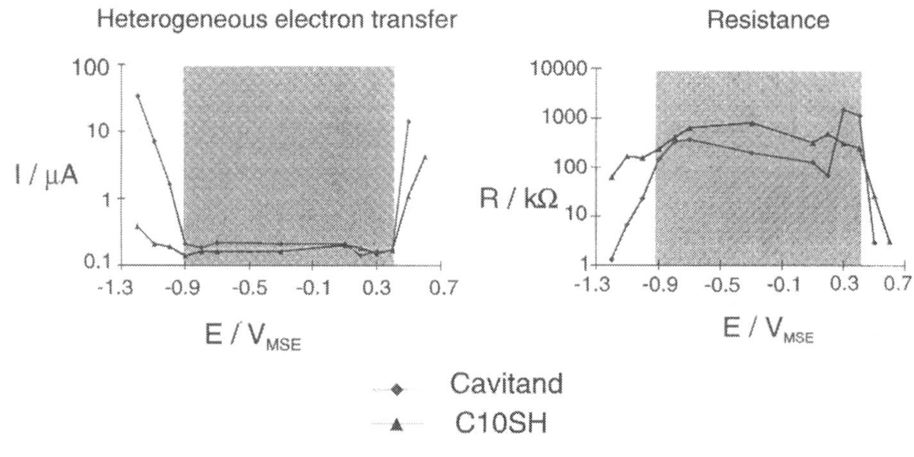

Figure 2. Potential window for decylthiol and cavitand monolayer.

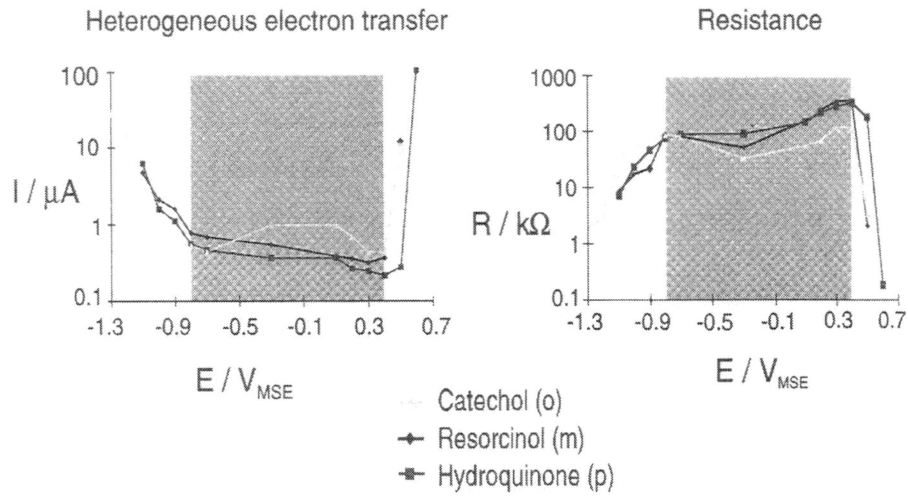

Figure 3. Potential window of differently substituted benzenes.

The potential window where self-assembled monolayers are electrochemically stable is thus from -0.8 to +0.4 V_{MSE}. The quality of the decylthiol and cavitand monolayer is higher than of the three substituted benzene derivatives.

2. pH-dependent response of mixed monolayers of ferrocene and acid.

Simple ferrocenethiol monolayers are electrochemically well-defined and easily prepared. Electrochemically-active self-assembled monolayers can be used for sensor-experiments in several ways. The receptor part can be present in the redox-molecule. A new approach uses mixed monolayers of an electrochemically-active adsorbate and a receptor adsorbate. Complexation at the receptor part of the monolayer causes a change in the surrounding of the redox-center and might lead to a shift in redox-potential. In this system there is a transfer of information from the receptor part to the other part of the mixed self-assembled monolayer, the redox center. Experiments were performed with a mixed monolayer of ferrocenethiol (Fc-C_6-SH) and an acid-thiol (HOOC-C_7-SH).

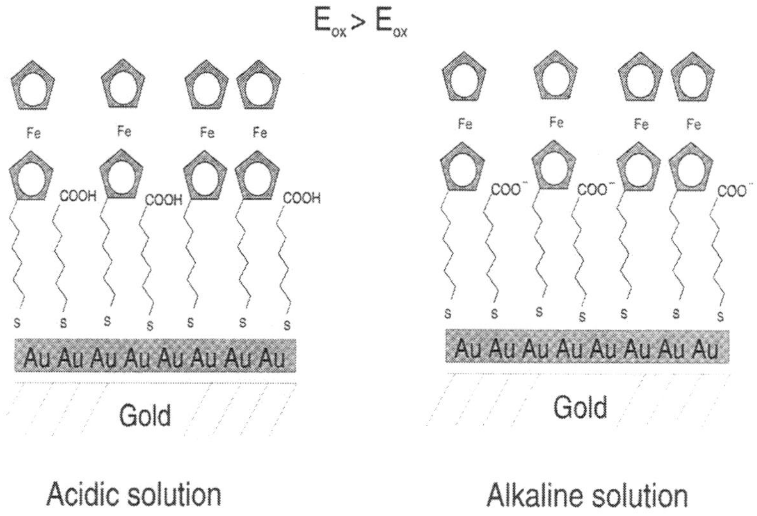

Figure 4. Mixed ferrocene/acid monolayer for sensor experiments.

Raising the pH of the electrolyte will deprotonate the acid function in the monolayer. The resulting negative charge can favor the ferrocene oxidation and shift it cathodically (oxidation at a lower potential). The mixed monolayer shows a clear pH-dependent response of the ferrocene (see Figure 5, mixed monolayer of 25% ferrocene and 75% acid in adsorbate solution). In $HClO_4$ (0.1M) the formal potential of the ferrocene has a value of -130 mV_{MSE} (see Figure 5). Changing the pH from 1 to 10 results in a broad peak at about -130 mV_{MSE} and an extra peak at about -300 mV_{MSE} is present. The shift of the redox peak (the deprotonation and protonation of the acid) is nearly reversible as changing the pH back and forward results in the same spectra. Information is transferred in a mixed self-assembled monolayer and measured electrochemically via the non-receptor part.

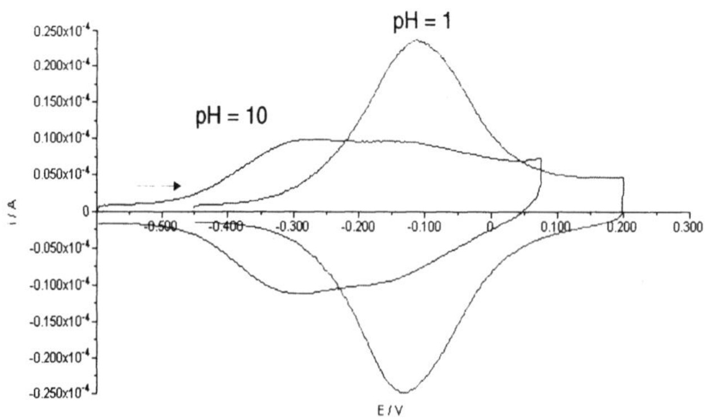

Figure 5. CV's of a mixed ferrocene/acidthiol monolayer at different pH's.

3. References

1) M.W.J. Beulen, B.-H. Huisman, P.A. van der Heijden, F.C.J.M. van Veggel, M.G. Simons, E.M.E.F. Biemond, P.J. de Lange, D.N. Reinhoudt, *Langmuir*, **1996**, *12*, 6170.

2) K.D. Schierbaum, T.Weiss, E.U. Thoden van Velzen, J.F.J. Engbersen, D.N. Reinhoudt, W. Göpel, *Science*, **1994**, *265*, 1413.

3) B.-H. Huisman, R.P.H. Kooyman, F.C.J.M. van Veggel, D.N. Reinhoudt, *Advanced Materials*, **1996**, *8*, 561.

BINDING OF METAL IONS TO SELF-ASSEMBLED MONOLAYERS OF CROWN ETHERS

Simon Flink, Frank C. J. M. van Veggel, David N. Reinhoudt
Supramolecular Chemistry and Technology, MESA Research Institute,
University of Twente, P.O. Box 217, 7500 AE Enschede, The Netherlands.

0. Abstract

The reversible binding of cations to self-assembled monolayers of (6-mercaptohexyloxy)methyl-12-crown-4 (**1**) and (6-mercaptohexyloxy)methyl-15-crown-5 (**2**) in aqueous solutions is reported. Binding of the electrochemically inactive cations can be monitored by cyclic voltammetry and impedance spectroscopy using $Ru(NH_3)_6^{2+/3+}$ as "reporter ions". As a result of the high surface density of crown ether ligands sandwich complexes are probably formed, which accounts for the observed selectivities of the monolayer.

1. Introduction

Self-assembled monolayers (SAMs) of sulfur-containing molecules, like thiols, disulfides, and sulfides are the subject of numerous investigations.[1] The ease of their preparation, their stability, and the possibility to introduce different functional groups makes them a facile way to create surfaces with tailor-made properties.[2,3] Self-assembly of receptor molecules onto gold enables their use in sensing devices. We have previously shown that self-assembled monolayers of resorcin[4]arene tetrasulfides can be used for the detection of organic compounds from the gas phase.[4] Only a few receptor adsorbates have been used to investigate the binding of guests in solution.[5] Rubinstein *et al.*[6] and Mandler *et al.*[7] have shown that the *voltammetric* detection of electroactive metal ions using self-assembled monolayers is a very sensitive and selective method. Recently, also impedance spectroscopy was used to detect the binding of guests to monolayers. Rubinstein *et al.*[8] reported changes of the monolayer capacitance due to the binding of Fe^{3+} by trishydroxamate SAMs. Henke *et al.*[9] used a cyclodextrin monolayer to bind an organic sulfonic acid which changed the charge transfer resistance of the SAM.
In this poster we describe the reversible and selective complexation of metal ions from aqueous solutions to self-assembled monolayers of 12-crown-4 and 15-crown-5 adsorbates, monitored by cyclic voltammetry and impedance spectroscopy.

2. Experimental

(6-mercaptohexyloxy)methyl-12-crown-4 (**1**) was synthesized by alkylation of 2-(hydroxymethyl)-12-crown-4 with 5 mol equivalents of 1,6-dibromohexane and NaH in DMF. This intermediate product was subsequently converted into the desired product **1** by treatment with thiourea in boiling ethanol, followed by hydrolysis with KOH in boiling water. The 15-crown-5 derivative **2** and the triethyleneglycol derivative **3** were synthesized accordingly from 2-(hydroxymethyl)-15-crown-5 and triethylene glycol monomethyl ether, respectively. The structure and purity of the products was verified by TLC, ^{1}H NMR, ^{13}C NMR, and EI-MS.

The monolayer preparation and the used substrates were applied as discribed before in ref 3c. Experimental details regarding the IR characterization were described in ref 3c. Advancing and receding contact angles were determined automatically on a Krüss G10 Contact Angle Measuring Instrument, equipped with CCD camera, during growth and shrinkage of the droplet by the drop shape analysis routine. Cyclic voltammetry and impedance spectroscopic measurements were performed with an Autolab PGSTAT10 in a three-electrode cell containing a gold working electrode (clamped to the bottom of the cell exposing a geometric area of 0.44 cm^2 to the electrolyte), a platinum counter electrode, and a mercurous sulfate reference electrode. All measurements were conducted in electrolyte containing 1 mM $Ru(NH_3)_6Cl_3$ as a redox couple and a total concentration of the background electrolyte of 100mM. Impedance spectroscopic experiments were conducted at the formal redox potential of $Ru(NH_3)_6^{3+/2+}$, in a frequency range of 10 kHz to 0.1 Hz, with an ac-amplitude of 5 mV. Analysis of the spectra was performed using the software package "Equivalent Circuit" see ref 14.

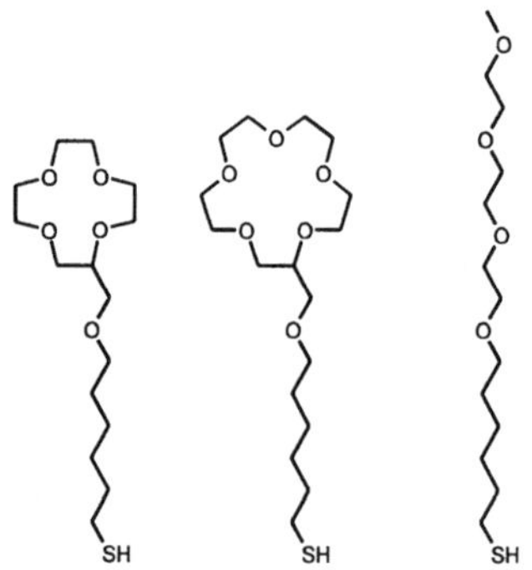

1 **2** **3**

3. Results and discussion

The monolayers of both crown ether adsorbates showed methylene C-H stretching vibrations at 2930, 2904, and 2859 cm^{-1}, which corresponds with the bulk spectra. The absorption at 1140 cm^{-1} was assigned to the C-O stretching mode of the ether groups.[10] The monolayer of the triethylene glycol adsorbate **3** showed a broad absorption centered around 2870 cm^{-1}, which was assigned to the various C-H stretching modes of the methyl and methylene groups. In addition a peak of the C-O streching vibration was present at 1115 cm^{-1}. Wettability studies of the three SAMs with water showed contact angles that are characteristic for a surface of low hydrophobicity. The self-assembled monolayers of 12-crown-4 **1**, 15-crown-5 **2**, and the triethylene glycol adsorbate **3** had advancing contact angles of 59, 55, and 58 ± 1°, respectively. In the case of both crown ether monolayers **1** and **2**, the hysteresis between the advancing and receding contact angle was 20°. SAMs of **3** showed a slightly smaller hysteresis of 15°, which indicates a more ordered layer.

Characterization of the crown ether monolayers by cyclic voltammetry in the presence of $Ru(NH_3)_6Cl_3$ showed that the heterogeneous electron transfer from the gold electrode to the redox couple in aqueous solution was hardly influenced by the

presence of the SAM. The large current density of the redox waves around -0.56 V_{MSE} indicates that the redox couple can easily access the electrode surface.
Addition of NaCl to the electrolyte leads to a *drastic decrease* of the faradaic current for the 12-crown-4 **1** modified electrode as shown in Figure 1. The 15-crown-5 **2** modified gold electrode showed a large attenuation of the redox current when KCl was added to the electrolyte (data not shown).

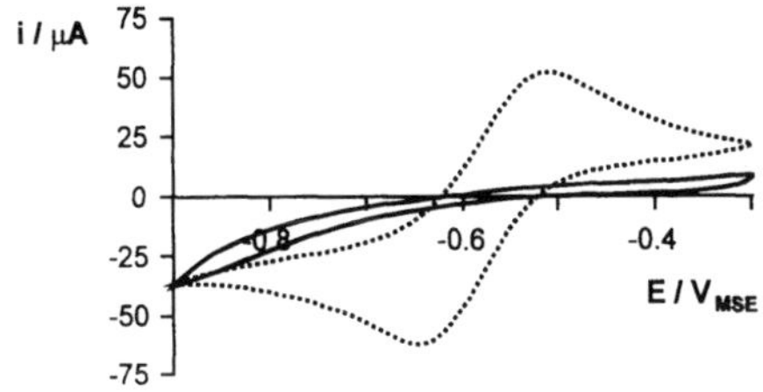

Figure 1. Cyclic voltammograms of a 12-crown-4 1 modified gold electrode in aqueous solutions. The dotted CV is recorded in 100 mM Et_4NCl + 1 mM $Ru(NH_3)_6Cl_3$; the solid CV is recorded in 14 mM NaCl + 86 mM Et_4NCl + 1 mM $Ru(NH_3)_6Cl_3$.

Similar effects have previously been observed by cyclic voltammetry of monolayers with amine or carboxylic acid headgroups which can be charged by protonation and deprotonation, respectively.[11] The influence of a charged monolayer on the electron transfer to a charged redox couple was attributed to the electrostatic forces acting between both, and the shift of the potential drop across the diffuse layer. Hence, complexation of metal cations by the SAM generates a repulsion towards the positively charged redox couple which attenuates its redox current.[11b]
This means that cyclic voltammetry allows the detection of metal cation complexation by the monolayer. However, the application of large potential sweeps (in the range of several hundred mV) might disturb the structure of the monolayer.[12] Impedance spectroscopy is a less interfering technique which enables the simultaneous determination of the electrochemical properties of the monolayer by using a wide frequency range.[13] The impedance spectra of the crown ether SAMs **1** and **2** were best described by the Randles equivalent circuit (Figure 2).[14,15]

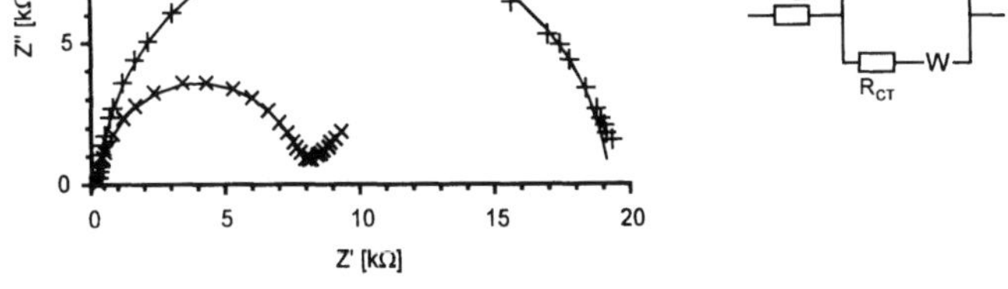

Figure 2. Impedance spectra of 12-crown-4 1 modified gold electrode in solutions containing 1 mM NaCl (x) and 2.5 mM NaCl (+). Fits to the spectra are indicated by the solid line. The Randles equivalent circuit, used to fit the spectra, consists of the electrolyte resistance (R_{EL}), the monolayer capacitance (C_{ML}), the charge transfer resistance (R_{CT}), and a diffusion element (W).

The dependence of R_{CT} on the concentration of sodium demonstrates that binding of metal ions to the monolayer of 12-crown-4 increases its resistance from 250 Ω to a limiting value of 145 kΩ at concentrations above 23 mM (see Figure 3). The initially linear increase of R_{CT} at low concentrations of sodium levels off at concentrations above 10 mM, and finally reaches a plateau value. These observations indicate that the binding sites become saturated at high guest concentrations.

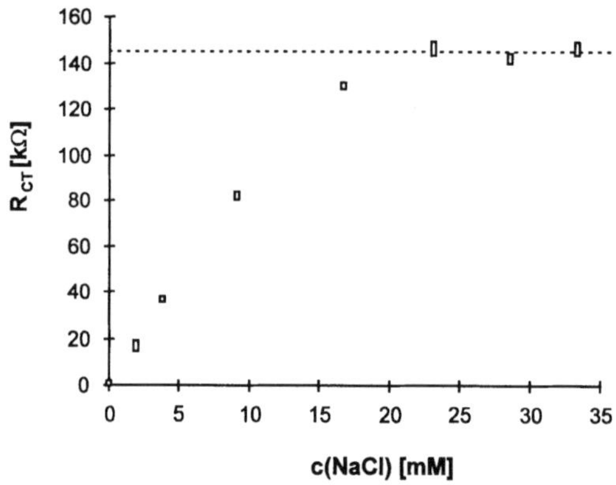

Figure 3. Charge transfer resistance (R_{CT}) of 12-crown-4 monolayer 1 as a function of the sodium concentration. For each concentration at least two impedance spectra were recorded within 30 minutes to establish the stability of the response.

Replacement of the solution by a sodium free electrolyte of 0.1 M Et_4NCl + 1 mM $Ru(NH_3)_6Cl_3$ reduced the charge transfer resistance to the initial value of 250 Ω and the addition of sodium resulted in the same changes of R_{CT} as before. This demonstrates that the binding of sodium to the monolayer is a *reversible* process. The monolayer capacitance (C_{ML}), determined simultaneously from the impedance spectra, shows a very weak response to the binding of sodium (6.4 $\mu F/cm^2$ at $[Na^+] = 0$ mM and 6.8 $\mu F/cm^2$ at $[Na^+] > 15$ mM). The slight increase of C_{ML} is probably caused by a change of the relative permittivity ε_r of the monolayer upon binding of cations.[8] The response of the monolayer (ΔR_{CT} vs. salt concentration) was also determined for several other alkaline and alkaline earth salts. Figure 4a shows the change of the resistance of the 12-crown-4 SAM at two guest concentrations (3.8 and 9.1 mM). It is clear that sodium gives the largest response. Also potassium does affect ΔR_{CT}, but the increase of the resistance is roughly a hundred times less than for the same concentrations of sodium. Other ions had even smaller effects on ΔR_{CT}. The selectivity of sodium over potassium makes it possible to *detect sodium concentrations down to 0.1 mM in the presence of 10 mM of potassium.*

A self-assembled monolayer of 15-crown-5 showed a completely different response pattern towards alkaline and alkaline earth cations (Figure 4b). This larger crown ether shows the highest response for potassium. Besides potassium also cesium shows a significant increase of the charge transfer resistance. It is remarkable that sodium gives rise to a signal which is less than one percent of that observed for the same concentration of potassium.

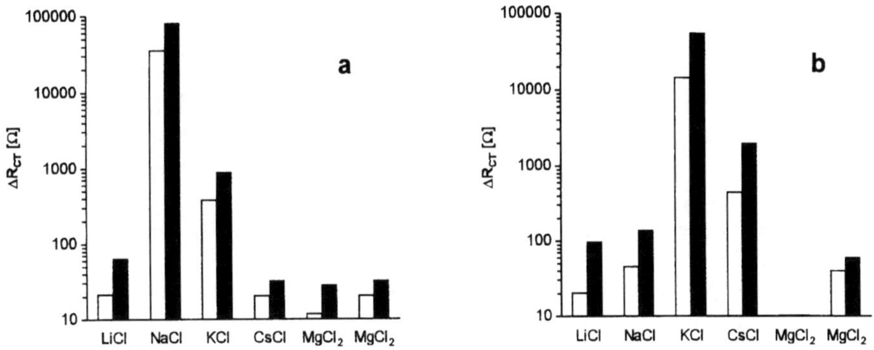

Figure 4. (a) ΔR_{CT} of 12-crown-4 monolayer in the presence of different salts (the white and the black column represent concentrations of 3.8 and 9.1 mM, respectively). (b) ΔR_{CT} of 15-crown-5 monolayer in the presence of different salts (the white and the black column represent concentrations of 2.0 and 9.1 mM, respectively).

Monolayers of the triethylene glycol adsorbate **3** showed no detectable change in both parameters (R_{CT} and C_{ML}) upon addition of alkaline metals. Hence, the binding of metal ions to the crown ether SAMs is a result of the macrocyclic structure of the receptor adsorbates. On the basis of the association constants determined for the 1:1 complexes of 12-crown-4 and 15-crown-5 with metal ions in aqueous solutions preferential binding of sodium and potassium to monolayers of 12-crown-4 and 15-crown-5 would not be expected.[16] The fact that the metal ions are bound to a monolayer with its high surface concentration of crown ethers makes the formation of sandwich complexes, in which one metal ion is complexed by two crown ethers, very likely. The influence of sandwich complexation on the selectivity has been studied using ion-selective electrodes.[17] Bis(12-crown-4) exhibits a much larger Na^+/K^+ selectivity than monomeric 12-crown-4. Similarly, bis(15-crown-5) has a higher K^+/Na^+ selectivity than 15-crown-5. The high resemblance of the selectivity reported for these ion-selective electrodes with the responses of the crown ether SAMs **1** and **2** strongly suggests that sandwich complexation in the monolayer occurs.

4. Conclusions

Modification of a gold electrode with self-assembled monolayers bearing crown ether groups enables the reversible binding of metal ions from water. The use of a positively charged redox couple in the electrolyte transduces the binding event into an electrochemically detectable signal. Moreover, the remarkable selectivity of the crown ether SAMs indicates the cooperative binding of one metal ion by two crown ether ligands.

References

(1) (a) Ulman, A. *An Introduction into Ultrathin Organic Films*; Academic Press: San Diego, CA, **1991**. (b) Finklea, H. O. in *Electroanalytical Chemistry* Vol. 19, Bard, A. J.; Rubinstein, I. (Eds.), Marcel Dekker, NY, **1996**.

(2) (a) Porter, M. D.; Bright, T. B.; Allara, D. L.; Chidsey, C. E. D. *J. Am. Chem. Soc.* **1987**, *109*, 3559. (b) Bain, C. D.; Troughton, E. B.; Tao, Y.-T.; Evall, J.;

Whitesides, G. M.; Nuzzo, R. G. *J. Am. Chem. Soc.* **1989**, *111*, 321. (c) Chidsey, C. E. D.; Loiacono, D. N. *Langmuir* **1990**, *6*, 682.

(3) Thoden van Velzen, E. U.; Engbersen, J. F. J.; de Lange, P. J.; Mahy, J. W. G.; Reinhoudt, D. N. *J. Am. Chem. Soc.* **1995**, *117*, 6853. (b) Huisman, B.-H.; Thoden van Velzen, E. U.; van Veggel, F. C. J. M.; Engbersen, J. F. J.; Reinhoudt, D. N. *Tetrahedron Lett.* **1995**, *36*, 3273. (c) Beulen, M. W. J.; Huisman, B.-H.; van der Heijden, P. A.; van Veggel, F. C. J. M.; Simons, M. G.; Biemond, E. M. E. F.; de Lange, P. J.; Reinhoudt, D. N. *Langmuir* **1996**, *12*, 6170.

(4) (a) Schierbaum, K.-D.; Weiss, T.; Thoden van Velzen, E. U.; Engbersen, J. F. J.; Reinhoudt, D. N.; Göpel, W. *Science* **1994**, *265*, 1413. (b) Huisman, B.-H.; Kooyman, R. P. H.; van Veggel, F. C. J. M.; Reinhoudt, D. N. *Adv. Mater.* **1996**, *8*, 561.

(5) (a) Spinke, J.; Liley, M.; Schmitt, F.-J.; Guder, H.-J; Angermaier, L.; Knoll, W. *J Chem. Phys.* **1993**, *99*, 7012. (b) Sastry, M.; Patil, V.; Mayya, K. S. *J. Phys. Chem. B* **1997**, *101*, 1167.

(6) (a) Rubinstein, I.; Steinberg, S.; Tor, Y.; Shanzer, A.; Sagiv, J. *Nature* **1988**, *332*, 426. (b) Steinberg, S.; Tor, Y.; Sabatani, E.; Rubinstein, I. *J. Am. Chem. Soc.* **1991**, *113*, 5176. (c) Steinberg, S.; Rubinstein, I. *Langmuir* **1992**, *8*, 1183.

(7) (a) Turyan, I.; Mandler, D. *Anal. Chem.* **1994**, *66*, 58. (b) Turyan, I.; Mandler, D. *Anal. Chem.* **1997**, *69*, 894.

(8) Gafni, Y.; Weizman, H.; Libman, J.; Shanzer, A.; Rubinstein, I. *Chem. Eur. J.* **1996**, *2*, 759.

(9) Henke, C.; Steinem, C.; Janshoff, A.; Steffan, G.; Luftmann, H.; Sieber, M.; Galla, H.-J. *Anal. Chem.* **1996**, *68*, 3158.

(10) Bruening, M. L.; Zhou, Y.; Aguilar, G.; Agee, R.; Bergbreiter, D. E.; Crooks, R. M. *Langmuir* **1997**, *13*, 770.

(11) (a) Cheng, Q.; Brajter-Toth, A. *Anal. Chem.* **1992**, *64*, 1998. (b) Takehara, K.; Takemura, H.; Ide, Y. *Electrochim. Acta* **1994**, *39*, 817. (c) Cheng, Q.; Brajter-Toth, A. *Anal. Chem.* **1995**, *67*, 2767. (d) Cheng, Q.; Brajter-Toth, A. *Anal. Chem.* **1996**, *68*, 4180.

(12) Sato, Y.; Ye, S.; Haba, T.; Uosaki, K. *Langmuir* **1996**, *12*, 2726.

(13) (a) Sabatani, E.; Rubinstein, I.; Maoz, R.; Sagiv, J. *J. Electroanal. Chem.* **1987**, *219*, 365. (b) Sabatani, E.; Rubinstein, I. *J. Phys. Chem.* **1987**, *91*, 6663. (c) Sabatani, E.; Cohen-Boulakia, J.; Bruening, M.; Rubinstein, I. *Langmuir* **1993**, *9*, 2974. (d) Finklea, H. O.; Snider, D. A.; Fedyk, J.; Sabatani, E.; Gafni, Y.; Rubinstein, I. *Langmuir* **1993**, *9*, 3660. (e) Nahir, T. M.; Bowden, E. F. *Electrochim. Acta* **1994**, *36*, 2347.

(14) (a) Equivalent Circuit version 4.55, 1996, Boukamp, B. A., University of Twente, Dept. Chemical Technology, Enschede, The Netherlands. (b) Boukamp, B. A. *Solid State Ionics* **1986**, *18-19*, 136.

(15) Sluyters-Rehbach, M.; Sluyters, J. H. in *Electroanalytical Chemistry*, Vol. 4, Bard, A. J., Ed.; Marcel Dekker: New York, 1970.

(16) (a) Izatt, R. M.; Pawlak, K.; Bradshaw, J. S.; Bruening, R. L. *Chem Rev.* **1991**, *91*, 1721. (b) Yurdakoç, M. K.; Erk, Ç. *Spectrosc. Lett.* **1987**, *20*, 983.

(17) (a) Tamura, H.; Kimura, K.; Shono, T. *Anal. Chem.* **1982**, *54*, 1224. (b) Moody, G. J.; Saad, B. B.; Thomas, J. D. R. *Analyst* **1989**, *114*, 15.

FUNCTIONALIZED CYCLODEXTRIN-CALIX[4]ARENE HOST MOLECULES FOR DETECTION OF ORGANIC ANALYTES

Jürgen H. Bügler, Menno de Jong, Johan F.J. Engbersen and David N. Reinhoudt

Laboratory of Supramolecular Chemistry and Technology,
University of Twente, P.O. Box 217,
NL-7500 AE Enschede,
The Netherlands

0. Abstract: Monofunctionalized β-Cyclodextrins were coupled to calix[4]arene derivatives bearing different fluorophores to give fluorescent cyclodextrin-calix[4]arene couples. The cyclodextrin-calix[4]arene couple **1** was investigated as a sensing molecule for the detection of several organic analytes in aqueous solution. Inclusion of a guest into the fluorescent sensing molecule leads to a decrease in fluorescence intensity. Compound **1** shows selectivity for organic analytes of a certain length. The fluorescent sensing molecules **2-6** were synthesized in order to investigate the influence of the different parts of the molecules on the sensing properties.

Keywords: Fluorescence Sensors, Cyclodextrins, Supramolecular Chemistry

1. Introduction. Cyclodextrins are a unique group of naturally occuring cyclic D-glucose oligomers. These molecules have the shape of a truncated cone and are capable of complexation of organic molecules in their cavity.[1] This host-guest complexation of cyclodextrins is of great interest in view of the development of sensors for specific molecules. Fluorescent cyclodextrins have been prepared by appending a fluorophore to one of the primary hydroxyl groups of β- and γ-cyclodextrin to give fluorescent probes for organic guest molecules.[2] Competition for the hydrophobic cavity between the guest molecule and the covalently linked optical reporter group results in a change in the optical properties of the reporter group. This change in properties is dependent on the concentration of the species to be detected.

A large number of organic analytes can only be detected by rather laborious methods, requiring sampling, pretreatment and analysis by relatively expensive techniques. Development of fast responding sensors is therefore of great use in environmental control, agriculture, food technology, etc. Good selectivity and transduction properties of appropriate sensing layers are a prerequisite for the construction of sensors. Cyclodextrin derivatives have good prospects to be used in such sensing layers, as they can accomodate analytes in their cavity. By appropriate functionalization of the wider, secondary face with groups that provide the cyclodextrin host with additional binding sites for the target molecule, the selectivity of the sensing layer can be increased. The main aim of this project is to combine both the advantages of cyclodextrins and calixarenes. Cyclodextrins offer good water solubility entailed by a promising host-guest chemistry. Calix[4]arenes are widely used as rigid platforms in supramolecular chemistry and can be selectively modified with functional groups.[3]

Preliminary experiments revealed, that calix[4]arene cyclodextrin couples showed increased binding abilities for the fluorescent probes 1-anilino-8-naphthalenesulfonate (ANS) and 2-p-toluidino-6-naphthalenesulfonate (TNS) in aqueous solution compared to cyclodextrin.[4,5] Herein we report the synthesis and the host-guest chemistry of the cyclodextrin-calix[4]arene couple **1** which bears a fluorophore as a reporter group for guest inclusion, as well as the synthesis of derivatives of **1**. Their properties might give closer insight in the sensing mechanism.

2. Results and Discussion. Compound **1** was obtained by reductive amination of a calix[4]arene derivative that bears an aldehyde functional group and a fluorophore opposite to each other with a silyl protected β-cyclodextrin having a (p-aminomethyl)tolyl spacer at one of the secondary hydroxyl groups (Fig. 1) followed by purification and deprotection. **1** was characterised by TLC, ^{1}H NMR and MALDI-TOF mass spectroscopy. The UV/VIS spectrum shows the expected absorption bands for the calix[4]arene-2-naphthylamine moiety. **1** is water soluble (soly. 0.2 mM) and is fluorescent in aqueous solution.

For the investigation of the sensing abilities of **1** the fluorescence spectra of an aqueous solution of **1** were measured while aliquots of a stock solution of an organic analyte were added succesively to the solution of sensor molecule **1**. The fluorescence intensity decreased upon addition of guest molecules (Fig. 2). A common parameter expressing the sensitivity of a fluorescent probe towards different analytes is the $\Delta I/I_0$-value.[6] As is shown in Figure 3, the new sensing molecule molecule **1** shows sensitivity for some terpenes and steroids. The remarkably high response for 1-adamantylbenzoate and norethindrone compared with the binding constants for 1-adamantanol or for the other examined steroids shows the high selectivity of **1** for guest molecules with a certain length. 1-adamantylbenzoate can be recognized by the new sensing molecule molecule with a selectivity of >200 over the other organic compounds. CPK models supply further evidence for the excellent fit of 1-adamantylbenzoate in **1**.

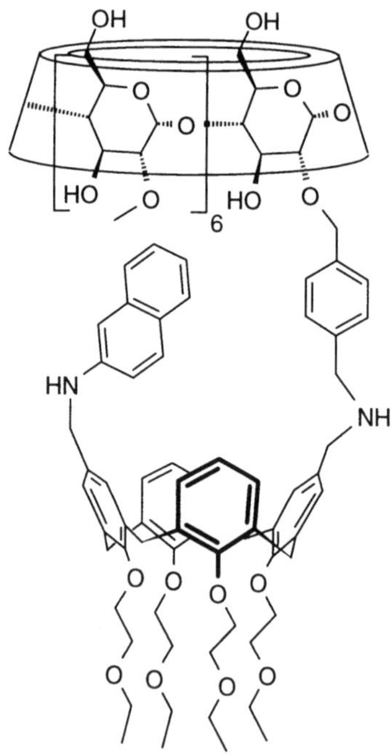

Fig. 1: Sensing molecule compound 1

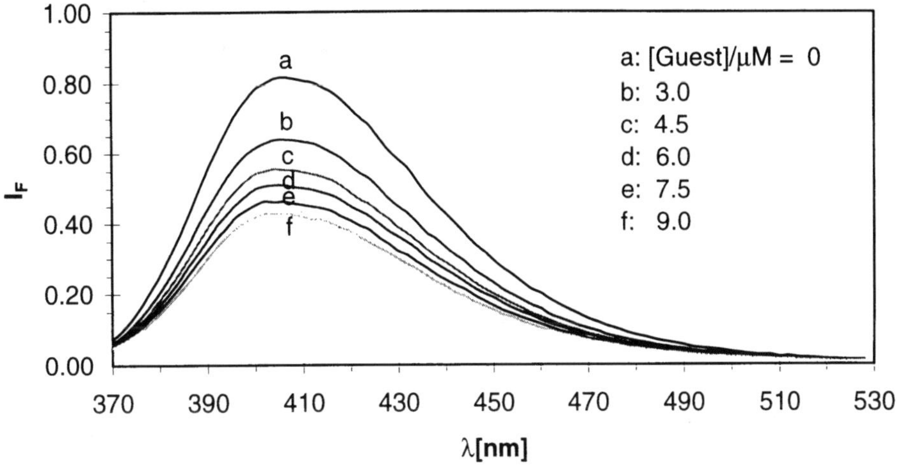

***Fig. 2:** Fluorescence spectra of **1** in aqueous solution ([**1**] = 2.9 μM, in phosphate buffer 0.01 mM, pH 7; λ(ex.) = 350 nm, λ(em.max.) = 408 nm). Guest: 1-Adamantylbenzoate.*

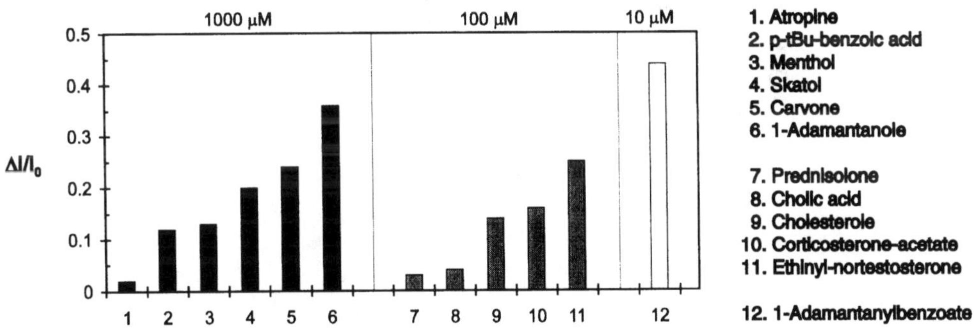

***Fig. 3**: Sensitivity factors $\Delta I/I_0$ for several neutral, organic analytes. [Analyte] = given above the scetch. [**1**] = 1.45 μM.*

To gain more insight in the factors governing the sensing properties, especially the sensitivity for certain analytes of cyclodextrin-calix[4]arene couples, several components of the molecule were varied (Fig. 4). Compounds **2** and **3** enable investigation of the influence of the spacer length and rigidity. Compounds **4** and **5** have been prepared to study the effect of different fluorophores. Therefore, the dansyl and the coumaryl fluorophore were chosen as they have rather high fluorescence quantum yields and comparably long excitation wavelengths. This might be important for future technical applications. In compound **6** the remaining upper rim hydrogens of the calix[4]arene have been substituted for groups which potentially attribute to the binding of suitable guest molecules. All couples were prepared by essentially the same procedure as that followed for the preparation of **1**. The binding and sensing properties of compounds **2-6** are currently under investigation.

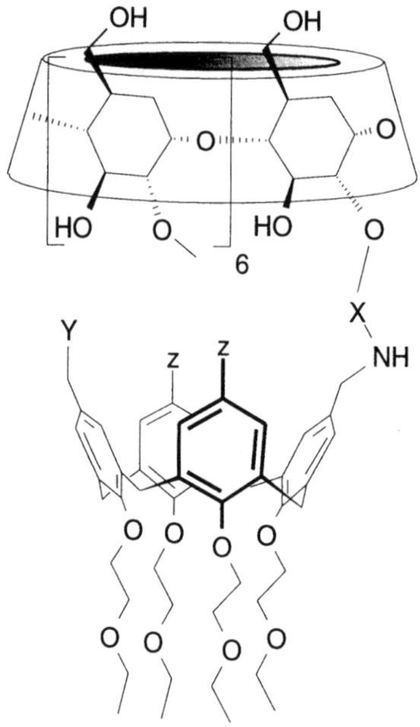

***Fig. 4**: Sensing molecule compounds **2-6** (see Tab. 1).*

***Table 1**: Substitution pattern of sensor molecules **1-6**.*

Compound	X	Y	Z
1	$-CH_2(p\text{-}C_6H_4)CH_2-$	2-aminonaphthyl-	H
2	$-(CH_2)_3-$	2-aminonaphthyl-	H
3	$-(CH_2)_6-$	2-aminonaphthyl-	H
4	$-CH_2(p\text{-}C_6H_4)CH_2-$	dansyl-	H
5	$-CH_2(p\text{-}C_6H_4)CH_2-$	coumaryl-	H
6	$-CH_2(p\text{-}C_6H_4)CH_2-$	2-aminonaphthyl-	NO_2

3. Conclusions. In summary, the new fluorescent cyclodextrin-calix[4]arene couple **1** is suitable for the sensing of organic molecules in aqueous solution. It shows remarkably high sensitivity for guest molecules with a certain length. Further efforts were made to optimize the binding properties of **1** by varying the different parts of the molecule. The sensing behaviour of the recently synthesized compounds **2-6** will allow further insight in the sensing mechanism and provide a basis for the design of further optimized, applicable sensor molecules.

Acknowledgements: This work was supported by a FAIR Programme of the EU (Project No. PL 95-0300) and by STW (Project No. 349-4213).

References:

[1] M.L. Bender, M. Komiyama, *Cyclodextrin Chemistry*, Springer Verlag, New York, 1978.
[2] A. Ueno, *Supramolecular Science*, **3**, 1996, 31, and references cited herein.
[3] C.D. Gutsche, *Calixarenes* in: *Monographs in Supramolecular Chemistry, Vol. 1*, J.F. Stoddart (Ed.), Royal Society of Chemistry, Cambridge, 1989.
[4] E. van Dienst, B.H.M. Snellink, I. von Piekartz, M.H.B. Grote Gansey, F. Venema, M.C. Feiters, R.J.M. Nolte, J.F.J. Engbersen, D.N. Reinhoudt, *J. Org. Chem.*, **60**, 1995, 6537.
[5] E. van Dienst, B.H.M. Snellink, I. von Piekartz, J.F.J. Engbersen, D.N. Reinhoudt, *J. Chem. Soc., Chem. Commun.*, 1995, 1151.
[6] A. Ueno, S. Minato, J. Suzuki, M. Fukushima, M. Ohkubo, T. Osa, F. Hamada, K. Murai, *Chem. Lett.* **1990**, 605.

SELF-ASSEMBLED MONOLAYERS ON GOLD FOR THE DETECTION OF MOLECULAR INTERACTIONS *VIA* SURFACE PLASMON RESONANCE

Arianna Friggeri, Frank C.J.M. van Veggel, David N. Reinhoudt

Supramolecular Chemistry and Technology, MESA Research Institute,
University of Twente, P.O. Box 217, 7500AE Enschede, The Netherlands.
e.mail: A.Friggeri@ct.utwente.nl

0. Abstract

Self-assembled monolayers, of receptor adsorbates, on gold have been used to detect the presence of small aromatic compounds, in water. Surface Plasmon Resonance (SPR) was used to monitor the *in situ* host-guest interaction processes occuring at the monolayer-water interface.

Keywords: Surface Plasmon Resonance, monolayers, resorcinarenes.

1. Introduction

Self-Assembled Monolayers (SAMs) on gold consist of one layer of molecules adsorbed, via sulfur coordination bonds, onto a gold (111) surface. The scope of this project is to investigate the possibility that SAMs on gold have, to act as detectors for molecular host-guest interactions [1]. So far, several receptor molecules with a variety of functionalities, have been adsorbed on gold substrates and their interactions with a series of aromatic guests has been investigated. These receptors belong to a class of molecules that can be synthesized starting from *resorcinarenes* and will be refered to as cavitands (Fig. 1). The monolayers formed by these adsorbates have been characterized by contact angle, Polarized Infrared External Reflectance Spectroscopy (PIERS), electrochemical measurements and X-ray Photoelectron Spectroscopy (XPS). These techniques show that the adsorbates form well packed monolayers, attached to the gold via the four sulfide moieties and with the head group extending towards the interface between monolayer and the air/water medium.

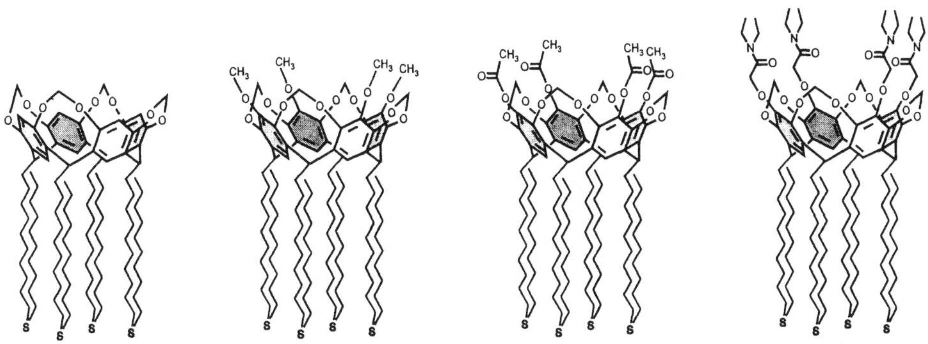

1 2 3 4

Fig. 1 Cavitand receptor adsorbates

2. Characterization: Results and Discussion

Contact angle results (Table 1) show that all monolayers are relatively hydrophobic. An evaluation of the packing of the molecules in the monolayers can be made by considering the hysterysis values: $\Delta\theta$. Taking into account that decanethiol layers have been reported to have $\Delta\theta$ values of around 17, then, monolayers of **1** can be defined as relatively well ordered, whilst monolayers of compounds **2**, **3** and **4**, are somewhat less well ordered. These results are to be expected considering that the substituents on the upper rim of these cavitand adsorbates are probably arranged in a disordered fashion.

Table 1: Advancing (θ_a) and receding (θ_r) contact angle data for water drops on monolayers of **1**, **2**, **3** and **4**.

	θ_a	θ_r	$\Delta\theta$
Decanethiol	108 ± 1	91 ± 1	17
1	107 ± 1	91 ± 1	16
2	98 ± 2	52 ± 2	46
3	109 ± 2	85 ± 1	24
4	107 ± 3	88 ± 3	19

Capacitance measurements performed on layers **2** and **4** are in the same range as the values reported for **1** and decanethiol layers, whilst **3** gives a slightly higher value. The heterogeneous electron transfer cyclic voltammograms show that all three layers (**2**, **3**, **4**) are able to block the Fe^{2+}/Fe^{3+} redox couple just as effectively as a layer of **1**. Therefore, these types of substituents on the upper rim of cavitands, do not seem to interfere with the overall permeability and packing of the molecules in the monolayers. This is also evident from the resistance values obtained by impedance measurements, shown in Table 2 along with the capacitance values.

Table 2: Capacitance and resistance values for monolayers of compounds **1**, **2**, **3**, **4** and decanethiol

Adsorbates	C ($\mu F \cdot cm^{-2}$)	R (ohm)
2	1.27	2.14e+6
4	1.43	1.08e+6
3	2.64	4.02e+5
1	1.14	3.75e+5
decanethiol	1.24	8.00e+5

Results obtained from Grazing Angle FT-IR are presented in Table 3. It is interesting to note that monolayers of **2** are liquid like, this can be seen from the shifted values of the CH_2 and CH_3 stretching vibrations, whilst monolayers of **3** and **4** are more crystalline.

Table 3: Grazing angle FT-IR data for monolayers of **2**, **3** and **4**.

	ν_{as} CH_3	$\nu_{as}CH_2$	ν_sCH_3	ν_sCH_2	ν C-H_{arom}	δ CH_2
2	2959	2924	2874	2855	3005	1474, 1456, 1423
3	2964	2921	2879	2850	-	-
4	2964	2921	2877	2851	-	-

3. SPR: Results and Discussion

All measurements were carried out in pure water and were repeated at least 3 times. It is important to note that the signals show <u>no</u> memory effect. The SPR [2] set-up used to make the measurements consents simultaneous detection of reflectivity changes on two different monolayers, in contact with the same solution of guest. One of the two layers is a reference layer (monolayer of $C_{18}SH$) used to take into account the change in reflectivity due to changes of refractive index in the bulk solution.

Explaination of some of the results is relatively easy, of others, very difficult, but some interesting conclusions can be made. Reading Table 4 in a vertical fashion (i.e. responses of one adsorbate to several guests) it can be observed that none of the guests show particular affinity for adsorbate **2** and **4.** The most probable explanation for this is the fact that both these layers resemble quite closely the reference layer: they both expose to the aqueous solution, alkyl groups. Furthermore, the shallow aromatic cavity of the adsorbate molecules is not available to the guests, therefore, no particular interaction is present due to the typical cavitand structure. Considering the results obtained for adsorbate **1**, the first observation that can be made is that all guests except *p*-methoxy phenol, show some interaction with the monolayer. Some guests seem to interact more favourably than others, but the reason for this cannot be pinpointed to singular features of the guest. For example, comparing *p*-toluic acid and benzoic acid, it cannot be said that a CH_3 group is necessary for interaction with the monolayer to occur. Of the guests studied, it can be said that the presence of a nitro group seems to favor interactions with receptor surface **1**, although it is not clear how the guest is placed upon the monolayer. Some interesting observations can be made if one considers the <u>solubility</u> and the <u>acidity</u> of the guests. Solubility wise, none of the results show that the more hydrophobic guests interact more strongly with the host monolayers. Looking at the pKa's of the guests, it is also true that there is no particular correlation between the acidity and the responses obtained with layer **1**.

Table 4: SPR results expressed as the difference between the SPR angle change of the receptor layer and reference layer.
*Sol. is the relative solubility in water. 1: insoluble, 2: slightly soluble, 3: soluble.

guests	**pKa**	**Sol.***	**SAMs**			
			1	**2**	**3**	**4**
p-CH_3 benzoic acid	4.36	1	0.06	-	0.05	-
benzoic acid	4.20	2	0.12	-	-	-
p-OH benzoic acid	4.48	2	0.04	-	-	-
p-NO_2 benzoic acid	3.41	1	0.08	0.10	-	0.07
p-cresol	10.17	2	0.04	-	-	-
p-NO_2 phenol	7.15	3	0.10	0.13	-	0.10
p-OCH_3 phenol	>10	3	-	-	-	-

It seems as though functionalization of the upper rim of these cavitand adsorbates leads to selectivity in the recognition of guests in water. Although this might not be a particularly surprising conclusion, it becomes quite important in view of the size of the guests involved and of the experimental medium: water.

4. Conclusions

This study provides *experimental evidence* that, in water, hydrophobic interactions, although present and perhaps the initial driving force of molecular association process, are not the main cause of recognition between host and guest. Unfortunately, because very few studies have been made regarding cavitand host-guest chemistry in water, and none have been carried out on the monolayer/water interface, it is very difficult to extrapolate quantitative data, as well as qualitative information on how the guest is placed with respect to the functionalized layer. From some of the values obtained from the measurements, it seems as though more than one layer of guests is present 'on top' of the adsorbate layer.

5. References

[1] Molecular Recognition by Self-Assembled Monolayers detected with Surface Plasmon Resonance; B.-H. Huisman, R. P. H. Kooyman, F.C.J.M. van Veggel, D.N. Reinhoudt; *Adv. Materials*, 8, 561-564, **1996.**

[2] Vibrating Mirror Surface Plasmon Resonance Immunosensor; R.P.H. Kooyman, A.T.M. Lenferink, R.G. Eenink, J. Greve; *Analytical Chemistry*, Vol. 63, No. 1, Jan. 1, **1991**.

BATTERY-LESS SENSORS FOR CONTINUOUS MEASUREMENT OF PHYSIOLOGICAL PARAMETERS OF ANIMALS

W.J. Eradus
DLO-Institute of Agricultural and Environmental Engineering (IMAG-DLO)
P.O. Box 43, 6700 AA Wageningen, The Netherlands
email: w.j.eradus@imag.agro.nl

0. Abstract

RF electronic animal identification, as used for tracing livestock, opens the possibility for the incorporation of sensing devices. Continuous measurements with such implanted devices requires a power supply battery. Because of the limited lifetime of this energy source, a prototype of battery-less implantable sensor system has been developed, using inductively coupled energy. Satisfying PTT regulations, an average energy flow of 80 μW could be realized with an inductive loop. This will be sufficient for the supply of the sensing system, built around an ultra low-power microcontroller. Besides this continuous way of energy supply, also a solution with a pulsed supply is discussed.

Keywords: **Sensors, powering, physiological parameters**

1. Introduction

In agriculture, RF electronic animal identification becomes widely used for tracing livestock, for tracing animals when there has been a disease outbreak, for preventing fraud and for farm automation. The latter application opens the possibility of management on an individual basis. The identification system consists of an active interrogator, energizing the passive transponder, implanted in the animal, which in turn replies with the transmission of the identification code [1]. To make more profit of this radio link with the animal, incorporation of sensing devices into the transponder is useful. During energizing, the measurement electronics will become active and add the measured value to the identification code. Continuous measurement or more frequent measuring of parameters such as temperature, heart rate and activity level, however, requires a power supply battery. This energy source has a limited lifetime and normally contains toxic components. Therefore, the idea of inductively coupled energizing is sometimes used. In literature, implementations are described, working at an operational distance of some centimetres [1][2]. For applications in livestock houses, where the cattle (dairy cows, pigs, etc) can freely move, this energizing distance is far too short.

2. Inductive loop powering

With the idea in mind of the inductive loop used to transmit audio signals directly to hearing-aids, the feasibility of such a loop for energizing implanted sensing devices has been studied. Basically, the working range is about the same as the area covered by the loop. The receiving coil has been made part of a resonator circuit, necessary to achieve a voltage level that is sufficiently high. The main factors determining the available power are the field strength, the dimensioning of the receiving coil and the quality factor of the resonant circuit. These factors in turn are dependent on the chosen frequency.

Other aspects are the environmental health aspects associated with the use of electromagnetic field energy. According the preliminary Dutch PTT regulations (1990) for inductive systems [4], the maximum field strength for distances up to 1 m is 240 mA/m for frequencies up to 16 kHz. At a distance of 3 m, this limit is 90 mA/m. For frequencies between 16 and 500 kHz the limits drop down with 3.5 dB/oct. as depicted in fig.1.

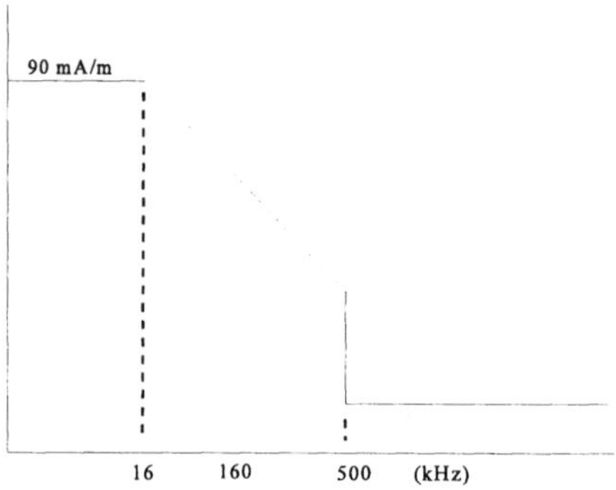

Fig. 1. Dutch PTT limits of field strength at distances larger than 3 m.

A prototype of a battery-less implantable sensor system for the measurement of activity level and temperature has been developed. At a frequency of 16 kHz, an average energy level of about 80 μW could be achieved. Apart from the problem of stabilizing the voltage in a non-homogeneous field, this energy level proved to be sufficient for the supply of the new generation of ultra-low power CMOS circuitry. The sensing system has been built around the PIC 16C54/LP microcontroller, which at 3V and 10 kHz clock frequency consumes about 40 μW. One of the problems is to stabilize the supply voltage. A main requirement is that the energy consumption of the stabilizing circuit will be very low. This has been done by means of a buffer capacitor in combination with a chain of silicon diodes switched in forward direction. At 3V this chain consumes only 1 μA, but at 5 V already 100 micro amperes. This will also contribute to the stabilization of the voltage and preventing the electronic circuits from damaging due to excessively high fields.

3. Discussion

Normally, the measured values stored in the implanted sensor system have to be transferred to an external database for further processing. As mentioned in the introduction, this can be done by communication with a transponder system integrated with the sensor [5]. During interrogation, the sensor/transponder is very close to the antenna of the external reading device (distances normally less than 50 cm), in order to transmit enough energy for the activation of the transponder.

In situations where the animal will stay a relatively long time in the vicinity of such an energy source, the excess of radiated power can be stored in an energy buffer. This occurs for instance in the milking parlour where cows remain two times a day for about 10 minutes. If the energy buffer, integrated with the sensor/transponder, contains enough energy, the sensor can be activated later on and the measurement data can be stored without need for a continuous inductive loop powering. This alternative can be referred to as pulsed loop powering. The advantage is that the movements of the animal are not longer restricted to the inductive loop areas. Also the associated cabling and continuous energizing field are not longer necessary to the same extend. Bottleneck however is the capacity and the volume of the energy buffer. Very tiny capacitors are available with a capacity of more than one Farad, but the energy/volume ratio of rechargeable batteries normally is higher. Therefore the performance of Lithium-based rechargeable foils is being evaluated for implementation in an implantable sensor/transponder system which will be energized by a pulsed loop powering approach.

4. Conclusions

Basically, powering implanted transponders for electronic animal identification with integrated sensors without the use of batteries proved to be feasible. Continuous inductive powering by means of an energizing loop provides an energy level of some tenths of microwatts in the implanted device. A precondition is that the animals are always located within the field of the energizing loop. To avoid this restriction, pulsed loop powering may be used. This necessitates a tiny internal energy buffer to brigde the time between two consecutive energizing periods

References

[1] Eradus,W.J. The development of standards for automatic animal identification, Proceedings of the XXV CIOSTA CIGR V Congress, Wageningen (NL), 10-13 May 1993, p.307-311.

[2] Piquard,J.F. and de Vel,O. Transmission of physiological parameters with an inductively coupled RF system, International Conference on Biomedical Transducers, Paris, 3-7 Nov. 1975.

[3] Thummim,A et al. High power electromagnetic energy transfer for totally implanted devices, IEEE Trans. on Magnetics, 2 June 1970, p.326-332.

[4] Limits for inductive systems, Preliminary regulation of the Dutch PTT, 4 March 1990.

[5] Kampers F.W.H., Rossing W., Eradus W.J. Technical concept of the standard for RF identification of animals, Proceedings of AgEng 96, Madrid, Sept. 1996, paper 96G-007.

Authors Index

Subject index

SPRINGER NATURE

GPSR Compliance

The European Union's (EU) General Product Safety Regulation (GPSR) is a set of rules that requires consumer products to be safe and our obligations to ensure this.

If you have any concerns about our products, you can contact us on ProductSafety@springernature.com

In case Publisher is established outside the EU, the EU authorized representative is:

Springer Nature Customer Service Center GmbH
Europaplatz 3
69115 Heidelberg, Germany

FSC
www.fsc.org
MIX
Papier | Fördert
gute Waldnutzung
FSC® C083411

Kolibri 360 GmbH
Ferdinand-Jühlke-Straße 7
99095 Erfurt, Deutschland
produktsicherheit@kolibri360.de